Active Safety and the Mobility Industry

Other SAE Books of Interest:

**Performance Metrics for Assessing Driver Distraction:
The Quest for Improved Road Safety**
(Product Code: R-402)

Automotive Safety Handbook, Second Edition
(Product Code: R-377)

Automotive Vehicle Safety
(Product Code: R-341)

For more information or to order a book, contact SAE International at
400 Commonwealth Drive, Warrendale, PA 15096-0001; phone 877-606-7323
(U.S. and Canada) **or 724-776-4970** *(outside U.S. and Canada);* **fax 724-776-0790;**
e-mail CustomerService@sae.org; website http://store.sae.org.

Active Safety and the Mobility Industry

Edited by
Dr. Andrew Brown, Jr.

Published by
SAE International
400 Commonwealth Drive
Warrendale, PA 15096-0001 U.S.A.

Phone: (724) 776-4841
Fax: (724) 776-5760
www.sae.org
April 2011

PT-147

SAE International

400 Commonwealth Drive
Warrendale, PA 15096-0001 USA

E-mail: CustomerService@sae.org
Phone: 877-606-7323 *(inside USA and Canada)*
       724-776-4970 *(outside USA)*
Fax: 724-776-1615

Copyright © 2011 SAE International. All rights reserved.
No part of this publication may be reproduced, stored in a retrieval system, distributed, or transmitted, in any form or by any means without the prior written permission of SAE. For permission and licensing requests, contact SAE Permissions, 400 Commonwealth Drive, Warrendale, PA 15096-0001 USA; e-mail: copyright@sae.org; phone: 724-772-4028; fax: 724-772-9765.

ISBN 978-0-7680-4766-0
Library of Congress Catalog Number 2011923419
SAE Order No. PT-147

Information contained in this work has been obtained by SAE International from sources believed to be reliable. However, neither SAE International nor its authors guarantee the accuracy or completeness of any information published herein and neither SAE International nor its authors shall be responsible for any errors, omissions, or damages arising out of use of this information. This work is published with the understanding that SAE International and its authors are supplying information, but are not attempting to render engineering or other professional services. If such services are required, the assistance of an appropriate professional should be sought.

Applicable to SAE International technical papers: The Engineering Meetings Board has approved these papers for publication. They have successfully completed SAE's peer review process under the supervision of the session organizer. This process requires a minimum of three (3) reviews by industry experts. Positions and opinions advanced in these papers are those of the author(s) and not necessarily those of SAE. The author is solely responsible for the content of the paper.

To purchase bulk quantities, please contact:
SAE Customer Service
E-mail: CustomerService@sae.org
Phone: 877-606-7323 *(inside USA and Canada)*
       724-776-4970 *(outside USA)*
Fax: 724-776-1615

Visit the SAE Bookstore at
http://store.sae.org

TABLE OF CONTENTS

Introduction

Active Safety and the Mobility Industry 3
Dr. Andrew Brown, Jr., P.E., FESD, NAE – President, SAE International, 2010

Special Contributions

**Advancing Safety in the Future: The Role of Technologies,
the Government, and the Industry** 13
Dr. Joseph N. Kanianthra, President of Active Safety Engineering LLC

Road Safety – A Better Way Forward 27
Patrick Lepercq, Michelin – Corporate Vice President Public Affairs and Chairman
of the Global Road Safety Partnership

Traffic Safety: International Status and Active Technologies for the Future 35
Dr. Dinesh Mohan, Volvo Chair Professor for Biomechanics and Transportation Safety
at the Transportation Research and Injury Prevention Programme of the Indian Institute of
Technology, Delhi

Addressing the Deployment Issues of Intelligent Cooperative Systems 41
Richard Harris, Director of Intelligent Transport Systems at Logica plc, English Language Secretary
of the World Road Association (PIARC) Technical Committee Network Operations,
and Technical Secretary of the PIARC/FISITA Joint Task Force on Intelligent Cooperative Systems

Active Safety Systems, Crash Sensing and Sensor Fusion

**Developing Integrated Vision Applications
for Active Safety Systems (2009-01-0158)** 49
Adam Prengler – NEC Electronics America, Inc.

**Successive Categorization of Perceived Urgency in Dynamic
Driving Situations (2009-1-0780)** 55
Gerald J. Schmidt, Ali Khanafer and Dirk BalzerAdam – Opel GmbH

Estimation of Vehicle Roll Angle and Side Slip for Crash Sensing (2010-01-0529) ... 61
Aleksander Hac, David Nichols and Daniel Sygnarowicz – Delphi Corporation

Sensor Data Fusion for Active Safety Systems (2010-01-2332) 75
Jorge Sans Sangorrin, Jan Sparbert, Ulrike Ahlrichs and Wolfgang Branz – Robert Bosch GmbH
Oliver Schwindt – Robert Bosch LLC

**Frontal Crash Testing and Vehicle Safety Designs:
A Historical Perspective Based on Crash Test Studies (2010-01-1024)** 83
Randa Radwan Samaha, Kennerly Digges and Thomas Fesich – George Washington Univ.
Michaela Authaler

Road Safety

Synthesizing a System for Improving Road Safety in China (2009-01-0592) 99
Hongtao Yu, Jiahe Zhang, Yuankui Meng and Shili Ni – CATARC Shanghai Operation
Weijian Han – Ford Motor Company
Peng Ren – Clemson University

**Structural Improvement for the Crash Safety
of Commercial Vehicle (2009-01-2917)** 107
Libo Cao, Zhonghao Bai, Jun Wu, Chongzhen Cui and Zhenfeng Niu
The State Key Laboratory of Advanced Design and Manufacturing for Vehicle Body

**Innovative Concepts for Smart Road Restraint Systems
to Provide Greater Safety for Vulnerable Users – Smart RRS (2010-36-0034)** 113
Arturo Davila and Mario Nombela – IDIADA Automotive Technology
Juan José Alba – Universidad de Zaragoza
Juan Luis De Miguel – Centro Zaragoza

Driver Assistance and Modeling

**Calibration and Verification of Driver Assistance
and Vehicle Safety Communications Systems (2010-01-0664)** 131
Mohammad Naserian and Kurt Krueger – Vector North America

**Driver Alcohol Detection System for Safety(DADSS).
Background and Rationale for Technology Approaches (2010-01-1580)** 139
Susan A. Ferguson – Ferguson International LLC
Abdullatif (Bud) Zaouk and Clair Strohl – QinetiQ North America

Drowsiness Detection Using Facial Expression Features (2010-01-0466) 151
Satori Hachisuka, Teiyuu Kimura, Kenji Ishida, Hiroto Nakatani and Noriyuki Ozaki
– Denso Corporation

**Modeling of Individualized Human Driver Model for
Automated Personalized Supervision (2010-01-0458)** 161
Xingguang Fu and Dirk Söffker – University of Duisburg-Essen

**Designing Reusable and Scalable Software Architectures
for Automotive Embedded Systems in Driver Assistance (2010-01-0942)** 171
Dirk Ahrens, Andreas Frey and Andreas Pfeiffer – BMW Group, Munich, Driving Dynamics
Torsten Bertram – Technische Universität Dortmund

Communications

Integration of Car-to-Car Communication into IAV (2009-01-1479) 187
Tae-Kyung Moon, Jun-Nam Oh, Hyuck-Min Na and Pal-Joo Yoon – MANDO Corporation
Enabling Safety and Mobility through Connectivity (2010-01-2318) 193
Chris Domin – Ricardo Inc.
Time Determinism and Semantics Preservation in the Implementation
of Distributed Functions over FlexRay (2010-01-0452) 203
Marco Di Natale, Scuola S. Anna and Haibo Zeng – General Motors
Exploring Application Level Timing Assessment
in FlexRay based Systems (2010-01-0456) 213
Sandeep U. Menon – Electrical & Controls Integration Lab, General Motors R&D

Market and Consumer Preferences

Driver's Attitudes Toward the Safety of
In-Vehicle Navigation Systems (2009-01-0784) 225
Andrew Varden and Jonathan Haber – University of Guelph
Consumer Attitudes and Perceptions about Safety and Their Preferences
and Willingness to Pay for Safety (2010-01-2336) 233
Veerender Kaul, Sarwant Singh, Krishnasami Rajagopalan and Michael Coury – Frost & Sullivan
Commercial Business Viability of IntelliDrive[SM]
Safety Applications (2010-01-2313) 243
Robert White, Tao Zhang, Paul Tukey and Kevin Lu – Telcordia Technologies
David McNamara – MTS LLC

Editor's and Special Contributors' Biographies

About the Editor 257
Dr Andrew Brown, Jr.
Special Contributors 258
Dr. Joseph N. Kanianthra
Patrick Lepercq
Dr. Dinesh Mohan
Richard Harris

INTRODUCTION

Active Safety and the Mobility Industry

INTRODUCTION

Our global societies, whether developed or emerging, face several challenges related to vehicles and transportation. Although not limited to, they mainly include:

- An aging population that will operate personal vehicles.
- Continued integration of electronic systems replacing mechanical systems.
- The need to reduce driver or operator distraction from external media.
- Balancing greater operational efficiency and safety implications by reducing traffic congestion.
- The need to reduce impaired driver/operator involvement.
- Aging air traffic control infrastructure.

These dominant trends provide distinct opportunities in all regions, focused on safety inside and outside the vehicle. Safety will continue to be driven by legislation and regulation, but it is also a clear consumer trend.

One of the goals of any society is to ensure its people are afforded the opportunity to use transportation systems that are safe to both the operator and those who interact with the vehicle. This pertains not only to the vehicles themselves but also their interaction with the infrastructure built to accommodate and serve them whether they are automobiles, trucks, off-road vehicles, personal watercraft, aircraft, or spacecraft. There are many challenges related to safety today and in the future.

To respond to the evolving demands of drivers, mobility industry OEMs have to pay attention to the world we all drive in. Popular trends, environmental challenges, and technological advances push the development of the latest automotive safety innovations. We are clearly witnessing acceleration from "passive," where products and technologies would only protect occupants in the case of an accident, to "active," where sophisticated systems prevent the accident from occurring.

Sensors in vehicle side panels combined with front and rear cameras, 76 GHz electronically scanned radar, and other features create a "cocoon of safety" around the vehicle that warns the driver of an impending impact. Other systems such as throttle, brakes, and steering respond together as needed, so a collision can be avoided with very little driver intervention.

An example of this kind of proactive safety feature is Delphi's Driver Alert System. It scans the driver's eyes using a monocular camera and infrared illuminators. If it senses fatigue or distraction, it generates an alert. For a half century, the mobility industry has focused its safety efforts on minimizing the damage or injury that a crash can cause. Today, Delphi and others are developing ways to help prevent the crash from happening at all.

So, how much further can these technologies really take us? Recently, SAE participated in the FISITA World Automotive Summit 2010 to help address this question. A FISITA workshop of global experts established these key messages on the subject:

- Vehicle technology has not reached its limit. New technologies will continue to increase road safety in the future. Benefits may also be enabled by technologies whose implementation is primarily driven by non-safety, comfort-oriented applications, such as C2C (C2X) systems. These may yield spinoffs for enhanced safety applications.

- Adoption rates of modern assistance systems need to be raised. Awareness must be increased for both the end-user and OEM personnel at the point-of-sale.

- Use rates of modern assistance systems can be increased by creating HMI concepts that are intuitive, i.e., easy to learn and understand.

- Harmonized regulatory requirements (both functional and legal) in all markets will enhance the development and rollout of new technology.

- Vehicle safety technology is only one element to improve road safety. Priorities on where to take the first and most effective steps will vary regarding their operating environments, which differ completely between developed and developing countries.

- Quick wins with great potential for road safety may lie outside the vehicle for developing countries, e.g., roadway separation for vulnerable road users and vehicles.

- Sound, data-driven evaluations of the problem are essential before finding solutions. Internationally harmonized in-depth accident research with a minimum, common data set is needed.

- A basic set of required safety equipment for all markets may comprise seatbelts, airbags, and ESC; optional "musts" depend on cost/benefit evaluations based on the above approach. These may also include "disruptive" rather than "evolutionary incremental" technology, e.g., mobile phone-based safety applications in developing countries.

In this publication, we propose to address "Active Safety and the Mobility Industry" through a series of new articles written specifically for it and with a collection of hand-picked SAE International technical papers. The dedicated articles discuss societal, technological, political, and consumer issues. The topics and authors are:

- *Advancing Safety in the Future: The Role of Technologies, the Government, and the Industry*, by Dr. Joseph N. Kanianthra, President of Active Safety Engineering LLC

- *Road Safety – A Better Way Forward*, by Patrick Lepercq, Michelin – Corporate Vice President Public Affairs and Chairman of the Global Road Safety Partnership

- *Traffic Safety: International Status and Active Technologies for the Future*, by Dr. Dinesh Mohan, Volvo Chair Professor for Biomechanics and Transportation Safety at the Transportation Research and Injury Prevention Programme of the Indian Institute of Technology, Delhi

- *Addressing the Deployment Issues of Intelligent Cooperative Systems*, by Richard Harris, Director of Intelligent Transport Systems at Logica plc, English Language Secretary of the World Road Association (PIARC) Technical Committee Network Operations, and Technical Secretary of the PIARC/FISITA Joint Task Force on Intelligent Cooperative Systems

We augment these articles with a series of 20 SAE technical papers published between 2009 and 2010, divided into five categories, as follows:

- **Active Safety Systems, Crash, and Sensor Fusion**

  - *Developing Integrated Vision Applications for Active Safety Systems*, 2009, by Adam Prengler
  - *Successive Categorization of Perceived Urgency in Dynamic Driving Situations*, 2009, by Gerald J. Schmidt, Ali Khanafer, and Dirk Balzer
  - *Estimation of Vehicle Roll Angle and Side Slip for Crash Sensing*, 2010, by Aleksander Hac, David Nichols, and Daniel Sygnarowicz
  - *Sensor Data Fusion for Active Safety Systems*, 2010, by Jorge Sans Sangorrin, Jan Sparbert, Ulrike Ahlrichs, Wolfgang Branz, and Oliver Schwindt
  - *Frontal Crash Testing and Vehicle Safety Designs: A Historical Perspective Based on Crash Test Studies*, 2010, by Randa Radwan Samaha, Kennerly Digges, Thomas Fesich, and Michaela Authaler

- **Road Safety**

  - *Synthesizing a System for Improving Road Safety in China*, 2009, Hongtao Yu, Jiahe Zhang, Yuankui Meng, Shili Ni, Weijian Han, and Peng Ren
  - *Structural Improvement for the Crash Safety of Commercial Vehicles*, 2009, by Libo Cao, Zhonghao Bai, Jun Wu, Chongzhen Cui, and Zhenfeng Niu
  - *Innovative Concepts for Smart Road Restraint Systems to Provide Greater Safety for Vulnerable Users – Smart RSS*, 2010, by Mario Nombela, Arturo Dávila, Juan José Alba, and Juan Luis de Miguel

- **Driver Assistance and Modeling**

  - *Calibration and Verification of Driver Assistance and Vehicle Safety*, 2010, by Mohammad Naserian and Kurt Krueger
  - *Driver Alcohol Detection System for Safety – Background and Rationale*, 2010, by Susan A. Ferguson, Abdullatif (Bud) Zaouk, and Clair Strohl
  - *Drowsiness Detection Using Facial Expressions Features*, 2010, by Satori Hachisuka, Teiyuu Kimura, Kenji Ishida, Hiroto Nakatani, and Noriyuki Ozaki
  - *Modeling of Individualized Human Driver Model for Automated Personalized Supervision*, 2010, by Xingguang Fu and Dirk Söffker
  - *Designing Reusable and Scalable S/W Architectures for Automotive Embedded Systems in Driver Assistance*, 2010, by Dirk Ahrens, Andreas Frey, Andreas Pfeiffer, and Torsten Bertram

- **Communications**

  - *Integration of Car-to-Car Communications into IAV*, 2009, by Tae-Kyung Moon, Jun-Nam Oh, Hyuck-Min Na, and Pal-Joo Yoon
  - *Enabling Safety and Mobility through Connectivity*, 2010, by Chris Domin
  - *Time Determinism and Semantics Preservation in the Implementation of Distributed Functions over FlexRay*, 2010, by Marco Di Natale and Haibo Zeng
  - *Exploring Application Level Timing Assessment in FlexRay Based Systems*, 2010, by Sandeep Menon

- **Market and Consumer Preferences**

  - *Driver's Attitudes toward the Safety of In-Vehicle Navigation Systems*, 2009, by Andrew Varden and Jonathan Haber
  - *Consumer Attitudes and Perceptions About Safety and Their Preferences and Willingness to Pay for Safety*, 2010, by Veerender Kaul, Sarwant Singh, Krishnasami Rajagopalan, and Michael Coury
  - *Commercial Business Viability of IntelliDrive Assistance and Vehicle Safety*, 2010, by Robert White, Tao Zhang, Paul Tukey, Kevin Lu, and David McNamara

Consumers are increasingly concerned about the future of the world they live in and keeping themselves and their loved ones safe. Governments, OEMs, suppliers, policy and technical experts, and professional societies such as SAE and FISITA are responsible for providing the solutions for the future. We expect this publication will be part of their foundation.

Dr. Andrew Brown, Jr., P.E., FESD, NAE
2010 SAE President
Executive Director & Chief Technologist
Delphi Corporation

SPECIAL CONTRIBUTIONS

Advancing Safety in the Future: The Role of Technologies, the Government, and the Industry

By Dr. Joseph N. Kanianthra

Dr. Joseph N. Kanianthra
President, Active Safety Engineering LLC,
Former Associate Administrator for Vehicle Safety Research (Retired),
National Highway Traffic Safety Administration,
U.S. Department of Transportation

Advancing Safety in the Future:
The Role of Technologies, the Government, and the Industry

By Dr. Joseph N. Kanianthra
President, Active Safety Engineering LLC,
Former Associate Administrator for Vehicle Safety Research (Retired),
National Highway Traffic Safety Administration,
U.S. Department of Transportation

Significant gains have been made in safety during the last four decades in reducing fatalities and injuries occurring in traffic crashes. However, the reduced levels that exist today are not sustainable once the economy improves and normal driving is resumed by the public. Approaches of mandated regulation of occupant protection systems have worked well in the past. But that alone may not suffice to bring casualties down to a substantially lower level. Other approaches are being attempted in other countries. Active safety technologies hold promise for preventing crashes if robust systems are introduced into the fleet in significant numbers. However, no new strategies for accelerated deployment of active safety technologies have evolved to date. The underlying reasons are discussed in this paper together with suggestions on a new approach to address the problem, pointing out the role of industry and government in bringing it to a reality.

Introduction

In over four decades, there has been a concerted effort to reduce the carnage on U.S. highways, which have brought fatalities and serious injuries in motor vehicle crashes. The data show that fatalities have been reduced from 54,589 in 1972 to 33,808 in 2009. Similarly, the number of injured in traffic crashes also declined appreciably from 3.4 million in 1988 to 2.2 million in 2009. In the same period, the number of estimated police reported crashes reduced from nearly 7 million in 1988 to 5.5 million [1]. These are significant accomplishments due to the collective efforts of many people from the industry, the government, and safety promoters such as the Insurance Institute for Highway Safety (IIHS), the Advocates for Highway Safety, and other similar organizations.

Certainly, the number of motor vehicle safety standards that were promulgated by the National Highway Traffic Safety Administration (NHTSA) has had a significant impact on these statistics. This impact still continues as the safety countermeasures that have been implemented and that are phased in proliferate throughout the vehicle fleet. In spite of the encouraging numbers, based on 2009 data, the fact remains that over 33,000 people are dying every year in motor vehicle-related crashes and many more are seriously injured in those events.

The historical data show two sets of information: the fatalities in motor vehicle crashes and fatality rates, calculated as fatalities per 100 million vehicle miles traveled (VMT). The fatality trends during the last four decades show that it is cyclical, starting from the early 1970s at an all-time high of nearly 55,000 lives lost to a low of 44,525 in 1975, another low of 42,589 in 1982, and an all-time low of 39,250 in 1992. Since then the number of lives lost steadily increased with very little fluctuation through 2007.

It is noteworthy that approximately at the end of every nine- or 10-year cycle, a low point for fatalities is reached. However, it is also evident that the peaks reached in each cycle after each of these lows is slightly lower than in the immediately preceding cycle. Thus, there is an overall decreasing trend in fatalities seen in the last four decades.

Since 2007, the fatalities have decreased substantially by 9 and 10% in 2008 and 2009, respectively. Similarly, when the fatality rates per 100 million miles traveled are examined, they dropped from 4.33 in 1972 to 3.17 in 1981, from 2.76 to 1.91 between 1982 and 1991, from 1.75 to 1.51 between 1992 and 2001, and from 1.51 to 1.13 between 2002 and 2009. The average fatality rates for the above four periods are 3.5, 2.38, 1.65, and 1.38 respectively, clearly showing a downward trend in those rates as well.

However, this trend has slowed down in recent years, indicating a shallower slope except in the last couple of years.

The cyclical behavior of the data, and its underlying reasons, can only be explained if it is analyzed thoroughly to identify the causes. One could surmise that there are many reasons for it, one of which could be the economic conditions and its relationship to selective use of vehicles by the motoring public. The sharper decline of fatalities in recent years may also be due to the many regulated and other safety countermeasures becoming available in most of the vehicles that exist in the fleet today.

It is recognized that some of these regulated countermeasures will continue to provide benefits until the entire fleet is so equipped. Several of those countermeasures are already present in a significant portion of the fleet. Examples of these are frontal airbags, dynamic side-impact protection, and upper interior head impact protection. If these are the likely reasons for the lower fatality numbers seen recently, then it is quite possible that as economic conditions improve, and as the full effect of the safety countermeasures of the last several years take hold, the new fatality and injury numbers may reach a new normal that is lower than that seen in the past but still higher than what it is today.

Much of the safety improvements during the last four decades have come about from the various safety countermeasures for occupant protection such as improved structures to preserve occupant compartment integrity, frontal and side crash protection, rollover protection, increased roof strength, seatbelt improvements and their increased use, and greater use of child restraint systems. Many of these have resulted from the regulatory approach and the safety campaigns of the government, the safety rating schemes promoted by IIHS and the government, and the improvements made by the automotive industry. It is claimed that airbags alone have saved 28,000 lives from their inception until now.

The obvious question that must be raised is whether we will be able to sustain the gains of the last 40 years for the future without some new strategies, and if so, what are those and where does the responsibility for them lie? It is clear that we have reached a point of diminishing returns as far as safety countermeasures such as the ones already promulgated for crash protection and the limited prevention systems are concerned.

There may still be further improvements possible in some of those items and in the features of seatbelts and airbags already found in today's vehicles. For example, belt use has not reached its highest possible level in spite of intense efforts. Improving belt fit, configuring belts for better comfort and flexibility in buckling from the left or right, and other approaches might improve their use and efficiency. Similar improvements could be tried in passive restraints as well. However, would that be sufficient to wipe out any increase in casualties that may result in the future? Don't we also need to drive the casualty numbers down even further to a reasonable level ultimately? These are the questions that must be raised, if the U.S. is to be the safety leader once again.

Safety goals for the future

The approach of the U.S. government's efforts in safety has always been to identify the safety problems in the real world, prioritize them, and find the necessary solutions that are feasible to implement in a cost-effective manner. This has worked well in the past, and many of the easy solutions have already been found. The remaining problems are far more complex to find answers for. However, many jurisdictions are going far beyond the above in establishing clearly defined safety goals that have the potential for periodic self-assessment.

For example, the "Vision Zero" program established by Sweden is based on the objective that no one should be killed or seriously injured in road traffic there by 2020 [2]. Similarly, Japan established in its 8th program for traffic safety from 2006 to 2010 that the number of fatalities were to be reduced to no more than 5,500 by 2010 [3]. Such goal setting may be important in light of the finding by TRB that the U.S. is lagging behind in reducing traffic deaths and injuries compared to other high-income nations [4]. Occasionally, goals are set in the U.S. as well but not seriously tracked. As safety gains are likely to come at a slower pace in the future, it is all the more important that thoughtful goals are set and tracked for measuring their success.

As mentioned earlier, due to improvement in the economy and the likely increase in exposure due to more driving, it is a possibility that an increase in fatalities and injuries will occur in another year or two. The early assessment figures for 2010 are already showing a 2.5 percent increase for the third quarter in comparison to the figures for the same quarter in 2009. Since many of the important crash protection measures are already available in many vehicles, the fatality and injury numbers are highly unlikely to reach the levels of 2005. However, it is hard to predict where those levels will be because of the uncertainties of economic impact, fuel prices, inflation, and other factors.
One thing is certain: continuing along the same path that we have been on for the last 40 years is not going to get us to a point where significant reductions in casualties can be expected.

Assuming that in the U.S., the lowest level of annual fatalities in its history of 31, 000 for 2010 (when that data become available) is reached – an optimistic figure –it is quite possible that approximately another 80 fatalities per year may be added, due to the assumed growth in vehicle miles traveled at the rate of 0.25 percent, and approximately 300 more lives lost as economic conditions improve, thus bringing the total to an average increase of nearly 400 lives annually. Therefore, by 2020, it is possible that we may see 35,000 fatalities per year.

The same would be true of injuries. If our goal is to keep the baseline number for 2010, namely, 31,000 fatalities even in 2020, it means we have to get an annual average reduction of 400 fatalities. This is an impossible task without new approaches to enhance safety.

Why active safety

Figure 1 shows the fatality rate per 1000 crashes and injury rate per 10 crashes calculated from vehicle occupant fatality numbers and total number of police reported crashes for the period from 1988 to 2009. Fatality rate from 1988 to 2007 is in the range of 4.93 to 5.38 per 1000 crashes with an average of 5.19 fatalities per 1000 crashes. For 2008 and 2009, these numbers have dropped to 4.61 and 4.45, respectively. Similarly, the number of crashes per million vehicle miles traveled was also calculated. This shows in 1988, the crash rate was 3.4 crashes per million miles. It declined steadily in the last 21 years, even though at a much slower pace, with the crash rate for most of this period hovering around 2 crashes per million miles, except in 2009 when it was 1.85. Though this decline has been slow, it is still noteworthy. Its reason can only be speculated that, overall, vehicles have generally improved their crash prevention performance due to braking, handling, and other characteristics along with additional improvements in crash protection.

In spite of all the safety improvements that have come about, most of which have been in the area of occupant protection, the fatality rate per 1000 crashes has remained nearly a constant for 20 years. During this period, significant crash safety protection devices have been incorporated into vehicle designs. This means that without these devices the fatality rates could have gone up. However, these and other future improvements in crash protection alone may not be sufficient to make a dent in the anticipated casualties in the future.

The question remains as to why there is a decline in the fatality rates in 2008 and 2009. One of the reasons is due to the overall reduction in the fatality numbers. The number of crashes also has decreased in those years. The reduction in fatalities for those two years is in the range of 10 to 12%, while crashes have reduced only about 4 to 5%. As mentioned before, once the economy improves and normal vehicle use is resumed, this rate may increase to a level somewhere between the current level of 4.45 and the previous level of 5.19. The same may be true of injury rate as well.

The injury rate trend, on the other hand, shows a decrease much earlier than the fatality rate, starting from 2000. The reason for this may be that fatalities occur in more severe crashes generally, and the safety countermeasure effectiveness in fatal crashes and less severe injury-producing crashes may not apply equally to both crash types. Therefore, even at a lesser penetration rate in the fleet, some countermeasures may have helped in reducing injury rates and not the fatality rate until a significant number of vehicles in the fleet are equipped with those countermeasures.

This means that if we are to have significant reductions in fatalities and injuries in the future, the only way would be by reducing the number of crashes. As vehicle miles traveled steadily increase and as normal driving habits are resumed, any significant reduction in total fatalities can occur only if new and bold initiatives are adopted for preventing crashes.

Role of technologies in crash prevention

It is a well-known fact that many of the causal factors in crashes are attributable to drivers failing to act to avoid a crash at the appropriate time. Driving is a benign activity until such time that a threat situation demands proper action from the driver. Until then, all that is required of the driver is to pay his/her full attention to driving, scanning the surrounding traffic, keeping eyes on the road, and keeping the vehicle under control at all times, poised to act when necessary. NHTSA's research on causal factors of crashes has shown that driver recognition errors (44%) were the leading cause of crashes. These include driver inattention, distraction, improper looking at surroundings, obstructed vision, roadway-related problems, and the like. Another factor is driver-decision errors (23%). They include misjudged speeds, tailgating, speeding, as examples. A driver's erratic actions and physical impairment combined (drunk and drugged driving, drowsiness, and illnesses) constitute another significant factor (23%) in crash causation [5]. This means that the primary causal factors in crashes are driver-related and, as such, human deficiencies in driving are the root cause of the problem.

If vehicle technologies are able to assist drivers in overcoming their own deficiencies in driving, then we have the potential to prevent crashes in many situations. NHTSA's research starting in the early 1990s under the Intelligent Transportation Research program was based on the hypothesis that advanced technological solutions can be found to assist drivers through appropriate and just-in-time warnings that prod drivers to act to prevent a crash and to help, if necessary, even in avoiding the crash through steering and braking assistance.

Such systems require the use of sensors, vision systems, computers, and algorithms that have to work in concert in sensing impending crash situations accurately, making appropriate decisions, and by providing the driver with the necessary "ammunition" to prevent a crash. It has to be done in such a way that the systems' decisions are flawless, actions are acceptable to the driver, and there are no unintended consequences. At the same time, they have to be affordable and must remain reliable and trouble-free throughout the life of the vehicle.

Fatality and Injury Rates

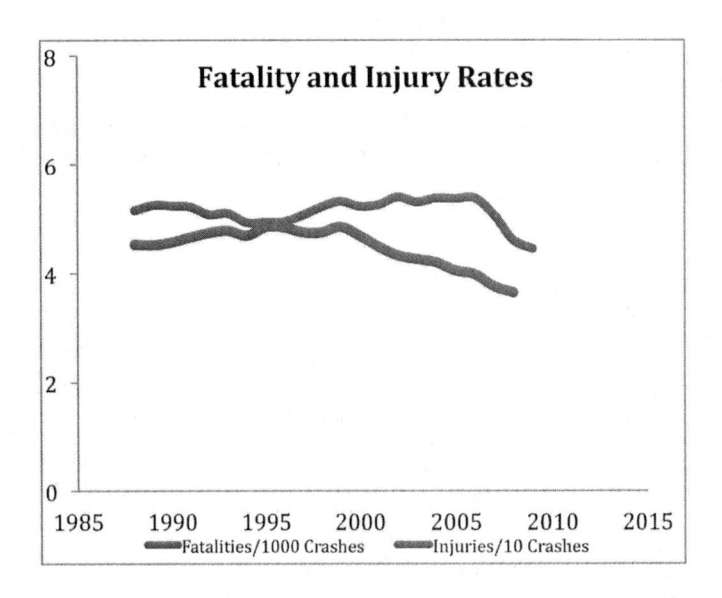

Figure 1   Source: NHTSA, Traffic Safety Facts

Their effectiveness will depend upon how successfully the driver-vehicle system can function in roadway and traffic environments to avoid crashes. One concern that is often raised by vehicle designers relates to the performance of these systems as they are placed in the hands of different drivers with varying driving behaviors, driving capacities, and levels of driver-vehicle operational skills and experience. Human factor issues become of paramount importance under these circumstances.

Therefore, research in the area of human factors is of great importance, and thorough evaluation of the technologies must be conducted. Simulator studies, test track testing, and short-term real-world operation of vehicles with and without active safety systems will yield the necessary data for determining the effectiveness of the system, proper functioning of the technologies, assessing false positives and negatives as well as customer acceptance issues.

Potential technologies described above are not only useful in preventing crashes but in enhancing protection as well. For example, sensors and vision systems could provide useful information from frontal crash prevention systems for enhancing the effectiveness of airbags and seatbelts by improving the precision in triggering airbag firing, and tailoring its inflation rate to suit the occupant and match likely crash severity. Thus, incorporating advanced technological devices in vehicle design, safety could be addressed not only in crash prevention but also in every other phase of the crash event in a continuous sweep. The effort could move in a step-wise fashion from crash prevention to severity reduction, to injury mitigation and medical attention, using advanced technologies to meet needs at every step. [6] However, the focus in this paper is on active safety and not on passive safety.

Integrated Safety: The continuous sweep approach

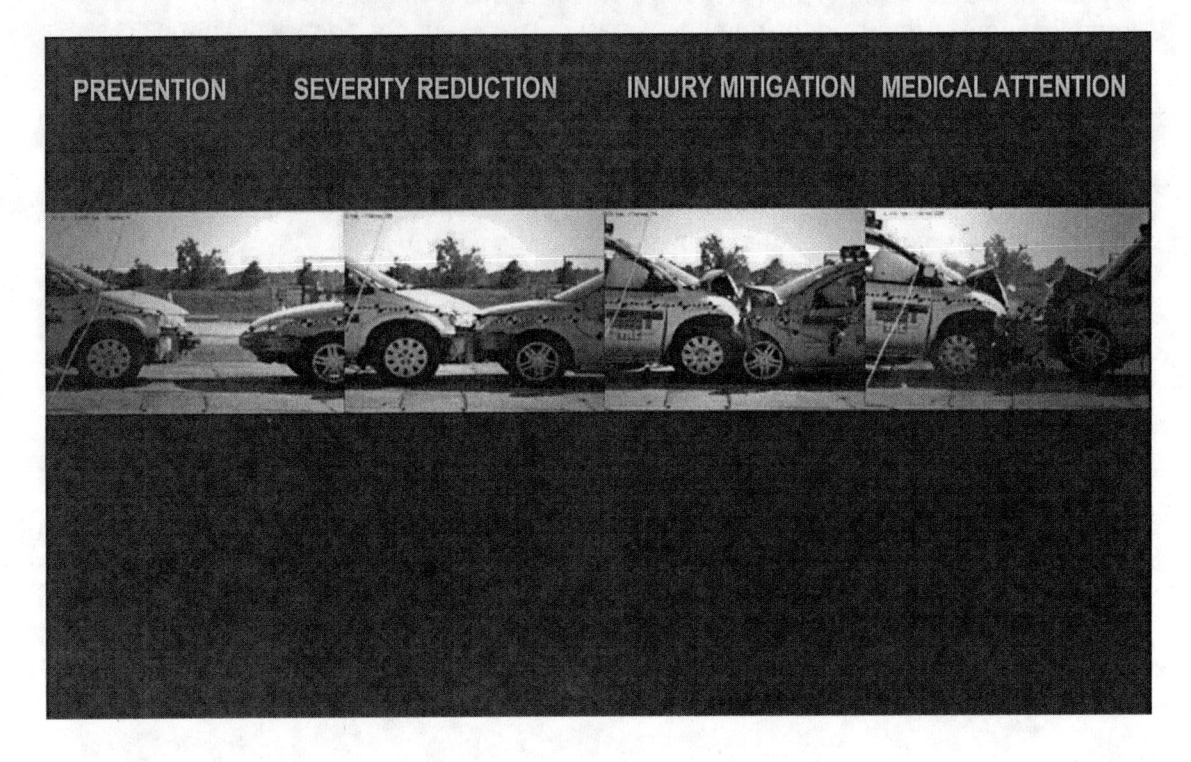

Figure 2   Source: NHTSA

While the concept of active safety technologies for safety enhancement is valid, it cannot be ensured that the system will perform as designed in every possible crash scenario on the road. However, once it has been attempted, even if one fails under some scenarios, even when crashes happen, its severity would have been reduced.

For more than 15 years, NHTSA has actively pursued research along these lines in collaboration with industry, automotive suppliers, and academic institutions. It has shown that the concept is sound, and the crash prevention systems installed in vehicles can indeed be effective, to some extent, in solving problems in many different crash situations. Automobile manufacturers have brought forward several of these safety technologies and crash prevention countermeasures in selected models. Yet, they have not widely offered such features in a large number of vehicles in the fleet in a large enough measure so as to have a significant impact on reducing crashes, fatalities, or severe injuries.

Active safety system deployment

In much of the work undertaken in NHTSA, though it was with the intent of safety improvements by reducing the number of crashes that occur, the agency did not have a clear deployment plan as it launched the research activities.

Traditional rulemaking approaches cannot be directly applied in active safety systems because of the difficulty in having to regulate "driver-vehicle" performance. In the case of crash protection, the regulations apply to vehicles only, and the burden of it is entirely on the manufacturer for meeting the requirements of such a regulation. Any type of regulation that is brought forward for crash prevention has to have a clearly defined associated driver model. In the absence of that, traditional regulatory approaches are not useful for developing regulations for active safety systems. There must also be proper test and evaluation procedures based on real-world crash scenarios. Developing an average driver model is not an easy task. However, it is important that researchers in the safety community begin the process of developing such model as well as test and evaluation procedures.

Meanwhile, to accelerate deployment of active safety systems, a New Car Assessment type Program (NCAP) is best suited for deploying active safety technologies. NHTSA has already started acknowledging the presence of certain crash prevention technologies based on its research and testing done under the Intelligent Vehicle Initiative. But it is only a start in the right direction and there is a long way to go.

It is noted that rating schemes have to meet certain requirements before they are applied to crash prevention systems. First, each system must have a safety problem focus such that it is demonstrably able to act as a countermeasure for each type of real-world crash occurrence. Second, they have to be technology neutral and not geared to any specific technologies for sensing, vision, and other systems. Third, they should have the highest reliability possible with very little or no unintended consequences or false positives and false negatives in their operation. Fourth, it should be possible to determine the system performance effectiveness in objective tests, with the ability to discriminate performances of similar systems that use different technologies to address the same problem. Additionally, these driver-vehicle systems must have "fail-safe" provisions to ensure that backup systems are available and will work under nearly all conditions. At the same time, they must be cost-effective.

At this time, there are very few systems available that are able to meet all the above criteria. Regulatory authorities have not yet found a way to develop satisfactory test methods that are objective and tailored for various crash scenarios that are found in real-world crashes of each type. Collaborative efforts with the industry are under way for this but its progress is too slow. Therefore, no rating system has yet been developed in the U.S. with the net result that no fast-paced deployment of active safety systems is occurring. On the other hand, Europe is moving ahead. While active safety technologies have been under development for nearly 20 years, the slow pace with which deployment is moving makes it difficult for automotive suppliers to invest in their further development without some added incentive. Manufacturers themselves are in a similar situation because the limited systems they have developed are not yet found to be very popular. The reasons vary: cost, customer unfamiliarity, lack of government endorsement through regulations or ratings, and the amount of perceived risk by vehicle owners.

Therefore, it is highly unlikely that active safety systems will find their way into the fleet in large numbers soon unless all stakeholders – the government, the automobile industry, and consumers – have a common interest in finding a way to promote their proliferation.

Role of the government

In spite of the risk involved without the guarantee of safety benefits in the real world, the only option the government has is to estimate the potential benefits based on any data NHTSA can generate through its own research efforts, and use it for rating. In order for it to rate vehicles, effectiveness and benefit estimates for each safety problem that is addressed by different technologies are needed. Those benefits could then be the rating criteria used.

NHTSA has already completed crash scenario breakdowns [7] based on the data the agency collects. Now it is a matter of developing a set of test procedures based on the crash scenarios. If a driver model could be developed, initial evaluations could be on the basis of simulator testing using the "average" driver and using computer simulations. Some limited test track testing could be used for confirmation purposes. To supplement these, there are the field operational test data that have already been obtained. Since it is a rating procedure, after observation of the performance of vehicles thus introduced into the fleet for a few years, should the rating turn out to be totally fallacious or only even borderline acceptable, these ratings could be withdrawn or changed to a different rating based on the latest information. NHTSA could offer preliminary guidelines regarding speed ranges in which certain systems should function, and items such as the level of false alarms permissible and redundancy for ensuring that the chance of failure of technologies is minimal. Since ratings are not federally mandated standards, withdrawal of the rating may be easier than changing or dropping a standard. It still may need notice and comment before changes are made.

Such a different process may require changes to the existing NHTSA statute itself, and public meetings and other processes are also available at NHTSA's disposal.

The public are the least informed about the potential of active safety technologies in preventing crashes. The government could play a role in educating the public about the benefits and some of the difficulties drivers may face without developing familiarity with the systems. Coming from the government, this would have greater credibility than the industry trying to educate the public alone.

Constant monitoring of the systems' performance in the field and potential flaws of such systems will provide the public necessary confidence in the system that will attract more and more people to the safety potential that active safety technologies have to offer.

Role of the industry

Automobile manufacturers and suppliers have a larger stake in promoting advanced technologies. They have to make sure that active safety technologies are thoroughly researched and their potential properly evaluated. That responsibility cannot be passed on to the suppliers alone. Such research should include human factor issues, customer preferences, customer acceptance, and consumer education of advanced technologies. Manufacturers should also join together with technology suppliers to develop realistic test procedures for testing technologies for various crash scenarios to determine their effectiveness. These would be at the precompetitive stage so that the proprietary nature of the countermeasures will not be compromised. Agreed-upon test procedures could be presented to NHTSA for their evaluation and determination of their suitability for rating purposes. If necessary, NHTSA may fine-tune those procedures as needed. This may shorten the time it takes for NHTSA to develop the test procedures on its own.

In order to assess potential benefits, manufacturers could equip a few vehicles with the advanced technologies they wish to evaluate and place them in the hands of volunteer drivers. These vehicles should have event data recorders that have the capability to acquire a lot more channels of information that are critical for evaluation. If they also are equipped with cameras (drive cam) and only severe crashes and near misses that would have been severe are captured, the volume of data acquired will be manageable for analysis purposes. Comparing these data points with other vehicles' data without crash prevention technologies could provide valuable information.

Concurrently, manufacturers should also seek the help of the insurance industry and other partners to offer discounted insurance rates for crash prevention systems, as they stand to benefit from reduced payout due to potential decrease in damage claims and injury claims. Manufacturers and suppliers could team up and conduct tests using procedures that are generic and not technology-specific, advertising the estimated benefits and expected ratings based on those tests.

Vehicle manufacturers have to make sure the customers accept the manufacturers' claim regarding the safety potential of their products. For this, sharing of their evaluation data with consumers would be necessary. If manufacturers have made a good faith effort to avoid potential flaws and shortcomings, customers will accept that initially. Their continued faith in the systems will very much depend upon the customers' own experiences. Incentivized offerings to cover guaranteed repairs in case of problems and insurance coverage for liabilities may also attract consumers to the products. Customer complaints must be attended to promptly, and their feedback sought and acted upon quickly.

Conclusions

Active safety technologies hold enormous promise to enhance traffic safety in the U.S. However, the pace at which technologies are being introduced into the fleet and the lack of urgency in the government and industry to move ahead faster will likely result in increases in fatalities and injuries once the economy improves.

If a serious attempt is made to deploy these systems with government's help, there is a good chance crashes will be prevented and public safety improved. It may even be possible to set ambitious goals for the future to substantially reduce the fatalities and injuries that are happening today. The approaches that are implemented have to be jointly developed by government and industry working together. This requires a paradigm shift from the customary process of mandated regulations of the past 40 years. The motoring public will be the ultimate beneficiary of the accelerated introduction of active safety technologies and, to use President Obama's words, a way to "win the future" in safety.

List of References
1. National Highway Traffic Safety Administration's Traffic Safety Facts 2009 (DOT HS 811 392 and 2001 (DOT HS 809 484)
2. Kevin E. McCarthy, The Swedish "Vision Zero" Program, OLR Research Report, 2007-R-0635, November 2007
3. Road Safety in Japan, International Association of Traffic and Safety Sciences, Japan
4. TRB Special Report 300. Achieving Traffic Safety Goals in the U.S. Lessons learned from other nations, 2010
5. Kanianthra, J., Mertig, A. International Symposium on Real World Crash Injuries, Leicestershire, U.K., 1997
6. Kanianthra, J. Re-inventing Safety: Do Technologies Offer Opportunities for Meeting Future Safety Needs? Paper Presented at the SAE Convergence Conference, Detroit, MI, 2006
7. Wassim, Najm G., et. al., Pre-Crash Scenario Typology for Crash Avoidance Research, DOT HS 810 767, 2007

Road Safety – A Better Way Forward

By Patrick Lepercq

Michelin – Corporate Vice President Public Affairs
and Chairman of the Global Road Safety Partnership

Road Safety – A Better Way Forward

By Patrick Lepercq
Michelin - Corporate Vice President of Public Affairs and GRSP Chairman

The number of road users is growing each year worldwide. Unfortunately, accidents are one of the heavy tolls that society pays for the rapid development of the modern world.

Today, according to the World Health Organization (WHO), motor vehicle crashes kill about 1.3 million people each year, the equivalent to the deaths caused by communicable diseases. Between 20 and 50 million people are severely injured in road accidents each year. The number of fatal victims is set to rise to 2 million by 2020 unless new safety measures are taken, making road traffic injuries the third largest cause of death and disability as well as the first cause of death among people aged 17 to 29.

These numbers don't tell the full story. They are too abstract to fully grasp, and so we try to make sense of the scale with comparisons. As Russian President Dmitry Medvedev told government ministers assembled for the first Ministerial Conference on Road Safety in Moscow in 2009, it's as if each year the equivalent of a midsized city quietly dropped off the planet.

Road crashes are a financial burden, costing up to US$500 billion annually – the same as the GDP of Switzerland.

Nearly 90% of fatalities occur in middle and low-income nations despite the fact they account for under half of the total number of vehicles. These are the countries where the number of vehicles will increase the most, and where the mix of vehicles and vulnerable road users also makes the risk greater. As traffic intensifies, the toll will get worse.

The scale of the problem is demonstrated by significant average increases in the rate of road deaths in developing regions between 1986 and 1995. During this period, it increased 40% in Asia Pacific, 26% in Africa, 36% in the Middle East and North Africa, and 16% in Brazil. In short, the situation is an escalating health, social, and economic disaster for developing and transition countries. The public health sector is creaking under the strain. In many developing countries, road crashes represent an unwelcome drain on medical resources. Crash victims occupy up to 10% of hospital beds, 10 times the rate of the U.K.

Road safety throughout the world

In general, even though today's roads still claim far too many victims, there remains reason for hope. For the last 10 years, a lot of initiatives have been introduced and numerous efforts have been made in order to enhance road safety in the world.

In May 2004, the WHO facilitated the development of a group of UN and other international road safety organizations – now referred to as the UN Road Safety

Collaboration (UNRSC). The UNRSC comprises over 40 UN and international agencies working on road safety, with a broad range of skills and experience from the transport, health, and safety sectors, and representing governmental and nongovernmental organizations, donors, research agencies, and the private sector.

The UNRSC gathers biannually, with meetings alternating between WHO headquarters and the UN Regional Commissions offices. One of the major projects of the UNRSC is the production of a series of "how to" manuals that will help governments implement some of the recommendations of the World report on road traffic injury prevention. The manuals are practical and user-friendly, providing step-by-step guidance on implementing specific interventions. Four of the six manuals focus on key factors identified in the World report on road traffic injury prevention, namely, helmets, seatbelts and child restraints, speed, and drinking and driving. The other two will address the establishment of a lead agency on road safety and traffic/injury data collection.

The first Global Ministerial Conference on Road Safety in November 2009 became a historical turning point in enhancing international awareness of the road safety issue. All stakeholders – ministries, specialized agencies, international bodies, associations, and others – met in Moscow for the conference organized by the WHO and the Russian Ministry of Transports.

They drafted a common declaration asking the United Nations General Assembly of March 2010 to pronounce a decade of action (2011–2020) for road safety, and made commitments to simple and proven solutions. On March 2, 2010, the UN General Assembly effectively proclaimed 2011–2020 the "decade of action for road safety," endorsing the conclusions of the Moscow conference and stating the necessity of responding to the needs of all road users. The goal was clear: stabilize and then reduce the forecast number of deaths from road accidents.

The UNRSC has developed a Global Plan for the Decade of Action for Road Safety 2011-2020, with input from many partners through an extensive consultation process. The plan, which provides an overall framework for activities, includes the following categories: building road safety management capacity; improving the safety of road infrastructure and broader transport networks; further developing the safety of vehicles; enhancing the behavior of road users; and improving post-crash care. Indicators have been developed to measure progress in each of these areas. Governments, international agencies, civil society organizations, the private sector, and other stakeholders are invited to make use of the plan as a guiding document for the events and activities they will support.

The "White Paper for Safe Roads in 2050" became a significant contribution to the action plan for the Decade of Action for Road Safety made by the Road Safety Task Force. This task force was composed of representatives of private companies, academia, government, and international organizations that collaborated during the global road safety round table and the plenary session of Challenge Bibendum Rio 2010. These experts took up the challenge of finding practical solutions for achieving zero work-related deaths by 2050. Addressing work-related safety as a springboard toward universal road safety, the task force drafted a global roadmap that centered on key goals and actions for the decades from 2020 to 2050. It indicates that, as employers, private companies have a vital role to play in improving road safety.

In 2009, the WHO founded Youth for Road Safety (YOURS), the first global youth nongovernmental organization (NGO) specifically focused on road safety. YOURS is a unique youth-led organization that acts to keep young people safe on the world's roads.

Partnerships in road safety

Partnerships are an effective way to tackle the road safety challenge. Progress in this issue is only possible through the participation of all involved: the public sector, the private sector, and civil society including the overall local community.

Within a country it is helpful to consider the three of them when examining roles in road safety: government, business, and civil society organizations – often termed nongovernmental (NGO) or not-for-profit organizations. In addition, at the international level are the multi- and bilateral development agencies and UN organizations, which provide advice and resources generally through a national government.

There is growing recognition that partnerships between business, government, and civil society organizations offer the possibility of more innovative and sustainable solutions to development issues, such as death and injury on the roads as a consequence of rapid motorization.

As with all aspects of public safety, government has the final responsibility for creating an appropriate legislative and social environment to enable road safety to be improved on a continuous basis. But other stakeholders can influence this – whether they are NGO pressure groups representing road crash victims or a business concerned about the health and safety of its employees. Government also plays the principal role in providing the basic road infrastructure and operational services such as the traffic police. Sadly, in many low-income countries, governments often lack the political will or the financial and professional resources to tackle road safety issues effectively on their own.

The Global Road Safety Partnership (GRSP), which brings together governments and governmental agencies, the private sector, and civil society organizations to address road safety issues in low- and middle-income countries, was launched in 2009. GRSP is a hosted program of the International Federation of Red Cross and Red Crescent Societies (IFRC), based in Geneva. It aims to find more effective and innovative ways of dealing with road safety in developing and transition countries. Through a comprehensive approach to road safety, GRSP partners collaborate and coordinate road safety activities. This approach aims to build the capacities of local institutions, and enhance the ability of professionals and communities to tackle safety problems proactively.

The Global Road Safety Initiative (GRSI), on the other hand, is funded by seven of the world's largest automotive and oil companies that have committed US$10 million to road safety. GRSI was launched as a five-year program in 2005, and the companies involved were Ford, GM, Honda, Michelin, Renault, Shell, and Toyota. A similar program has been extended for another 5 years (2010-2014) and is being implemented as well by GRSP.

Good examples of the effectiveness of a partnership approach are the diverse road safety projects that have been implemented in the city of São José dos Campos, in the state of São Paulo, Brazil. The most effective ones are related to road safety education, engineering, and enforcement. Proactive actions in these areas led to a decrease of the main rate (deaths + serious injury per 10,000 motor vehicles) from 14.04 in 2007 to 12.49 in 2008. This improvement is the result of the partnership between São José dos Campos and the GRSP, which supported the implementation of the Proactive Partnership Strategy (PPS). Such a strategy also enabled the municipality to gather a systemic model of crash data and services related to their prevention.

Joint efforts of public and private sectors

Roads are generally publicly owned assets managed and controlled by central or local government. Although roads are mainly used by private individuals and businesses, traditionally road safety has been seen as a responsibility of government, linked to their ownership of the infrastructure. This traditional view, however, is changing. The understanding that collaboration in road safety is a win-win business case, both for public and private sectors, is now growing.

Governments are interested in committing resources to this issue, in cooperation with the business sector for the following opportunities:
- Ability to gain political credibility and to improve relationships with communities.
- Greater transparency.
- Learn of more appropriate and replicable technical, social, and institutional solutions.
- Leverage funding.
- Turn lessons into legislation.
- Disseminate knowledge and new skills.

The main reasons for business sector participation in road safety are:
- Cost savings: large fleet operators can save substantial sums by reducing the number of times their vehicles are involved in road crashes. They may also reduce vehicle operating costs and insurance premiums as a result of safer driving practices.
- Market development and branding: some businesses are willing to demonstrate the value of their safety products to persuade others to use them. Likewise, some businesses use their reputation for safety as part of their brand identification.
- Legal requirements under occupational health and safety laws: many high-income countries have laws and regulations that treat business vehicles as part of the workplace. The vehicles therefore have to be of safe design, well maintained, and safe to use. Many businesses voluntarily go beyond the health and safety requirements imposed by national legislation. They may even adopt the goal of having no work-related accidents, including road crashes, and apply this to all their employees and subcontractors.

- Corporate social responsibility: most global businesses now recognize that they have social responsibilities in countries where they operate. Road safety activities often take place under this heading.
- Company reputation and quality assurance: some businesses view fleet safety as part of total quality management. This involves influencing the need for road transport, how road transport is implemented, and the choice of vehicles and equipment.

NGOs are of many kinds, from local to international, and with objectives ranging from promoting single issues to broadly based humanitarian organizations such as the Red Cross/Red Crescent movement, or associations such as the International Road Transport Union (IRU), which represents the road haulage industry at the international level.

Opportunities for NGOs include:

- Strengthened and expanded access to influence public or private policy.
- Capacity building and fostering enabling environments for communities to voice their needs.
- Implementation of projects that may not otherwise have been completed.
- Scaling up, replicating, and expanding existing programs.
- More sustainable, long-term income.

GRSP and Challenge Bibendum

GRSP is one of the most significant contributors to the struggle against the worldwide crisis of millions of vulnerable people being killed and injured in traffic accidents in the developing world. It has already established effective platforms in several countries that support sustainable projects, based on known successes and owned by local stakeholders. After fewer than three years in operation, this is a significant achievement. The main element of this progress is the fact that GRSP's approach between business, civil society, and government has tapped into new capacities, and combined them as a credible alternative response to this road safety crisis.

During the partnership's initial phase, resources were concentrated on raising awareness so support could be mobilized to tackle what is a "silent" developmental disaster that attracts only fragmented international attention. But now GRSP has moved ahead and is coordinating a global response, which it is piloting in selected developing countries through projects best suited to local conditions. By testing various models for road safety partnerships, GRSP is helping partners adopt solutions that best fit their needs. The GRSP program in Ghana, for instance, is quite different from the one in Bangalore. Despite structural and operational distinctions between countries, the whole GRSP program is united toward one goal: the building of working alliances that reduce road deaths and injuries.

GRSI, which is being implemented by GRSP, focuses on the critical road safety issues identified in the World report on road traffic injury prevention (2004, WHO and World Bank). These include pedestrian safety, drinking and driving, helmet use, speed management, and seatbelt use. GRSI will build upon the good practice guides on these issues being developed by GRSP, WHO, World Bank, and FIA-Foundation under the auspices of the UN road safety collaboration. It will provide training to road safety professionals in developing countries and seed money to support pilot projects to improve road safety in these areas.

In addition, GRSI will help build the capacity of developing countries to reduce traffic fatalities and expand GRSP's capability to deliver road safety improvements in line with the recommendations of the World report on road traffic injury prevention.

GRSI targets three regions or very large countries with substantial road safety problems. The first to be confirmed was the ASEAN region, building on the regional and national road safety plans that have been developed with the support of the Asian Development Bank. China is the second and Brazil is the third.

Another fine example of a joint effort to change the situation in road safety for the better is Challenge Bibendum, which is a forum to address all the issues of a multi-stakeholder sector such as road mobility. The goal of this event, organized by the Michelin Group and partners, is to push transformations and invent more sustainable road mobility. Challenge Bibendum combines technical tests and evaluations, demonstrations, and test drives to measure the progress of vehicles, energies, and technologies. Technologies and trends are debated in forums and conferences. Private and public sector opinion leaders and decision-makers gauge the advances and future sustainable energy paths for a better mobility.

The scale of the complex change needed at a global level is massive. Its undertaking will require time to implement so we must start now. We have the technologies, and we have the ability to envision new realities and to craft effective strategies.

Traffic Safety: International Status and Active Technologies for the Future

By Dr. Dinesh Mohan

Volvo Chair Professor for Biomechanics and Transportation Safety at the Transportation Research and Injury Prevention Programme of the Indian Institute of Technology, Delhi

Traffic Safety: International Status and Active Technologies for the Future

By Volvo Chair Professor for Biomechanics and Transportation Safety
at the Transportation Research
and Injury Prevention Programme of the Indian Institute of Technology, Delhi

The WHO released the "Global Status Report on Road Safety: Time for Action" in July 2009. This report is the first broad assessment of the status of road safety in 178 countries. The data were obtained from national governments using a standardized survey form. The status report shows that low-income and middle-income countries on an average have higher road traffic fatality rates (21.5 and 19.5 per 100,000 population respectively) than high-income countries (10.3 per 100,000) and that over half of those who die in road traffic crashes are pedestrians, bicyclists, and users of motorized two-wheelers (MTW). Here we analyze the data reported, which include a vast majority of the middle- and low-income population of the world, to understand the injury trends by national income and modal shares of traffic in different societies. These data are used to propose road safety countermeasures and policies that may be necessary to accelerate the reduction in road traffic injuries (RTI) in the future.

Current status

Figure 1 shows the country reported and WHO estimates for RTI fatality rates per 100,000 persons plotted against national per-capita income for 177 countries. Only 85 (48%) of the countries have reported fatality rates within 5% of the WHO estimates, and 80 countries (45%) reported much lower numbers with 55 reporting numbers less than half that estimated by the WHO. For seven countries, the WHO estimates were lower than those reported. The WHO status report also gives 90% confidence intervals for fatality estimates. More high-income countries seem to have reported rates close to WHO estimates than low-income countries. However, it is interesting that both low-income and high-income countries can have underreporting and realistic reporting. For example, both a high-income region such as United Arab Emirates (UAE) and low-income Kenya seem to have significant underreporting, and both low- and middle-income countries (e.g., Vietnam, Iran, South Africa, and Argentina) and high-income countries (e.g., Netherlands, Qatar, U.S., and U.K.) report fatality rates close to the WHO estimate. Therefore, it appears that it's not necessary to have high-income levels to develop reliable RTI reporting systems as commonly assumed.

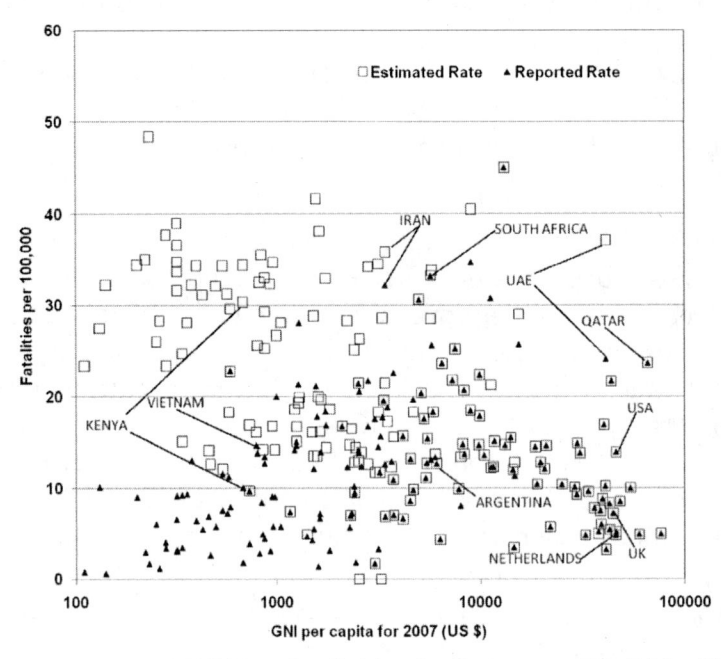

Figure 1. Reported and estimated (WHO) road traffic injury fatality rates per 100,000 for different countries.

Figure 1 also shows that national RTI fatality rates per 100,000 persons (reported or WHO estimates) do not have a high correlation with national incomes. The WHO estimates seem to have a lower correlation than the rates reported by individual countries. Some high-income countries such as United Arab Emirates (UAE), Qatar, and the U.S. have higher rates than middle- and low-income countries such as Argentina and Vietnam. At all income levels, countries with the highest fatality have values three to five times higher than the ones with low rates. For example, among high-income countries, the U.S. has a fatality rate that is three times greater than those with the lowest rates (e.g., Netherlands).

This suggests that higher national incomes do not necessarily produce better road safety policies. This is contrary to the widely held belief that RTI rates are highly dependent on per capita incomes (Kopits, Elizabeth & Cropper, Maureen, 2005). This is probably because all earlier analyses depended on official fatality rates as reported by individual countries. The WHO estimates for low-income nations are generally higher than country reports, whereas for high-income countries the two estimates are generally closer (with some exceptions).

Vulnerable road users

The "Global Status Report" indicates that the number of fatalities involving vulnerable road users (sum of pedestrian, bicycle, and motorized 2/3-wheelers) is greater than 20% in all counties except one, whereas 14 countries report less than 20% share for motor vehicle fatalities. Countries such as Thailand and Malaysia (middle-income countries) report very low pedestrian fatality proportions and very high 2/3-wheeler proportions. However, most countries report vulnerable road user fatalities in excess of 50% of the total. This is partly because some of these countries also have a high proportion of motorcycle use.

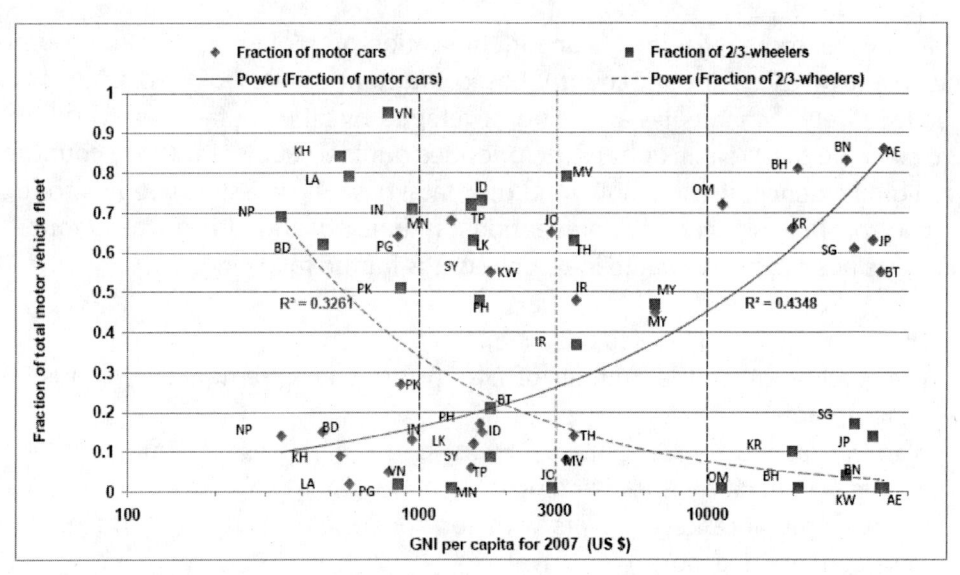

Figure 2. Proportion of 2/3-wheelers and motor cars in vehicle fleet vs. per-capita income in Asian countries

In order to propose safety policies for the future, it is necessary to have some idea about how vehicle fleet distributions change with increases in income, especially motorcycle ownership. Figure 2 shows that generally car proportions increase and 2/3-wheeler proportions decrease with increases in per-capita incomes. Here again there are large variations at similar levels of income. The correlation by income is weak for both over the whole range of incomes. Proportion of car ownership becomes greater than 2/3-wheeler ownership for all cases only when incomes exceed US$10,000 per capita per year. Incomes double every 10 years at growth rates of 7% per year. Since most countries are below US$3,000 income levels at present, it is unlikely that many countries' annual per-capita incomes will exceed US$10,000 in the next two decades. At present, Japan is the only high-income country in Asia that has a large population. It is also interesting that Japan and Singapore have a relatively high motorcycle ownership level but only Japan has low fatality rates both overall and for 2-wheelers.

Keys to the way forward

- Vulnerable road user fatalities constitute the majority of all RTI fatalities in all large Asian countries including the high-income countries. These groups need special attention in all road safety activities.
- In large low- and middle-income countries in Asia, 2/3-wheeled vehicles constitute a high proportion of all vehicles. This is not likely to change over the next two decades.
- Both overall and road user specific fatality rates do not have a high correlation with income levels. The reasons for this are not known.
- It appears that factors other than income levels, car and road design, and policing influence fatality rates per capita for each country. Much more work will have to be done in this area before the variability in crash rates can be explained satisfactorily for Asian countries.

Role of active safety

In the absence of more reliable data and identification of risk factors for each country, it is not possible to give very specific country-based countermeasures for road safety. However, since active safety technologies are being developed by all manufacturers, it is important that we examine their role in light of the priorities outlined above for those countries that have a high incidence of vulnerable road user fatalities. These issues need us to focus on speed control, braking, and the prevention of drinking and driving as priorities. The technologies need to be evaluated in these settings for the following:

- The willingness of drivers to use them.
- The level at which the number of false-positive or false-negative alarms become unacceptable.
- Warnings at too frequent intervals (e.g., pedestrian presence) may irritate drivers who may then ignore such warnings.
- Behavior modification - drivers with too much faith in the systems may be less observant or drive more aggressively.
- Testing effectiveness of these technologies in high-density pedestrian environments and heterogeneous traffic will be very useful.

Technologies with promise include:
- Adaptive and generalized pre-crash braking
- Speed limiting systems
- Alcohol interlock for cars
- Mandatory airbags for small cars
- ABS and electronic stability control

Research agenda

- Development of street designs and traffic-calming measures that suit mixed traffic with a high proportion of motorcycles and non-motorized modes.
- Highway design with adequate and safe facilities for slow traffic.
- Design of lighter helmets with ventilation.
- Pedestrian impact standards for small cars, buses, and trucks.
- Evaluation of policing techniques to minimize cost and maximize effectiveness.
- Effectiveness of pre-hospital care measures.
- Traffic calming measures for mixed traffic streams including a high proportion of motorized 2-wheelers.

List of References
Kopits, E. & Cropper, M. (2005). Traffic fatalities and economic growth. Accident Analysis & Prevention, 37, 169-178.
WHO (2009). Global Status Report on Road Safety: Time for Action. Geneva: World Health Organization.

Addressing the Deployment Issues of Intelligent Cooperative Systems

By Richard Harris

Director of Intelligent Transport Systems at Logica plc,
English Language Secretary of the World Road Association (PIARC)
Technical Committee Network Operations,
and Technical Secretary of the PIARC/FISITA Joint Task Force
on Intelligent Cooperative Systems

Addressing the Deployment Issues of Intelligent Cooperative Systems

By Richard Harris
Director of Intelligent Transport Systems at Logica plc,
English Language Secretary of the World Road Association (PIARC)
Technical Committee Network Operations,
and Technical Secretary of the PIARC/FISITA Joint Task Force
on Intelligent Cooperative Systems

ITS (Intelligent Transport Systems) is still a relatively new discipline, and implementation, operation, acceptance, and take-up varies between regions, countries, and authorities. Indeed the very term ITS is open to many interpretations. No matter what the interpretation or expectation, the key element of any ITS service is the final outcome, the benefit to the user, shipper, customer, and society.

It is the ever increasing computing and data storage capabilities, combined with widely available and reliable communications that provide the key for ITS applications. ITS can make our transportation safer, surer, smoother, and smarter. They support policy and business outcomes and are enabled by technology rather than being about technology.

The challenge with transportation is that it is everywhere and includes everyone. While we may have conventions to bring order to mobility, we are dealing with free-thinking individuals who are making decisions in real time and who are able to reduce even the most efficient and high-capacity network to a miserable assembly of inefficient and polluting vehicles.

Driving wide-scale ITS deployment

In recent years, the big successes in ITS have been those available "off the shelf" such as nomadic navigation devices. These have changed the way people travel and how they decide which route to follow. There is no doubt that such units provide opportunities for increased benefits for individuals. However, we have not been quite so successful in unlocking the wider community-based benefits for society as a whole.

Various ITS route maps, actions plans, and strategies have been developed and are being implemented to try to address this issue.

The two groups driving wide-scale ITS deployment and operation are the public authorities and the business sector.

Public authorities normally focus on the policy drivers of safety, efficiency, mobility, the environment, and the economy. Some more advanced authorities may also include quality of life, health, and inclusion within their policy drivers. The focus and priority may switch between these areas but generally they remain the key considerations.

The business sector has a different focus and is interested in compliance, competiveness, service levels, scalability, customer loyalty, market share, and stakeholder perception.

But ITS-based information technology advances continue to outpace our ability to coordinate, develop, define, fund, and deploy systems and services within either policy framework or within sustainable business models. This means that there is a real danger of significant safety and efficiency advances continuing to be interesting demonstration schemes, pilot projects, or field operational trials.

ADAS (Advanced Driver Alert Systems), CVHS (Cooperative Vehicle Highway Systems), IntelliDrive (formerly VII), or ICS (Intelligent Cooperative Systems), as we will refer to them in this article, have become the latest challenge for coordinated development and deployment. Seen as a breakthrough in automotive safety and efficiency, they are based on vehicle-to-vehicle, vehicle-to-infrastructure, and infrastructure-to-vehicle communications.

Regional development and cooperation

There are three distinct centres working on ICS: Europe, the U.S., and Japan.

In Europe, ICS is seen as a way to achieve the vision of connected road network operations. This would provide services including network control and advisory systems, traffic management services, network monitoring, and enforcement and information systems as varying levels (e.g., linked to micro-level technology, available to individual road users, built into cars and trucks). Applications include collision avoidance, driver support monitoring, intelligent speed adaptation, longitudinal and lateral vehicle control, convoy or platooning driving, floating car data, and ultimately fully automated highways.

In the U.S., the IntelliDrive program focuses on safety technology for situational awareness around the vehicle, autonomous technology, vehicle-to-vehicle connectivity, and services that will initially inform the driver and ultimately prevent crashes. Applications include prevention of the running of red lights and stop signs; gap-assisted signalization and stop control; speed warning for curves, school zones, and work zones; commercial vehicles services including wireless roadside inspection, universal truck identification, virtual weigh station, and parking information; and pedestrian applications including transit and alert systems.

Japan has Smartway, a road system that can exchange various types of information among cars, drivers, pedestrians, and other roadway users. Interestingly, the objectives are both to realize integrated ITS to provide safe, smooth road transportation and positive environment benefits. It also aims at supporting the foundation for affluence and comfort in life and society for people, goods, and information (including realizing comfortable living spaces and building infrastructure that provides safety and security). Smartway is already achieving impressive results. Rear-end type collisions have been reduced by 60% on the Expressway at Sangubashi Curve by warning drivers of the obstacle ahead. Japan is moving forward with national deployment with 1600 sites selected for ITS Spots (providing infrastructure-to-vehicle communications) spaced about every 10-15 km on the intercity expressways and about every 4 km on the urban expressways.

International development is already happening between the U.S. DOT and European Commission (Directorate-General Information Society), and between the U.S. DOT and Japan; an agreement between Europe and Japan is expected shortly.

Understanding the issues

ICS is such a far-reaching concept that it actually includes numerous aspects for consideration, which are listed below.
- Policy issues: role of government, uniform architecture and common regulations, minimum set of standards, public funding, and public policy targets.
- Standards: common standards for V2V and VSI Short-range communications, commercial 3G and 4G networks, and the specific needs and standards connected with electric vehicles.
- Business case: should the roadside infrastructure for DSRC be provided by government? A viable sustainable business case is a critical requirement, and dealing with road space shared with non-equipped vehicles.
- Technology: the three stages of safety applications, advisory (e.g., signing, speed management, traffic, and incident management), intersection safety (signal phase timing and information and crash avoidance), and V2V (the "here I am" message, a local broadcast); wireless communication, DSRC, and commercial cellular networks.
- Security and resilience: high-integrity data, attack proof, certification systems, confidence and protection from interference.
- Privacy: more important in some countries than others, confidential data, payment systems, integrity.
- Navigable databases: high-quality, attribute-rich, up-to-date mapping; ownership issues.
- Road operations: dealing with implementation, sensor-friendly roads and infrastructure.

As with all ITS services, applications, and deployment, the very introduction of the new approach means that the way in which we work has to change. This remains one of the biggest challenges to operational systems and services. The role of the road operator is key to connected vehicles services. Examples of operational challenges faced by road operators include understanding the impact on non-equipped vehicles and dealing with platoons of trucks, which may mask road signs from car drivers or make joining or leaving a motorway challenging. However, the added value for road operators includes a greater ability to influence drivers and routing; increased safety and smoother traffic flow; more and better quality traffic data and intelligence, road condition alerts, early notification of incidents and accidents, reduced signage costs, increased journey time reliability, improved enforcement (e.g., speeding, overweight vehicles); and great public relations opportunities.

To enable a debate on ICS between the motor industry and road operators, a joint task force was established in 2008.

PIARC/FISITA Joint Task Force

The World Road Association (PIARC) is a non-political and non-profit association. It was granted consultative status to the Economic and Social Council of United Nations in 1970. Current membership includes 118 governments. PIARC is the world leader in the exchange of knowledge on roads and road transport policy and practices within an integrated sustainable transport context. The PIARC technical committee on network operations considers aspects of ITS among its area of remit.

The International Federation of Automotive Engineering Societies (FISITA) is the world body for automotive engineers with a membership of more than 155,000.

In 2007, a Memorandum of Understanding was signed between PIARC and FISITA, which committed both organizations to work closely together. This proved to be a breakthrough for communications between road authorities and automotive engineers.

A joint task force (JTF) was established to consider the issues associated with ICS. The mandate was clear in that the JTF would not deal with technical or standardization aspects that were already being addressed elsewhere. The role of the task force is to consider the softer issues, those dealing with operations, funding, deployment rollout, and business cases. The objective is to inform stakeholders and to help accelerate deployment.

Conclusions

Road operators need evidence of benefits, a stable systems approach, access to data, to understand liability and privacy issues, change how they operate, and have control and understanding of the cost involved in operating and maintaining the systems.
The automotive sector needs to engage with road operators, seek public investment, require roadside equipment, rely on standards, understand liability and privacy aspects, and secure the necessary investment.

Deployment of inter-connected cooperative systems will require:

- A commercial case for investment.
- A public case for deployment, fitting in with current road network operations practice.
- Proof of consumer and societal benefits.
- A workable "connected vehicle" operations model.

- The means to manage deployment challenges.
- Political, financial, and operational justification.

The JTF report of its findings will be published in mid-2011. Those involved from the road operations and automotive engineering sides are committed to accelerated deployment to unlock the potential of Intelligent Cooperative Systems as they help us save lives, reduce accidents, improve information, and unlock the community-wide benefits of Intelligent Transport Systems.

ACTIVE SAFETY SYSTEMS, CRASH SENSING AND SENSOR FUSION

2009-01-0158

Developing Integrated Vision Applications for Active Safety Systems

Adam Prengler
NEC Electronics America, Inc.

Copyright © 2009 SAE International

ABSTRACT

Current image-processing solutions are limited with respect to being simultaneously flexible, scalable, high-performing and efficient. As vision-based safety systems increase in functionality and become more widely adopted, system makers will require flexible and scalable solutions to accommodate the needs of an expanding market. To support the highly complex embedded automotive vision systems of the future, therefore, engineers are now turning to dedicated vision processors in place of standard off-the-shelf solutions.

This paper will describe how to develop highly integrated image processing systems for active safety applications using the unique capabilities of a highly parallel, reconfigurable SIMD-MIMD processor architecture that offers the ability to handle both single- and multi-core designs. This architecture will enable safety systems to execute multiple applications simultaneously to provide more comprehensive driver assistance information. Several sensor sources will be discussed, including radar, vision, light detection and ranging (LIDAR), in addition to common advanced safety applications such as lane tracking and obstacle detection. The paper will also describe how the reconfigurable architecture allows for more robust and efficient algorithm development, as the unique performance of the SIMD-MIMD architecture supports real-time processing of images even when using highly complex algorithms.

PAPER

According to the National Highway Traffic Safety Administration's (NHTSA) *Motor Vehicle Traffic Crash Fatality Counts And Estimates of People Injured for 2007 Motor Report*, there were over 6 million highway crashes in the U.S. alone in 2007, resulting in 2,491,000 traffic-related injuries and 41,059 fatalities.

Current technology in vehicles is focused on protecting drivers and passengers in the event that an accident occurs, and these systems have proven successful at reducing the number of injuries and fatalities. The 2008 National Transportation Statistics report also states that an estimated 18,604 lives were saved in 2006 due to passive safety systems such as safety belts, child restraints and airbags. While these numbers are notable, reaching zero traffic-related fatalities is impossible using only these mitigation technologies. More widespread implementation of crash-avoidance technologies is required to make significant progress to this goal.

There are many ways to try to avoid a vehicle. In considering the most common causes of accidents—distracted drivers, driver fatigue, drunk driving, speeding, aggressive driving and weather—all except weather are directly related to driver behavior and actions.

Creating a comprehensive crash avoidance system means having to address each of these possible causes. However, developing an electronic system that can address each one is very challenging. A great deal of technology is available; however, nearly all of the possible accident causes must be monitored and processed in a different way. For example, a system designed to reduce the possibility of driver distraction might have to process a number of variables—from eye focus, to driver posture, to driving style and even driver behavior. A camera could be used to monitor a driver's face for signs of eyelids closing or a tilted head, but the camera alone could not identify all circumstances in which a driver might become distracted nor would a camera be 100% reliable. Combining a camera with something that could monitor steering angle, speed, and other feedback would create a more robust system that could better evaluate a driver's current condition. Further combining that system with navigational information and GPS capabilities could create a system that learns, and can predict and react to driver behavior. This example illustrates how complex these systems can be, but in no way should complexity be a deterrent. Even the simplest systems can have a positive effect on driver safety.

Automakers have started adopting these crash-avoidance technologies and their penetration rate is growing rapidly. Some reports state that by 2015, more than 14 million active safety systems and driver-assistance systems will be in production. However, recent NHTSA reports state that consumers still do not understand the benefits of these systems, and therefore are not as willing to pay extra for them.

To achieve expected penetration levels, OEMs will have to find ways to introduce these systems cost-effectively while at the same educating consumers about their value, that is, their life-saving capabilities. Unfortunately, many active safety systems require a significantly larger investment than traditional automotive electronic control units (ECUs). Many advanced safety systems rely on the processing of images, and while there are a number of experts in the field of image processing, the time spent developing robust applications can result in significant costs. With volumes still relatively low, and consumers still not convinced of their value, it has been difficult for automakers to increase adoption rates for advanced image-processing systems.

However, as adoption becomes more widespread, the benefits of these systems are likely to become more apparent. Additionally, many Tier 1 automotive suppliers are beginning to promote capabilities and benefits of their systems more and more. In the future, we also expect a big push from special interest and consumer advocacy groups to make certain types of active safety systems a requirement.

On the OEM side, the focus will continue to be on creating the most cost-effective system. The current approach is to create low-cost individual systems. For mainstream lane-departure and forward-collision warning applications, among others, OEMs are hoping that volumes will increase quickly enough to enable the OEMs to recoup their initial investments. However, as technology advances are made and new systems emerge, there will be a major focus on finding ways to create integrated systems. Many of the modules in development now use the same resources—cameras, RAM, video displays— so combining systems would be a natural approach to achieving optimal cost.

When considering how to develop integrated safety systems, it is important that we understand how the embedded automotive environment compares with the non-embedded one. Many current automotive image-processing solutions are based on consumer solutions. Vision processing has been performed in many industries on a variety of processors, for example Pentium-class machines used for monitoring assembly lines. In such cases, the algorithms and overall image-processing approach are very similar to automotive requirements: high resolution and high contrast are crucial, numerous methods are used to define edges and search for objects, and template matching and verification techniques are used for classification. Therefore, both types of image-processing systems require high-performance, real-time processing solutions. The difference, however, is that industrial and consumer requirements are much different than those for embedded automotive systems. The latter require low power consumption, a small physical area, and optimized cost. It is not acceptable, for instance, to add items such as cooling fans or heat sinks to dissipate heat in an embedded automotive system. Likewise, real estate in a vehicle is limited, and it is usually not possible to allocate large amounts of space to processing functions.

In addition, the quality levels required by automotive systems are much greater. Because these systems are used to determine potential danger so as to alert drivers, product failure is unacceptable. Stringent reliability levels, combined with the high-volume shipments associated with automotive electronics—totaling millions of units per year—have very difficult requirements that only the highest-quality semiconductor companies can meet.

Requirements for hardware in the automotive industry are vastly different than in other industries, as are requirements for software. Many software companies see active safety applications such as vision processing as good business, but they don't consider the resources available to the integrator. For example, developing an image-processing algorithm on a computer with Matlab® functions has no value if the algorithm cannot be implemented in an embedded controller that can be used in an automotive module. Many companies have very robust image-processing applications, and do not consider performance and memory size requirements as limiting factors.

As we consider how to develop active safety systems, it is important to understand how these systems will interact so as to create a comprehensive safety system. Active safety systems can consist of many different inputs. There are generally two groups of sensors or inputs: ranging systems and camera systems. Ranging systems—or distance-based inputs such as radio detection and ranging (radar), light detection and ranging (lidar), and ultrasonic systems—can be used to detect objects and to calculate their distance and velocity. Images from cameras can be used to confirm the ranging data, to identify items not detected by the ranging data, and to perform recognition or classification of data. With these inputs, automakers can create systems that use ranging sensors to detect obstacles and vehicles; vision sensors to detect lane makers, signs and signals; and a combination of the two to detect and recognize pedestrians and other objects. A truly integrated active safety system would employ all of these input technologies so as to acquire the most reliable data about driving conditions.

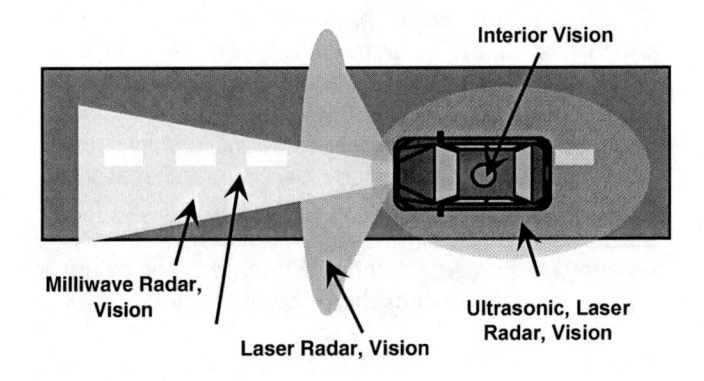

Figure 1. Autonomous Sensing

To ensure that an integrated system receives the most reliable data possible, it is important to understand the advantages and disadvantages of each input type. Camera systems yield the most data about driving conditions and offer superior sensing of vision-focused targets for lane tracking, traffic sign recognition and traffic light detection. Any instance in which image recognition is necessary, such as for recognizing a red light or a stop sign, requires visual data to be captured and processed by the system. In a single frame, an image processor could search for lane markers, identify objects in the roadway, and look for signs on the sides of the road. The challenge of working with cameras is that the processing required to search an image is very complex and requires algorithms that are bandwidth-intensive. For example, the processing of one video frame involves tens of thousands or even hundreds of thousands of pixels, each of which is treated as a piece of data. Each frame is typically preprocessed to correct distortion or perform scaling. A system might then binarize an image to create a black-and-white representation that can be used to look for edges or shapes, or alternatively, to produce a color histogram that can be used to search for specific colors. After that, systems often go through a correlation step to determine if objects match requirements or templates. Ideally, each of these and other steps are performed on every image frame, at rates up to 33 frames per second. Tracking algorithms can even combine data from multiple frames, further increasing a system's complexity.

Finally, while images can provide superior data about driving conditions, they are limited in their ability to provide distance information and suffer in situations with limited or poor visibility. This is where ranging sensors can help compensate for the limitations of cameras. Ranging technologies such as radar, lidar, and ultrasonic systems do not yield as much information but are easier to implement and can be an excellent complement to vision-based systems in non-ideal driving conditions.

Perhaps the most common ranging sensor is radar. Available in short- and long-range versions, radar is based on the reflection of radio signals at objects. The reflected signal is processed to detect objects and their distance. Radar is fairly immune to weather conditions and can provide range as well as relative velocity information. Additional benefits of using radar include the wide availability of radar sensors and an abundance of knowledge on radar processing.

Lidar is similar to radar but is based on the reflection of laser pulses from objects rather than radio signals. While algorithms used for lidar systems are simpler than those used for radar processing, lidar is based on light and therefore is more affected by environmental conditions. A small buildup of dust or dirt on a lidar sensor can keep the system from functioning. Additionally, lidar is newer in the market, so cost is still somewhat higher and there is not as much expertise available compared to radar.

Ultrasonic systems are based on sound waves being reflected by objects. The reflected sound waves can be used to detect distance and/or relative speed of objects. Like radar, ultrasonic systems are widely used in many industries and there is a lot of expertise on the subject.

With any of the ranging technologies, correct positioning and tuning of the system is critical to get successful data. A well-tuned ranging system offers a simpler and easier way to get distance and velocity information, and can be used to develop systems that detect objects and measure distance. In an integrated system, the ranging technology can be used as the first step to identify objects, and then the image processor can be used to confirm and classify them.

A robust and accurate active safety system would take advantage of multiple different sensor types and combine them, resulting in a so-called "sensor fusion" system. While a low-end sensor fusion system could rely on local processing of the sensor signals, sharing only the analysis results, a high-end integrated system would benefit from having all processing data available within one processing unit. However, this can only be achieved if the processing unit is capable of handling multiple different sensor interfaces. Flexible, yet efficient sensor interfaces are required to process the incoming data streams without loss of data, a challenge that requires thorough analysis and planning during the early phase of any device development. Devices such as the NEC Electronics IMAPCAR2® processor combine multiple capture interfaces with a high-performance, flexible parallel-processing architecture that provides an ideal solution to this problem.

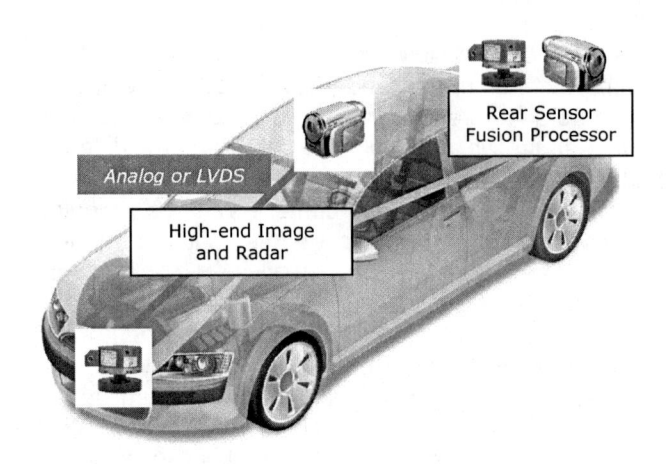

Figure 3. Integrated System

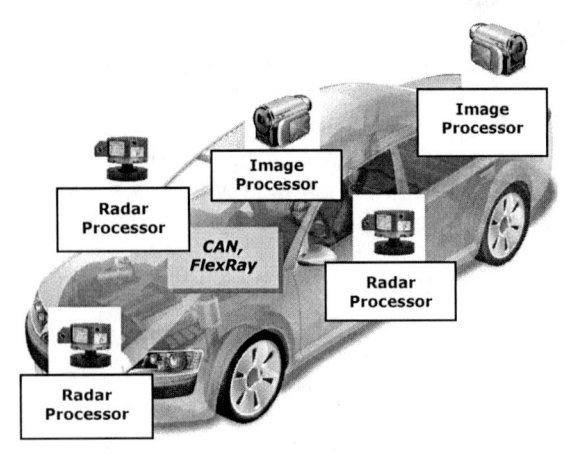

Figure 2. Low-End Systems

Decentralized systems execute local processing and exchange results through the automotive network

There are several challenges to realizing a combined or integrated solution. First, automakers must consider the layout of the system. A system with multiple ECUs located in proximity to the sensors can use the standard automotive networks such as CAN or FlexRay® to transmit the sensor results. With an integrated solution, it is necessary to transmit data from the radar sensors and cameras over a long distance, requiring the use of more expensive wiring such as analog cables or low-voltage differential signaling (LVDS) wiring. LVDS is currently one of the only options that can offer the required speed with low power consumption. In addition, an ideal combined system would require the processor to have multiple sensor interfaces and the bandwidth to process multiple inputs at one time.

Processing radar and image sensor information in a single module provides cost advantages on system level

Benefits of the combined system would include a reduction in the number of ECUs required, in addition to a system that does not have the latency of waiting for sensor data from other modules. With all data in one ECU, decisions about obstacle information and driving conditions could be made even faster, an important factor for systems where real-time functionality is critical. If you consider that a study by Daimler-Benz showed that an extra half second of early warning can prevent 60 percent of rear-end accidents, and 1.5 seconds will prevent 90 percent of them. Therefore, removing any latency from the system should be the target.

Creating a single module capable of processing these multiple inputs requires dedicated processing solutions that have the required sensor interfaces. For example, consider an integrated forward system that employs two cameras and a radar antenna array. A typical system such as this would require a minimum of two 8-bit video interfaces and another 8- to 16-bit capture interface for a radar chipset. To capture the multiple, simultaneous inputs, the processor would have to have built-in hardware for transferring the camera and sensor data to memory. Devices such as the NEC Electronics IMAPCAR2 processor have multiple capture interfaces, up to four channels on some devices, each with its own buffer to capture the image or radar frame. An on-chip direct memory access (DMA) controller then transfers the captured data to external memory, which can then be worked on by the individual processors. Also, with a flexible and fast memory controller, the IMAPCAR2 processor has the ability to read and write to external memory in 32- or even 64-bit increments, further increasing a system's ability to efficiently transfer data from the capture interfaces to the memory workspace.

Once the data is captured by the system, the challenge becomes processing it in real-time. While a radar frame and image frame are very similar with regard to spatial orientation of data relative to the receiver, the processing of the data is very different. The main use of the radar system is to measure Doppler Effect, or subsequent Doppler shift of the returned echoes from any objects in order to determine their distance and velocity. Processing of this data requires traditional signal processing, mostly using fast Fourier transforms (FFTs) to find shifts in the transmitted and received frequencies.

The image data is processed much differently than the radar. An image frame consists of thousands of pixels, which can represent a great deal of information to the system. Depending on what information the processor is looking for in a frame, different processing techniques are chosen. Generally the first step is to scale the image and isolate areas of the image that are considered regions of interest. For lane tracking and obstacle detection, this is obviously the area of the road directly in front and to the near sides of the vehicle. Traffic signs would be different areas, to the sides of the road. After scaling the image and making corrections, the system can take a number of approaches to perform detection and recognition. Many of the tasks for these systems involve searching for obstacles. A first approach to distinguishing objects is to search for their boundaries or edges, which can be done using algorithms such as those used in the Canny edge detector. From there, additional functions can search for and identify obstacles in the frame. Using various methods, such as symmetrical scanning and other verification techniques, the system can further classify the objects into vehicles, obstacles, and even pedestrians.

In functions that rely on color, producing histograms in the frequency domain can provide data that shows where certain colors appear in a frame. From there, additional processing can determine if areas of color have any additional characteristics that would help to identify them as a known object, such as a stop sign. Functions for lane detection could employ either a color approach or an edge detection approach. Using both can create an even more robust solution, such as the IMAPCAR2 processor.

The processor employs NEC Electronics' new IMAPCAR-XC® core, a reconfigurable core capable of single instruction/multiple data (SIMD) operation, multiple instructions/multiple data (MIMD) operation, or a combination of the two. In SIMD operation, the device behaves like a single processor that can handle multiple data points simultaneously. This is the solution of choice for applying filers to entire frames, for processing large amounts of data very efficiently, or for performing parallel operations such as FFTs. In MIMD operation, the device operates like a multiprocessor architecture with a shared memory base. This would be an idea processing solution once areas of interest have been identified. Depending on the original analysis, each MIMD element could process for a specific target using a unique function or operation. With up to 128 parallel cores available, and the ability to reconfigure operating mode on the fly, the IMAPCAR2 processor has the flexibility and performance to take on the processing challenge of an integrated system.

For radar systems, a pure SIMD configuration is most ideal. By running FFTs on each of the 128 SIMD cores, the system can process an entire radar frame efficiently. Image data processing can take advantage of both SIMD and MIMD capabilities. As stated, the first step is focused on preprocessing, such as filtering or scaling. SIMD operation is advantageous here, since it can process multiple data points in parallel. Once initial processing is complete, configuration of the cores depends on the particular application or subsequent algorithm being processed. Some algorithms can be designed with a highly parallel structure, while others might have a dependency on other data in the image and therefore require a sequential structure, which can be supported by configuring the cores for MIMD operation.

IMAPCAR2 devices make an excellent platform for developing an integrated solution for both hardware and software reuse. Based on the previous-generation IMAPCAR processor, the IMAPCAR2 lineup will consist of four derivatives starting with the highest-performing device capable of executing over 270 giga, or billion, operations per second (GOPS) and supporting three additional derivatives with a scaled number of processors to support varying customer requirements. Three of the devices will be available with the same packaging and pinouts, allowing customers to create a single hardware platform and then choose the IMAPCAR2 derivative for a given application's needs. This means that an automaker could develop a forward safety system based on IMAPCAR2, and add and remove features with only software modifications.

Active safety systems are becoming more popular and automakers worldwide are continuing to invest and develop even more advanced and innovative ways to protect drivers, passengers, and even pedestrians. In addition, many of these systems are used by automakers as ways to differentiate their vehicles from the rest of the market. As these systems increase in quantity, automakers will have to look at ways to reduce the overall system costs. One approach will certainly involve the creation of integrated modules. While the physical challenges required to combine these systems (along with high-bandwidth processing requirements) have been considered roadblocks in the past, advancements such as the IMAPCAR2 processor are providing automakers with real, scalable platforms on which to base designs that can be flexible, yet cost-effective.

REFERENCES

1. Ankrum, D.R. (1992) "Smart Vehicles, Smart Roads". *Traffic Safety* 92(3): 6–9
2. Imou, K., M. Ishida, Y. Kaizu, T. Okamoto, A. Sawamura, and N. Sumida. "Ultrasonic Doppler Sensor for Measuring Vehicle Speed in Forward and Reverse Motions Including Low-Speed Motions". *Agricultural Engineering International: the CIGR Journal of Scientific Research and Development.* Manuscript PM 01 007, Vol. III.
3. Strategy Analytics. Automotive Electronics Strategy Advisory Service. May 2008. *System Demand 2006 to 2015.*
4. U.S. Department of Transportation (DOT), National Highway Traffic Safety Administration (NHTSA). 2008. *New Car Assessment Program* (model year 2010), docket no. NHTSA-2006-26555.
5. U.S. Department of Transportation (DOT), National Highway Traffic Safety Administration (NHTSA). September 2008. *Motor Vehicle Traffic Crash Fatality Counts and Estimates of People Injured for 2007.* Based on the Fatality Analysis Reporting System (FARS) and the National Automotive Sampling System, General Estimates System (NASS GES). DOT-HS-811-034.
6. U.S. Department of Transportation (DOT), Research and Innovative Technology Administration, Bureau of Transportation Statistics. 2008. *National Transporta-tion Statistics.*

CONTACT

Adam Prengler is a senior technical marketing engineer in the Automotive Strategic Business Unit at NEC Electronics America, Inc. He can be reached by e-mail at adam.prengler@am.necel.com or by phone at +1-214-262-7873.

DEFINITIONS, ACRONYMS, ABBREVIATIONS

1. Canny edge detector: uses a multi-stage algorithm to detect a range of edges in images
2. DMA: direct memory access
3. ECU: electronic control unit
4. FFT: fast Fourier transform
5. GOPS: giga-operations per second
6. GPS: global positioning system
7. IMAPCAR: integrated memory array processor for the car
8. LIDAR: light detection and ranging
9. LVDS: low-voltage differential signaling
10. NHTSA: National Highway Traffic Safety Administration
11. MIMD: multiple instructions/multiple data
12. RADAR: radio detection and ranging
13. SIMD: single instruction/multiple data

2009-1-0780

Successive Categorization of Perceived Urgency in Dynamic Driving Situations

Gerald J. Schmidt, Ali Khanafer and Dirk Balzer
Adam Opel GmbH

Copyright © 2009 SAE International

ABSTRACT

The timing of warnings in Advanced Driver Assistance Systems is crucial for a fast and correct reaction, as well as the driver's acceptance. Knowledge about the human urgency perception is needed to match this timing. We developed a new technique to measure the perceived urgency in dynamic situations and evaluated it in a test track study.

Eight participants drove an equipped test car with constant speeds (30 – 130 kph) on a test track approaching a preceding vehicle that was either standing or moving with a lower constant speed. Similar to the last second braking method [1], the participants were required not to brake until a collision seemed inevitable. Approaching the vehicle, the participants judged the increasing perceived urgency with successive presses on a steering wheel mounted button to indicate a change of urgency category (3 ascending urgency categories).

We manipulated the speed of the participant's car and the speed difference. When the participants chose a certain urgency category the relative velocity had systematic effects on the distance and Time-To-Collision (TTC). Higher relative speeds lead to earlier urgency judgements compared by TTC. In addition, the variance of distance and TTC per category between the drivers decreased as the urgency category increased.

INTRODUCTION

More than 5,000 people were killed in German traffic in 2007 [11] and more than 41,000 people are killed on U.S. roadways every year in motor vehicle crashes [9]. Of all motor vehicle crashes in the U.S., 1/3 is reported as rear-end collisions. Of those rear-end collisions, it has been estimated that more than 60% are caused by driver inattention [7]. In addition to the well-engineered passive safety systems more and more active safety systems are currently under development or already introduced to the market. The collision avoidance systems (CAS) preventing rear-end collisions try to focus the driver's attention in the direction of the hazard and evoke a braking reaction by some sort of alert (i.e. tone, light, haptic etc.). The timing of warnings in CAS is crucial for a fast and correct reaction, as well as the driver's acceptance. Thus, knowledge about the human urgency perception under full attention is needed to match this timing.

MAIN SECTION

The most common approach to determine the timing of a rear-end collision warning is the Stop-Distance-Algorithm [2]. The stop distance algorithm takes into account the preceding and own vehicle absolute speed and deceleration. The driver is represented in the formula by two values: assumed brake reaction time and assumed deceleration (often independent of approach scenario). The formula results in an online stopping distance. If a certain value is exceeded the driver is warned. This algorithm seems appropriate to compute the timing warning to evoke a last second hard braking maneuver based predominantly on physical data.

A more driver oriented way to find this point are the studies on last-second braking behavior [1,4,5,6]. The test procedure in [1] was that participants drove towards a stationary Styrofoam object at different relative speeds.

The instruction was to either start hard braking (but without locking the wheels) at the latest moment the participants think they are able to stop in front of the object, or start normal braking at the latest moment they think they can stop safely in front of the object. The van der Horst study [1] revealed that the Time-to-Collision[1] (TTC), when participants start to brake, increases with speed, but less than could be expected on the basis of a constant deceleration model. Kiefer et al. [4,5] conducted the Crash Avoidance Metrics Partnership (CAMP) and added moving and braking vehicles to the [1] study design. They further enhanced the study by an instruction to brake comfortably hard. Interestingly participants showed distinct timing differences between the normal and hard braking trials, but no difference between hard and comfortably hard braking could be found. Thus with this instruction it is not possible to diagnose the development of the situation from the perspective of the driver, which may occur over a very short period of time. On the one hand with one drive it is only possible to collect one data point and on the other hand only two points in time are reasonable to investigate with this method.

Our intention was to collect data for the approaching process as a whole to understand the development of the situation from the viewpoint of the driver. We intended to collect a reasonable amount of data points in one drive without discontinuing the process.

METHOD – Eight participants (3 female, 5 male) between the age of 22 – 56 years (mean=33.75, median=31) took part in this study. We conducted our study on a test track in Germany. The vehicle driven by the participants was an Opel Vectra sedan. It was a fully equipped test car, able to record vehicle CAN-data plus information about the environment via a LIDAR (Light Detection and Ranging) sensor and a mono front camera system. LIDAR uses active sensing for the detection of objects and has its benefit in measuring distances and velocities similar to a RADAR. It uses time-of-flight measurement to detect objects and to measure their distance. In contrast to RADAR, LIDAR uses a calculated velocity only. Thus, with this equipment we had knowledge of the kinematic conditions of the test car and any preceding car. A confederate vehicle pulled a collapsible trailer as a target simulator. This was a safety procedure to avoid harm to the participants and the involved cars. To enhance the optical realism in reference to height and occlusion we mounted a balloon car to the trailer as can be seen on Fig. 1. With the balloon car we could assure that realistic looming could occur [10].

The target simulator was either stationary or driven with constant, lower speeds than the participant's test vehicle (PV). The primary task of the participants was to follow the target simulator with a certain constant speed. The speed of both vehicles was maintained by cruise control. The PV's speeds were 30, 50, 80, 100, and 130 kph.

[1] TTC: The time required for two objects to collide if they continue at their present speed and on the same path.

These velocities were chosen because they represent the standard speeds on German roads (urban, rural, and autobahn). The relative velocities between PV and target simulator were 10, 20, 30, 40 and 50 kph. Forty and 50kph speed difference were of course not applicable at

Figure 1. Preceding target simulator

30kph PV's velocity. Similar to the "last second braking studies" the participants had to brake in the "last second to avoid a collision. However unlike the original last second braking study by van der Horst [1], we did not instruct our participants to avoid blocking the brakes. We wanted their definite judgement of the "last second". Although there was plenty of room to swerve beside the target simulator, the participants were clearly instructed to avoid a collision by braking only.

Figure 2. Steering wheel button

During the approach towards the slower target simulator the participants had to judge the subjective criticality of the situation on a scale (cf. Fig. 2). This scale has been derived in design from the one used by [8].

Usually workload measurements and judgments on perceived security are used to collect subjective data on situation criticality. The advantage of this scale over the aforesaid two solutions is the clear-cut subjective tolerance limit definition. The participant is unambiguously instructed that everything less than 'dangerous' is acceptable for traffic safety, but 'dangerous' and above is not.

We instructed the participants to press a button on the steering wheel (cf. Fig. 3) in the instant they perceived a change in level on the criticality scale. This means, while the participants are far away from the target simulator they consider the driving situation as a 'free driving' Situation. When they were approaching the target simulator and their perception of the situation switched

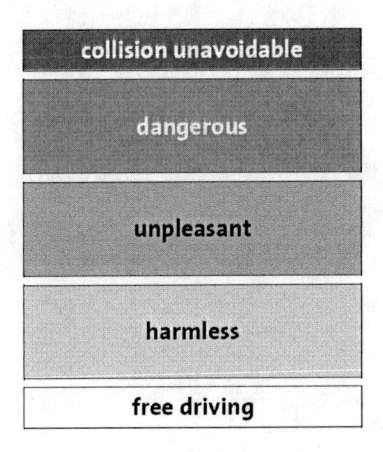

Figure 3. Situation criticality scale

start and ended by the first movement of the brake pedal. The last section ('collision unavoidable') can be determined by a collision itself or – more relevant for a CAS – the physically last chance to avoid the collision.

A main advantage of the method is to gather subjective data at exact points of time. With traditional methods a participant can only be asked to judge a situation at a certain time, but it is hard to determine the exact time. To do so one would have to use complicated methods like occlusion glasses to stop a situation for exact judgment. With our method the participant is aware of the situation and can decide the time to judge on his own. Additionally the push of a button does most likely not alter the driving behavior.

from 'free driving' to a following situation they should press the button the first time. Beginning with the first actuation of the button on they judged the situation as a 'harmless' following scenario. At the moment they first thought of the situation as being 'unpleasant', they should press the button again. As stated above this still means they evaluated this situation as acceptable for traffic safety. When they started to judge the situation as 'dangerous', hence as unacceptable, the button should be pressed the last time. Their last action in each approach was to brake hard in the second before they think a collision is inevitable. With this technique it is possible to define five sections of vehicle data corresponding to the subjective development of the scenario. The start of the first section ('free driving') is limited only by the range of used sensors and ended by the first push of the button. The middle sections are closed by the pushes of the button. The section 'dangerous' is defined with the press of a button on the

RESULTS – With the data of CAN, camera and LIDAR at hand we focused on evaluating the distance and the TTC between PV and target simulator. As we were interested in devising a method to map the subjective progress of criticality in an approaching situation, it is indispensable to show that the participants truly distinguished the categories of the scale in Fig. 3. On the other hand we wanted to see if our results match the work that has been done before. Kiefer et al. [5] found different behavior in relation to a moving and non-moving target. Therefore we analyzed both situations separately. In the following inferential statistics' analyses we only included trials with moving target simulators. For the PV at speeds of 30 and 50kph not all relative velocities could be realized. Thus, we also restricted the data we used to PV's velocities to 80kph and up. We did this to be able to run an Analysis of Variance (ANOVA) for repeated measures for all factors: 'Category changes', 'PV speed', and 'Relative Velocities'. To get an overview of all the data compare Figure 4 and 5. On the axis of ordinates in Figure 4 the distance between the two vehicles and in Figure 5 the TTC can be viewed. The PV's speed is on the abscissa, the single lines mark the speed difference, and every column represents a push of the button.

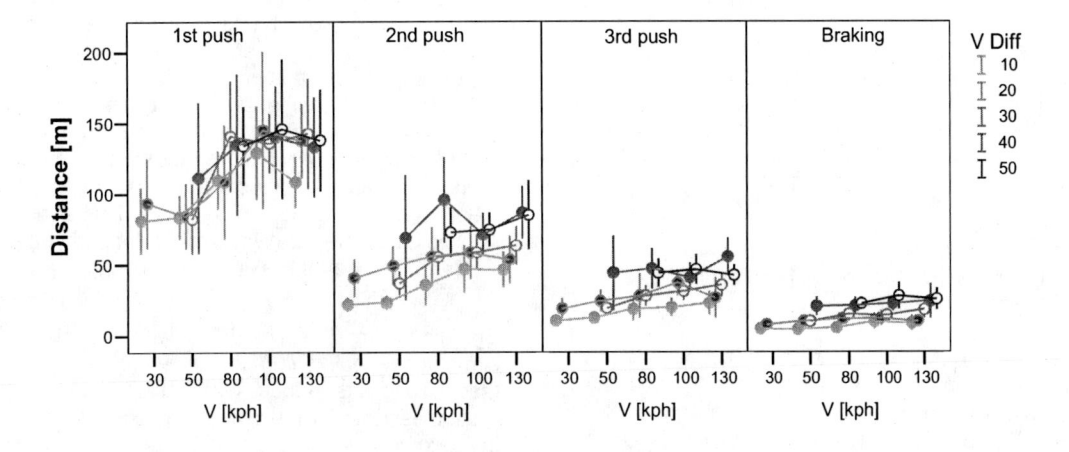

Figure 4. Distance criticality scale change with moving target simulator

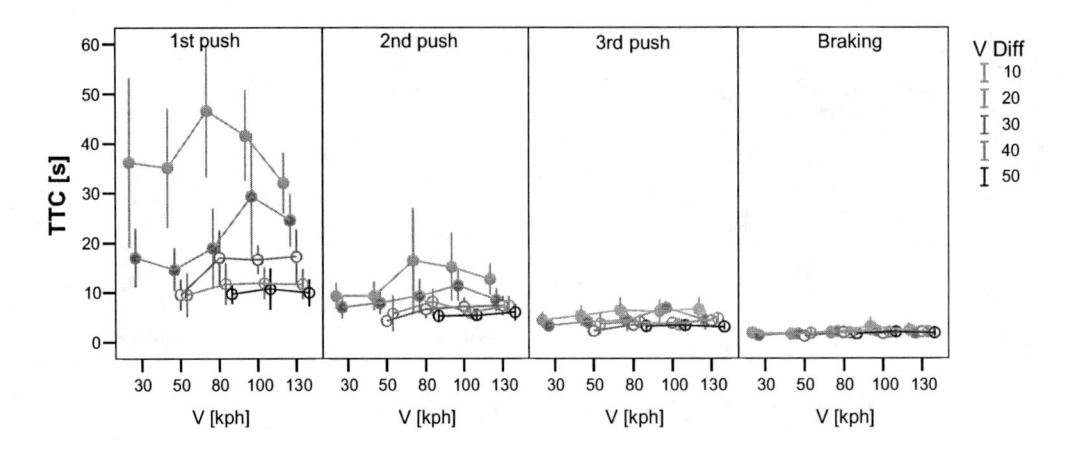

Figure 5. TTC (+/- 2 SD) at criticality scale change with moving target simulator

Category changes – The dependent variables distance and TTC showed clear differences between the four category changes. In the ANOVA the factor "category" (including the moment of braking) is highly significant for the distance (df=3, F=215.09, p < .001) as well as the TTC (df=3, F=6047.14, p < .001). Thus our participants were able to indicate three different category changes plus the last second to avoid a crash. These assumptions are enforced by the post-hoc pairwise comparisons which reveal that each category is significantly different from every other (p < .05). In addition to that statistical analysis the participants stated, that they had no problems evaluating the situation with the pushes of button. The participants also felt no disturbance of their driving behavior.

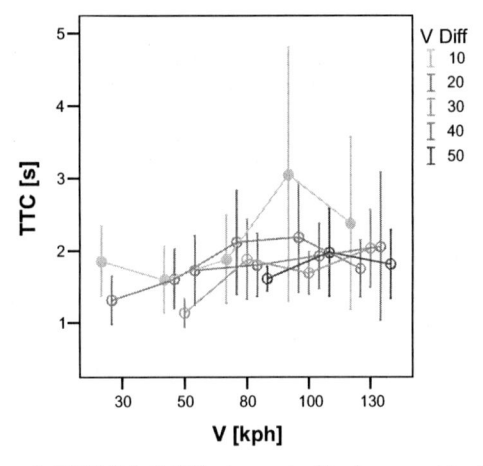

Figure 6. TTC (+/- 2 SD) at onset of last second braking with moving target simulator

PV's speed – The speed as a main factor has no significant impact on the distance (df = 2, F=2.00, p> .15), but shows an approaching significance on the TTC (df = 2, F=2.40, p> .10), at which the participants press the button (cf. Table 1 for overall means).

Regarding the overview in Figure 4 it seems reasonable that the speed has an impact on the distance - including the 30 and 50kph conditions. But remember, this ANOVA includes only the PV's velocities of 80, 100 and 130 kph. From the previous work done in [4,5] PV speed should have a significant impact on the last second braking behavior. Looking at the last second braking behavior only on Figure 6, there seems to be an upward trend, when the two lower speeds are included, which supports the approaching significance. But with the lowest mean at the highest speed the picture is blurred.

	80kph	100kph	130kph
Mean Distance [m]	59.16	63.32	63.15
Mean TTC [s]	9.04	9.50	8.53

Table 1. Means of Distance and TTC for PV Speed

Relative Velocity – The speed difference between the PV and the target simulator has a significant effect on the distance (df = 4, F = 26.45, p < .001) as well as on the TTC (df = 4, F = 36.08, p < .001). The higher the difference of speed is, the further away the participants push the button or hit the brakes. Overall this is also true for the TTC: the higher the speed difference the higher is the TTC when participants feel the change of category. When analysing each category change alone, the above stated significant relation of TTC and relative speed is true. The individual analyses for the onset of braking reveals a non-significant result for the effect of relative speed (p > .16). No matter which condition the participants are in, they start to brake at about a TTC of 1.9s (mean = 1.89s, sd = 1.22). But again, in other studies [4,5] the relative velocity had an effect on the TTC at the onset of braking.

Standing Target – The trials with the standing target simulator revealed very similar results. It does not seem, that participants judged substantially different due to the motionlessness of the object. The results are presented in Figure 7.

For both situations (moving and standing target) it is clear to see that the variance of the TTC to the time of judgement decreases as perceived urgency increases.

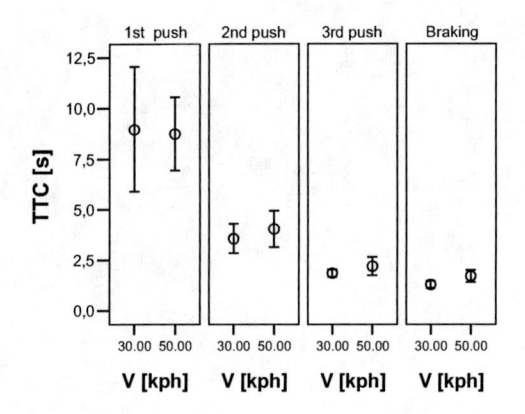

Figure 7. TTC (+/- 2 SD) at criticality scale change with standing target simulator

This is a good result for developing a warning system. The closer the point in time the system has to give out the warning, the more the participants agreed upon the time of the category change. Therefore it seems to be feasible to design one warning criterion (based primarily on relative speed) for a wide range of drivers. But it has to be kept in mind, that this kind of study is limited to tell at which point in time an alert driver is not willing to accept the situation anymore. If there is knowledge about the driver's state, adjusted warning timings are of course preferable.

CONCLUSION

This research examined a new categorization method to evaluate the criticality of a dynamic driving situation. The applied setting was the approach to a rear-end collision situation. The dependent variables discussed are the distance between the two vehicles and the TTC.

The analysis of the participant study shows, that our method is highly applicable for vehicle approach scenarios. The use of the adapted situation criticality scale [8] was easy to use by the participants. The use of our method with this scale gives the opportunity to define a hard cut objective criterion (i.e. TTC) based on subjective assessment (3rd push of the button).

The participating drivers judged the criticality of the situation primarily on the relative speed. We could not find a clear influence of the PV velocity. Regarding our last second braking results alone, even the relative velocity did not affect the corresponding TTC. These finding contradict the results of [4,5]. We assume that the reason for this discrepancy could be based on the different target simulator and therefore the perception of the target.

The result of a study with the successive categorization method can be fed into an algorithm evaluating a situation system for an active safety system. In our case this would be a Forward Collision Warning (FCW) algorithm. An example for an application of our method is described in [3].

As compared to other methods, our approach is easy to implement, driving is not interrupted, and multiple data points can be gathered during one drive. Finally, the technique may be readily adapted to other scenarios, and thus provide a highly accurate framework to assess perceived urgency by the driver in dynamic driving situations. Examples of other applicable situations could be lane departure, closing of following traffic, the beginning of an overtake maneuver with oncoming traffic or blind spot situations. Of course in all these cases the last second braking has to be substituted by the respectively adequate action.

REFERENCES

1. Horst, R. v. d. (1990). A time-based analysis of road user behaviour in normal and critical encounters. Unpublished doctoral dissertation, Delft. University of Technology, Delft.
2. ISO/TR 16532 (2005). Road vehicles - ergonomic aspects of in-vehicle presentation for transport information and control systems - warning systems. International Organization of Standardization.
3. Khanafer, A., Balzer, D. & Isermann, R. (in press). A rule-based Collision Avoidance System – Scene Interpretation, Strategy Selection, Path Planning and System Intervention, Rep. No. 09AE-0026. SAE World Congress, 2009.
4. Kiefer, R., Cassar, M., Flannagan, C., Jerome, C., & Palmer, M. (2005). Surprise braking trials, time-to-collision judgments, and "first look" maneuvers under realistic rear-end crash scenarios (Tech. Rep. No. DOT HS 809 902). National Highway Traffic Safety Administration.
5. Kiefer, R., LeBlanc, D. J., & Flannagan, C. A. (2005). Developing an inverse time-to-collision crash alert timing approach based on drivers' last-second braking and steering judgments. Accident Analysis & Prevention, 37, 295-303.
6. Kiefer, R., LeBlanc, D., Palmer, M., Salinger, J., Deering, R., & Shulman, M. (1999). Development and validation of functional defnitions and evaluation procedures for collision warning/avoidance systems (Tech. Rep. No. DOT HS 808 964). Department of Transportation.
7. Knipling, R.R., Mironer, M., Hendricks, D.L., Tijerina, L., Everson, J., Allen, J.C., and Wilson, C. Assessment of IVHS Countermeasures For Collision Avoidance: Rear-End Crashes. NHTSA technical report, Publication Number DOT HS 807 995, May, 1993.
8. Neukum, A., Lübbeke, T., Krüger, H.-P., Mayser, C. & Steinle, J. (2008). ACC-Stop&Go: Fahrerverhalten an funktionalen Systemgrenzen. In: M. Maurer & C. Stiller (Hrsg.) 5. Workshop Fahrerassistenzsysteme - FAS 2008. S. 141-150. Karlsruhe: fmrt.

9. NHTSA. Traffic Safety Facts - A Brief Statistical Summary, Tech. Rep. No. DOT HS 811 017, August 2008.
10. Summala, H., Lamble, D., Laakso, M. (1998). Driving experience and perception of lead car's braking when looking at in-car targets. Accid. Anal. Prev. 30, 401-407.
11. Statistisches Bundesamt Deutschland. Zahl der Verkehrstoten in 2007 kaum verändert, Pressemitteilung Nr. 492, retrieved Sept. 25th, 2008: http://www.destatis.de/jetspeed/portal/cms/Sites/destatis/Internet/DE/Presse/pm/2007/12/PD07__492__46241.psml

CONTACT

Gerald J. Schmidt
Adam Opel GmbH
IPC K1-08,
65423 Rüsselsheim – Germany
Email: gerald.schmidt@de.opel.com

SAE International

Estimation of Vehicle Roll Angle and Side Slip for Crash Sensing

2010-01-0529
Published
04/12/2010

Aleksander Hac, David Nichols and Daniel Sygnarowicz
Delphi Corporation

Copyright © 2010 SAE International

ABSTRACT

Estimation of vehicle roll angle, lateral velocity and side slip angle for the purpose of crash sensing is considered. Only roll rate sensor and the sensors readily available in vehicles equipped with ESC (Electronic Stability Control) systems are used in the estimation process. The algorithms are based on kinematic relationships, thus avoiding dependence on vehicle and tire models, which minimizes tuning efforts and sensitivity to parameter variations.

The estimate of roll angle is obtained by blending two preliminary estimates, each valid in different conditions, in such a manner that the final estimate continuously favors the more accurate one. The roll angle estimate is used to compensate the gravity component in measured lateral acceleration due to vehicle roll or road bank angle. This facilitates estimation of lateral velocity and side slip angle from fundamental kinematic relationships involving the gravity-compensated lateral acceleration, yaw rate and longitudinal velocity. The results of simulations and vehicle tests in a variety of maneuvers demonstrate accuracy of the proposed estimation methods and the use of side slip information in rollover discrimination.

INTRODUCTION

In recent years, electronically controlled brake systems and occupant protection systems, each using a distinct sensor set, have become common in production vehicles. Concurrently, networking within vehicles has improved dramatically, facilitating access to various signals, including the ones measured within other sub-systems. This creates opportunities for exchanging information to improve performance of these systems by using previously unavailable signals.

In this paper, a specific example of such integration is considered. It involves the Electronic Stability Control System (ESC) and the occupant protection system (side air curtains) designed to deploy during rollovers and side collisions. Each of these systems uses a separate set of measured and estimated signals. The ESC systems typically incorporate lateral acceleration, yaw rate, steering angle and wheel speed signals, from which vehicle longitudinal velocity, lateral velocity and side slip angle are estimated. The systems governing deployment of side air curtains typically utilize roll rate and lateral acceleration sensors and estimates of roll angle. In this paper, the possibility of sharing sensor information between both systems and integrating the estimation tasks is explored to improve the estimation accuracy. More specifically, yaw rate and longitudinal velocity signals are used to improve the estimate of vehicle body roll angle, while the estimated roll angle is used to remove the gravity component from the measured lateral acceleration, thus enabling one to simplify and improve the estimation of lateral velocity and side slip angle.

A number of methods for estimating the side slip velocity and side slip angle have been proposed for application in ESC systems, but the most common methods are estimators based on dynamic handling models [1, 2] of vehicles and estimators using primarily kinematic relationships [3]. The former methods have many performance advantages, but they also tend to be complex, require knowledge of many vehicle and tire parameters or characteristics (with associated sensitivity issues), and often demand access to additional signals, such as brake master cylinder pressure or brake caliper pressures [1]. The estimators based on kinematic relationships are much simpler, but have not been successful in production vehicles primarily because of insufficient robustness on banked roads due to inability to discriminate between the effect of change in lateral velocity and the effect of bank angle on the measured lateral acceleration. More recently, a solution to this dilemma has been proposed [4], but at the expense of adding four inertial sensors (rotational rates and accelerations in all three directions are measured) and algorithm complexity.

In addition, the requirements regarding accuracy of estimates are different in crash sensing applications than in controlled brake system. In ESC applications accurate estimates are required for side slip angle magnitude in the range of approximately 0 to 30 degrees, while in crash sensing it is desirable to be able to track the side slip angle of much larger value, even beyond 90°.

Information about roll angle of the vehicle body is important in predicting and sensing rollovers. The most common methods of estimating roll angle are based either on a dynamic model of vehicle (e.g. [5]) or on the kinematic relationship [6]. In the former case, a vehicle model with associated parameters, and often additional signals, are necessary. In the latter case an estimate of (the absolute) roll angle is often obtained by some form of integration of roll rate. This method may be sufficient for rollover sensing, when the roll angle changes fast, but it is inadequate during normal driving. It is required here that the estimate be accurate in the entire range of vehicle operation, including normal driving, driving on banked roads, emergency handling maneuvers and rollovers.

In this paper, a method based on simple kinematic equations is proposed to overcome fundamental limitations of the estimator. The approach uses only a roll rate sensor in addition to the sensors available within ESC system. The roll rate sensor provides additional independent information about the roll angle, thus helping differentiate the gravity component sensed by the lateral acceleration from the change in lateral velocity. At the same time presence of yaw rate and longitudinal velocity signals facilitates improvements in estimation of roll angle, especially in steady-state turns, when the estimates obtained via integration of roll rate are inaccurate. In the proposed algorithm, a preliminary estimate of roll angle, generally reliable in nearly steady-state conditions, is first determined from a kinematic relationship involving lateral acceleration, yaw rate and longitudinal velocity. The final estimate of roll angle is then obtained by blending the preliminary estimate with a second estimate based on the bias-corrected measure of roll rate. In the blending process, the relative weighting between two preliminary roll angle estimates depends on their frequency and on the driving conditions so that the final estimate is always close to the more accurate of the preliminary estimates. The roll angle estimate is then used to compensate the gravity component in measured lateral acceleration arising due to vehicle roll and/or road bank angle. The lateral velocity is determined from a simple observer based on a fundamental kinematic relationship involving gravity-compensated lateral acceleration. The side slip angle is subsequently determined. Information about the lateral velocity and side slip angle is then used in the algorithm controlling deployment of counter-measures in rollover events.

The results of simulations and vehicle test data in a variety of maneuvers performed on high friction and slippery surfaces and on banked roads demonstrate the accuracy of the proposed method. The role of lateral velocity information to help discriminate between rollover and near-rollover situations in tripped rollover events is also illustrated using vehicle data.

ALGORITHM OVERVIEW

In this section the proposed algorithm for estimation of roll angle, lateral velocity and side slip angle is briefly described. The algorithm's block diagram is shown in Figure 1.

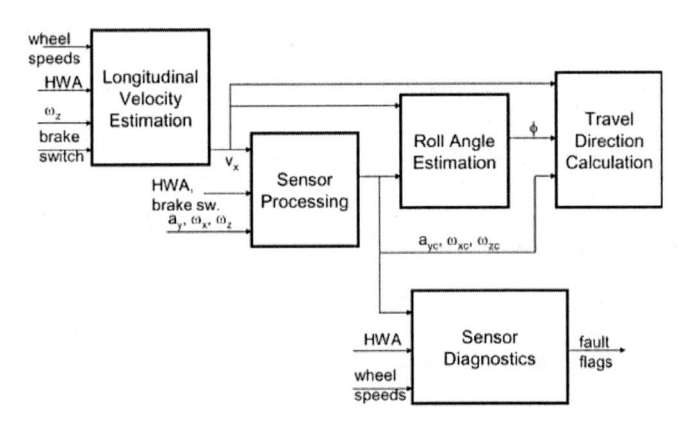

Figure 1. Functional Diagram of the Estimation Process

In the *longitudinal velocity estimation* block, an estimate of the longitudinal component of velocity at the vehicle center of mass, v_x, is determined. The primary signals used in this process are the four vehicle wheel speeds. Vehicle yaw rate and steering angle are also used to compensate for the differences in the linear velocities of the wheels and of the vehicle center of mass during turning via kinematic corrections. Brake switch signal may also be used to influence the logic between driver braking and non-braking situations. This part of the algorithm is very similar to that used by brake controllers in vehicles equipped with ESC systems. This block can therefore be omitted when the longitudinal velocity signal is available from a brake controller.

Within the *sensor diagnostics* block health monitoring of the sensors is performed and sensor faults are detected and isolated. Out of range faults are easily detected by observing individual sensor outputs. Sensor faults, which remain within normal operating range, are detected and isolated by continuously comparing the sensor outputs with those predicted from models using other sensors. A fault is declared when the differences between the measured and predicted signals (the residuals) exceed threshold values for some time. The thresholds for individual sensors depend on driving situation. The thresholds tend to be small in steady-state driving within the linear handling range (when the correlations among signals are strong and hold relatively precisely) and larger in transient or non-linear driving conditions. This adaptation of the diagnostic strategy to driving conditions is fundamentally similar to that used in ESC and other chassis systems, as reported for example by Fennel and Ding [7] and D'Silva et al. [8].

In the *signal processing* block, the outputs of yaw rate, lateral acceleration and roll rate sensors are filtered to attenuate noise, the lateral acceleration signal is compensated for the sensor location being away from the vehicle center of mass (if applicable), and the signal bias values are determined and compensated for. Yaw rate and lateral acceleration bias values are determined and slowly removed when vehicle is stopped or is in nearly straight driving conditions. These centering strategies are similar to those used in ESC systems. Large gravity components of measured lateral acceleration, such as contributed by significant bank angle of the road, are compensated in a separate process discussed in later sections. The bias and other unwanted components of the measured roll rate signal are determined and compensated for when the vehicle is in nearly steady-state driving conditions, including steady-state turning. This portion of the algorithm will be discussed in some detail.

In the *roll angle estimation* block, the value of the absolute (i.e. measured with respect to an inertial frame) roll angle of vehicle body is determined. Two preliminary estimates of roll angle are determined, one reliable in steady-state driving conditions, another in transient maneuvers. Then both estimates are blended in such a manner that the final estimate always favors the more accurate of the two.

In the *travel direction calculation* block, the estimates of the lateral velocity and side slip angle are obtained. Using the roll angle just determined, the measured lateral acceleration signal is compensated for the gravity component due to body roll. The lateral velocity estimate is then determined from a simple observer based on the fundamental kinematic relationship between the gravity-corrected lateral acceleration, yaw rate and longitudinal velocity. The side slip angle estimate is subsequently determined from the basic relationship.

In the next sections, the most relevant portions of the algorithm, namely those associated with centering of roll rate sensor, roll angle estimation and travel direction determination are described in more detail.

FUNDAMENTAL RELATIONSHIPS

By their very nature inertial sensors used in ground vehicles measure components of absolute lateral acceleration and rates of rotation, that is accelerations and rates with respect to the inertial (Earth-fixed) frame. The sensors, however, are attached to the vehicle body, hence they measure respective quantities along axes fixed to the vehicle body. Since during driving the vehicle body undergoes both translational and rotational motions relative to the inertial frame, the relationships among measured signals may become quite complicated.

Let us define the orientation of the vehicle body relative to the inertial frame by the angles the body longitudinal and lateral axes form *with the horizontal plane*, that is by the roll angle, ϕ, and the pitch angle, θ, respectively. Note that according to these definitions, the roll and pitch angles include contributions of road bank and fore-aft inclinations, respectively, which may differ from commonly employed definitions of these angles [9]. Then the following relationships hold [10]

$$\dot{\phi} = \omega_x + \sin\phi \tan\theta \, \omega_y + \cos\phi \tan\theta \, \omega_z$$
$$\dot{v}_y = -\omega_z v_x + \omega_x v_z + a_y + g \sin\phi \cos\theta$$
$$\dot{v}_x = \omega_z v_y - \omega_y v_z + a_x + g \sin\theta$$

(1)

Here ω_x, ω_y, and ω_z are the rates of rotation about the longitudinal (x), lateral (y) and vertical (z) axes of vehicle body (that is roll, pitch and yaw rates), v_x, v_y and v_z are the components of the velocity vector at the vehicle center of mass along the same axes, a_x and a_y are the components of acceleration at the center of mass along the x and y axes, *including the gravity components*, g denotes acceleration of gravity and a dot above a symbol designates a derivative with respect to time.

In nearly all operating conditions considered here, it can be assumed that

• the pitch angle, θ, is small (< 0.2 rad), that is $\cos\theta \cong 1$, $\sin\theta \cong \tan\theta \cong \theta$,

• the roll angle, ϕ, is small to moderate (< 0.5 rad), that is $\cos\phi \cong 1$, and

• the vertical component of velocity, v_z, is small.

Under these assumptions, equations (1) can be written as

$$\dot{\phi} = \omega_x + \theta\omega_z$$
$$a_y = \dot{v}_y + \omega_z v_x - g\sin\phi$$
$$a_x = \dot{v}_x - \omega_z v_y - g\sin\theta$$

(2)

The term $\sin\phi\tan\theta\omega_y$ in the first equation is neglected as a small value of higher order. Since the accelerations a_x and a^y include gravity components, they correspond to measured accelerations if measurement errors are negligible. The following observations can be made from equations (2):

• When the product of the pitch angle and yaw rate is small, then the roll angle can be determined in principle (but not in practice) by integrating the measured roll rate;

• When the vehicle is in a steady state condition, the time derivative of v_y is small and the roll angle can be determined from the second equation using measured lateral acceleration a_y, yaw rate ω_z, and estimated longitudinal velocity, v_x.

• When the roll angle, ϕ, is known (e.g. estimated), then the derivative of lateral velocity, $\dot{v}_y$, can be determined from the second equation using known signals. Lateral velocity may then be obtained in principle by integrating $\dot{v}_y$.

• When the longitudinal acceleration, a_x, is measured, then the pitch angle can be determined in principle from the last equation when the product of yaw rate and lateral velocity is small, since the time derivative of longitudinal velocity can be estimated by differentiating the estimated v_x.

Finally, it is noted that the product of the pitch angle and yaw rate is small in the great majority of driving conditions, but it can be significant, for example during driving on a spiral ramp with significant inclination angle (e.g. in parking garages). As discussed later, the effect of this term on roll rate is mostly compensated for by the centering algorithm.

ROLL ANGLE ESTIMATION

Estimation of roll angle involves some form of integrating the measured roll rate signal. In order to avoid unacceptable compromises in the estimation process, any slowly varying error in the measurement, such as sensor bias, must be removed. Since the estimated roll angle is used to remove the gravity component sensed by lateral acceleration sensor caused by body roll (including road inclination), the estimate must be reasonably accurate in all operating conditions, including slowly varying and rather small roll angles. Bias removal is therefore critically important and is discussed first.

ROLL RATE BIAS REMOVAL

The purpose here is to determine and then to remove a constant or slowly changing error in measured roll rate. This error may be the sensor bias or slow drift in the sensor output, or a result of slowly varying component of other angular rotation vector, such as indicated in the first of equations (2). The sensor's steady state error is estimated by first determining estimates of roll rate using two other methods relying on other sensors. If the roll rates obtained by these two methods are consistent with each other (e.g. sufficiently close to each other) but not consistent with the measured roll rate and the magnitudes of measured and estimated values are below thresholds, then the difference between the measured roll rate and one of the estimated roll rates contributes to the calculated sensor bias. The calculated bias is passed through a low pass filter and subtracted from the measured roll rate yielding a compensated roll rate. The bias calculations are performed only during normal, nearly steady-state driving (when the magnitudes of measured and estimated roll rates are below threshold values). Otherwise, the previous value of estimated bias is held. The estimated bias is subtracted from the measured roll rate, yielding the bias-compensated roll rate, ω_{xc}. Note that this process compensates not only for sensor bias, but also the product of pitch angle and roll rate in steady state driving. More accurate compensation for the latter product is possible when the longitudinal acceleration signal is available.

The two estimates of roll rate used in this process are determined as follows. At first, two roll angle estimates are calculated. The first estimate, ϕ_{eay}, is obtained using measured and filtered lateral acceleration:

$$\phi_{eay} = -R_{gain}a_{yf}$$

(3)

Here R_{gain} is the roll gain of vehicle, which relates the roll angle of vehicle body to lateral acceleration in steady-state cornering. The filtered lateral acceleration, a_{yf}, is obtained by passing the measured lateral acceleration through a low-pass filter to reduce the effect of measurement noise. The second estimate of roll angle is obtained from the second of equations (2), under the assumption that $\dot{v}_y$ is small (vehicle is in steady-state). Since these calculations are performed only for small roll angles (less than about 3 degrees), $\sin\phi$ is replaced with ϕ, yielding the following simplified equation

$$\phi_{e2} = \frac{\omega_z v_x - a_y}{g}$$

(4)

The estimates of roll rate are then determined by differentiating the estimated roll angles from equations (3) and (4), respectively. In practice differentiation is replaced by passing the estimated roll angle through a high pass filter.

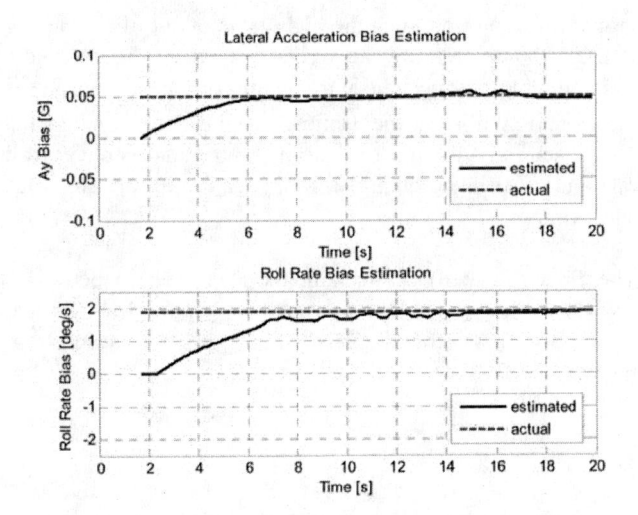

Figure 2. Illustration of Sensor Bias Determination

In Figure 2 an example of lateral acceleration and roll rate bias determination during straight driving is shown. The actual bias values are shown as horizontal lines. It is seen that the estimated bias values converge to the actual ones within seconds.

In spite of the best efforts in bias compensation, small sensor errors may remain; thus the estimation algorithm needs to posses some robustness with respect to these errors.

ROLL ANGLE ESTIMATION

Conceptually, the estimate of roll angle is obtained by blending together two preliminary estimates, each valid in different range of operation. In the blending process the weighing of the estimates is such that the final estimate is always close to the one which is more accurate.

In principle, an estimate of roll angle can be obtained by integrating the measured and compensated roll rate, that is

$$\phi_{e\omega} = \int_0^t \omega_{xc}(\tau)d\tau$$

(5)

Unfortunately, pure integration has infinite sensitivity to nearly constant errors, such as bias, because the error is integrated over time. In practice, integrator is often replaced by a pseudo-integrator, that is a low pass filter with a low cut off frequency (in the Laplace domain, an integrator $1/s$ is replaced by a filter $1/(s+b)$). This reduces sensitivity to bias, but also causes distortions of the estimated signal since the filter attenuates low frequency and DC components of the signal itself. For example, in a steady-state turn with a constant roll angle, an estimate thus obtained tends to decay exponentially. Therefore other means of determining roll angle is such situations are necessary.

As discussed earlier, the estimate of roll angle, which is valid in steady-state conditions, can be obtained from the second of equations (2) under the assumption that the time derivative of lateral velocity is small. Ignoring this term yields the following estimate from the kinematic relationship:

$$\phi_{ek} = \sin^{-1}\frac{\omega_z v_x - a_y}{g}$$

(6)

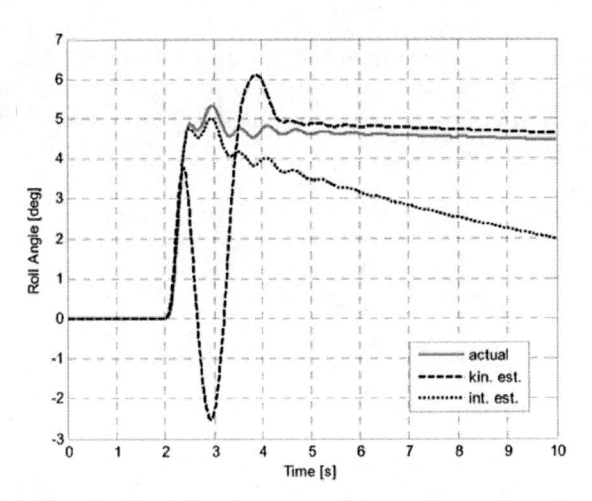

Figure 3. Roll Angle and the Simple Estimates in a Step Steer Maneuver

Estimates of roll angle obtained from pseudo-integration with cut off frequency of 0.1 rad/s and from equation (6) in a simulated step steer maneuver are shown in Figure 3. As illustrated in Figure 3, none of the above methods by itself can provide a reliable estimate of roll angle in all operating conditions. The estimate obtained from the kinematic relationship (6) tends to be accurate in steady-state conditions, when the ignored derivative of lateral velocity term is small, but it is invalid during quick transient maneuvers, when the ignored term can become large. Conversely, the estimate obtained via pseudo-integration of roll rate tends to be accurate during quick changes of roll angle, since then the signal distortion is minimal and the roll rate is large in comparison to the error (bias); the estimate is often poor during steady-state conditions. Thus both estimates are complimentary in the sense that when one of them is accurate, the other is not and vice versa.

It is therefore desirable to combine the two estimates in such a manner that the final estimate is always closer to the one which is more reliable in the given conditions. The blending process used here can be described by the following differential equation

$$\dot{\phi}_{ebl} + b_{bl}\phi_{ebl} = b_{bl}\phi_{ek} + \omega_{xc} \quad (7)$$

Here ϕ_{ebl} is the final (blended) estimate of the roll angle, ϕ_{ek} is the preliminary estimate obtained from equation (6), ω_{xc} is the compensated roll rate and b_{bl} is the blending filter parameter, which varies between two distinct values depending on operating conditions, as explained below. Note that the estimate obtained from integration is not explicitly used in equation (7). As shown in the Appendix, equation (7) can be derived from Kalman-Bucy filter theory. However, it has a straightforward intuitive interpretation, which becomes more obvious when equation (7) is viewed in the Laplace (or frequency) domain.

In Laplace domain equation (7) can be written as

$$\phi_{ebl} = \frac{b_{bl}}{s + b_{bl}}\phi_{ek} + \frac{1}{s + b_{bl}}\omega_{xc} \quad (8)$$

or on account of equation (5) as

$$\phi_{ebl} = \frac{b_{bl}}{s + b_{bl}}\phi_{ek} + \frac{s}{s + b_{bl}}\phi_{e\omega} \quad (9)$$

Denoting by $w(s)$ a frequency-dependent weighting function, equation (9) can be written as

$$\phi_{ebl} = w(s)\phi_{ek} + [1 - w(s)]\phi_{e\omega}, \quad w(s) = \frac{b_{bl}}{s + b_{bl}} \quad (10)$$

Hence, the final estimate of roll angle can be interpreted as a weighted sum of the estimate obtained from the kinematic relationship, ϕ_{ek}, and an estimate obtained by integrating the measured (and centered) roll rate, $\phi_{e\omega}$. The weight $w(s)$ depends on frequency and on the filter parameter b_{bl}, which varies with conditions of motion. At low frequencies ($s \cong 0$) $w(s) \cong 1$ and $[1 - w(s)] \cong 0$, so primarily the estimate obtained from the kinematic relationship, ϕ_{ek}, is used. Conversely, at high frequencies ($s \to \infty$) $w(s) \cong 0$ and $[1 - w(s)] \cong 1$, so predominantly the estimate obtained from the integration of roll rate, $\phi_{e\omega}$, is used. The latter estimate is not explicitly calculated in the algorithm, since the final blending of estimates is determined directly from equation (7) without the intermediate step of calculating $\phi_{e\omega}$.

The value of the blending filter parameter, b_{bl}, that is the cut off frequency of the filter, depends on conditions of vehicle motion. If the vehicle is at or near steady-state condition in terms of roll motion, then the high value of cut off frequency is used, thus increasing the weight placed on the kinematic estimate, ϕ_{ek}. During transient conditions, the lower value is selected. Essentially, the vehicle is considered to be in a steady-state condition when the measured and centered roll rate and its estimate obtained from equation (6) are small for some time.

The final (blended) estimate in the step steer maneuver is shown in Figure 4. A comparison with Figure 3 shows that the final estimate continuously follows the more accurate of the two simple ones.

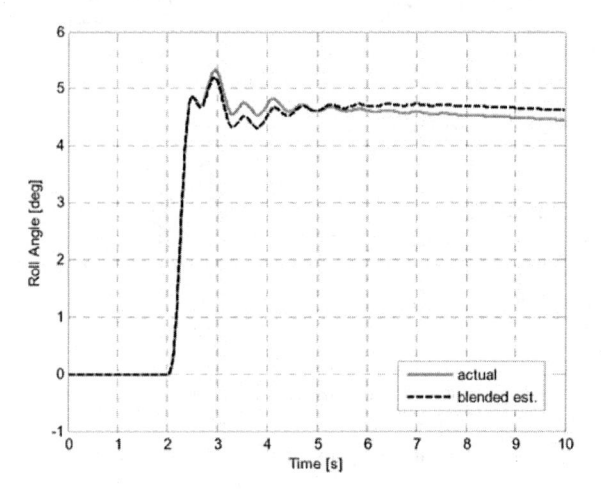

Figure 4. Roll Angle and the Blended Estimate in a Step Steer Maneuver

SIDE SLIP ESTIMATION

Determination of lateral velocity and side slip angle is based on the fundamental kinematic relationships, which are satisfied for any vehicle. The derivative of lateral velocity is obtained from the second of equations (2) as follows

$$\dot{v}_y = a_y - \omega_z v_x + g\sin\phi \quad (11)$$

All the variables on the right hand side of this equation are either measured directly (lateral acceleration, a_y and yaw rate, ω_z) or estimated (longitudinal velocity, v_x, and vehicle roll angle, ϕ). Therefore, the derivative of lateral velocity is determined from equation (11) with the measured and estimated values of the signals replacing the actual ones.

In principle, lateral velocity can be obtained as an integral of the lateral velocity derivative. Unfortunately, as discussed earlier, simple integration exhibits infinite sensitivity to nearly constant errors in the integrated signal. Viewing it from another perspective, an integrator is an observer with a zero eigenvalue; such observer is marginally stable but not asymptotically stable. Therefore, pure integration is replaced by pseudo-integration, which is essentially low pass filtering. In the time domain, the lateral velocity estimate, v_{ye}, is determined as a solution of the following equation:

$$\dot{v}_{ye} = -b_f v_{ye} + a_y - \omega_z v_x + g \sin\phi$$

(12)

Here b_f is the cut off frequency of the filter. In the Laplace domain, the pseudo-integrator can be represented as a low pass filter, $1/(s+b_f)$. Such filter can be interpreted as a combination of a high pass filter, $s/(s+b_f)$, (which attenuates nearly constant errors) and an integrator, $1/s$. If a DC component of the error in calculating a term on the right hand side of equation (11) is Δ, then the steady-state value of the error in v_y estimation is Δ/b_f. Therefore, a large value of b_f is desirable to reduce sensitivity of the estimator to nearly constant errors. On the other hand, a large value of the cut off frequency b_f causes signal distortion, because low frequency components of the actual integrated signal below the filter cut off frequency are attenuated along with the DC component of the signal. This trade off is illustrated in Figure 5.

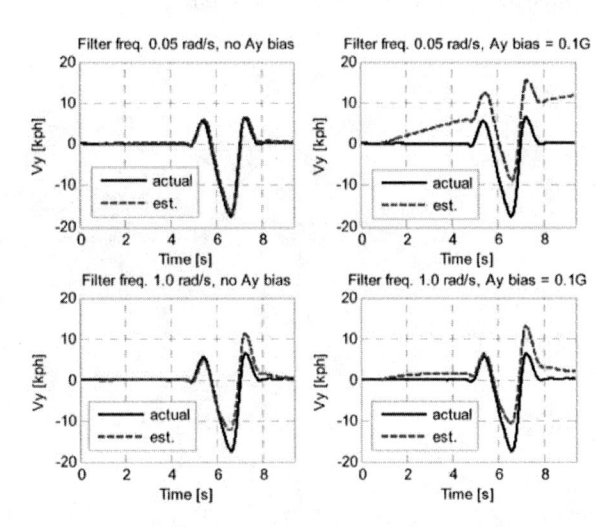

Figure 5. Estimate of Lateral Velocity in a Double Lane Change Maneuver with Different Filter Parameters

In Figure 5, the estimated values of lateral velocity are shown in a double lane change maneuver for two different values of the filter cut off frequency in two cases: no error in calculation of $\dot{v}_y$ (the left hand side) and 0.1G of error due to lateral acceleration bias. It is seen that as the filter constant decreases, the tracking performance for the case without error improves, but the estimate becomes more sensitive to DC errors. Obviously, a trade off exists between the estimator's ability to track slowly varying signals and the robustness with respect to nearly constant errors. This trade off is resolved by

- using sensor centering algorithms, which remove sensor bias, thus reducing errors in determination of $\dot{v}_y$ and allowing the use of low cut off frequencies and

- adaptively adjusting the cut off frequency depending on the operating conditions.

In general, the cut off frequency assumes a higher value when the vehicle is in steady-state conditions with respect to yaw motion. The value is lower, making the filter approximate an integrator, when vehicle is in a transient condition. Essentially, the steady state is declared when the derivatives of measured signals, such as yaw rate and steering angle, are small in magnitudes for some time.

The estimated longitudinal vehicle velocity, v_x, is generally indicative only of the velocity magnitude, since most wheel speed sensors presently used do not provide the sign of wheel speeds. In order to track the side slip angles at magnitudes above 90°, the sign of the longitudinal velocity component must be determined during a spin out condition. For this purpose, the estimate of side slip angle from the previous iteration is used. In order to make this determination more robust, a different method of calculating side slip angle for the purpose of determining the sign of v_x is used when the magnitude of side slip angle is near 90°. It can easily be shown that the rate of change of side slip angle, β, is

$$\dot{\beta} \cong -\omega_z - \frac{a_x}{v_y} + \frac{\dot{v}\cos\beta}{v_y}$$

(13)

Here $v = \sqrt{v_x^2 + v_y^2}$ is the velocity magnitude. When the magnitude of side slip angle, β, is close to 90° and the lateral velocity, v_y, is large, the last two terms in equation (13) are small and

$$\dot{\beta} \cong -\omega_z$$

(14)

Therefore, in these conditions the side slip angle is determined by integrating yaw rate. The signed estimate of longitudinal velocity is used in calculation of the lateral velocity estimate, equation (12), as well as in determination of side slip angle described below.

Vehicle side slip angle, β, is determined from the following trigonometric relationship:

$$\beta = a\tan 2(v_y, v_x)$$

(15)

Atan2 is the arctangent function with the domain including all four quadrants, that is the angle $-180° \leq \beta \leq 180°$, as opposed to the narrower range of $-90° \leq \beta \leq 90°$ for the atan function. The function (15) has a singularity when the velocity vector is zero since the side slip angle is then indeterminate; the function is very sensitive to errors when both estimates of v_x and v_y are small in magnitude. Therefore the magnitude of longitudinal velocity is limited from below when the vehicle is stopped or rolling slowly, but is not limited when the longitudinal velocity passes through zero, for example during vehicle spin out, in order to enable the estimator to track the side slip angle in these conditions. The estimate of lateral velocity is used to differentiate between these two conditions.

The estimates of lateral velocity and side slip angle are used by the rollover discrimination algorithm to help discriminate between rollover and near rollover events, thus improving decisions concerning deployment of countermeasures.

RESULTS OF SIMULATION AND VEHICLE TESTING

In this section selected results of vehicle testing and simulations are presented to illustrate performance of the estimation algorithms and the use of side slip information to improve rollover discrimination.

ROLL ANGLE AND SIDE SLIP ESTIMATION RESULTS

Performance of the proposed estimation algorithms has been verified using full vehicle simulation and vehicle tests. In addition to production grade sensors, test vehicles were equipped with laboratory grade IMU/GPS unit providing accelerations, rates and angles of rotation with respects to all three vehicle axes as the reference signals. Only production grade sensor data was used in the estimation process. Selected results are presented below. In each figure three signals and their estimates are shown: absolute roll angle of vehicle body, lateral velocity and side slip angle. Solid and dashed lines represent the measured and estimated signals, respectively.

In Figure 6, the results obtained in an aggressive double lane change maneuver performed at 80 kph on dry surface are shown. In this maneuver the roll angle and lateral velocity vary quite quickly and the side slip angle reaches a peak magnitude of 18 degrees. All estimates track the actual signals faithfully. A period of straight driving prior to maneuver is used to center the sensors.

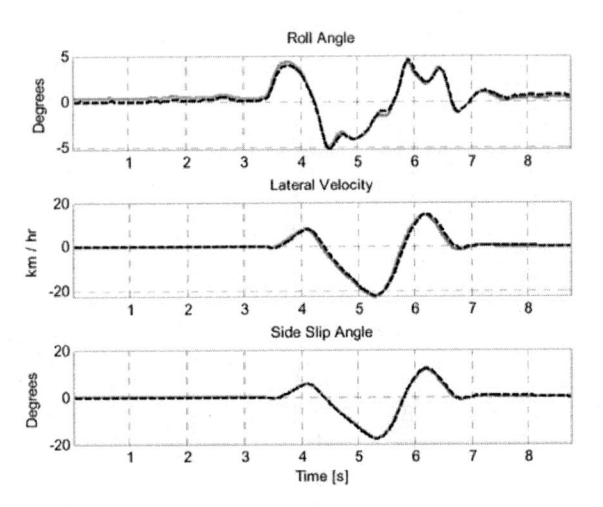

Figure 6. Estimated (dashed) and Measured (solid) Signals in a Double Lane Change Maneuver

In Figure 7 the results from a Fishhook test at 80 kph are shown. This maneuver is more challenging from the estimation point of view because the roll angle remains relatively constant in the second part of maneuver, while the side slip velocity and angle vary rather slowly.

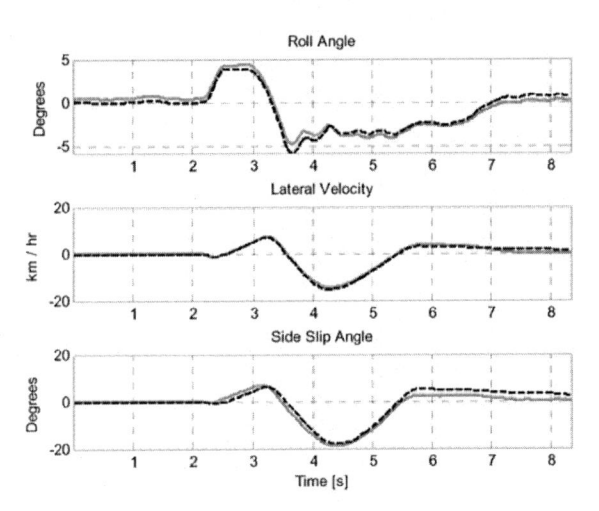

Figure 7. Estimated (dashed) and Measured (solid) Signals in a Fishhook Maneuver

The next maneuver, shown in Figure 8 is a prolonged slide on a very slippery surface (wet tiles). This maneuver is extremely challenging for any estimation method because of difficulties in estimating longitudinal speed (due to tire skid, motions of some tires are partially decoupled from vehicle motion); in addition, for model based estimation methods, the tire is out of the range of model validity for a significant period of time. Still, reasonably good estimates are obtained.

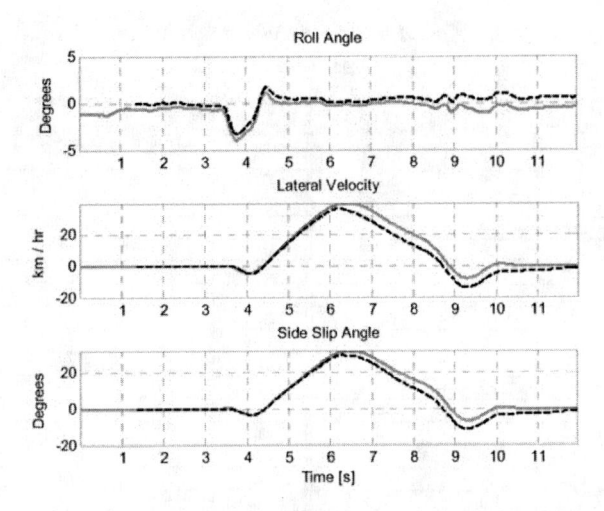

Figure 8. Estimated (dashed) and Measured (solid) Signals in a Slide Maneuver on Slippery Surface

In Figure 9, the estimation results during driving on a severely banked road are shown. The peak bank angle reaches 25 degrees. The estimate of roll angle is quite accurate; consequently, the contribution of the bank angle to lateral acceleration signal is compensated for accurately and the side slip velocity and angle estimates are good. Without this compensation the estimates would have drifted away resulting in a large error.

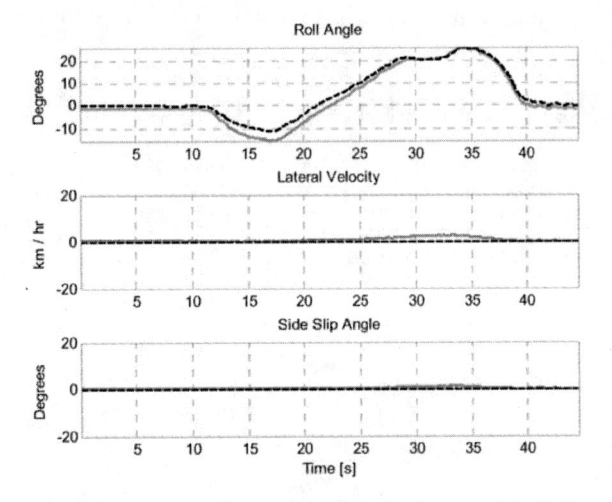

Figure 9. Estimated (dashed) and Measured (solid) Signals in Maneuver Performed on Banked Road

In Figure 10, the estimation results in a simulated fishhook maneuver at 120 kph leading to vehicle rollover are shown. The estimate of the roll angle is accurate, but the errors in estimation of lateral velocity and side slip angle increase in the last phase of rollover. This is primarily because at this stage of rollover, the assumptions used in the kinematic relationships (2) are violated and more complex equations (1) apply. For example, the product $\omega_x v_z$ in the second of equations (1) is not negligible, since both the roll rate ω_x and the vertical component of velocity at the vehicle center of mass v_z are significant. In addition, some accuracy is lost in estimating longitudinal component of velocity from wheel speeds after the two wheel lift off. However, it is typical to latch the lateral velocity and side slip angle at the onset of the rollover event to represent the initial conditions of the event. This latching occurs well before the accuracy loss.

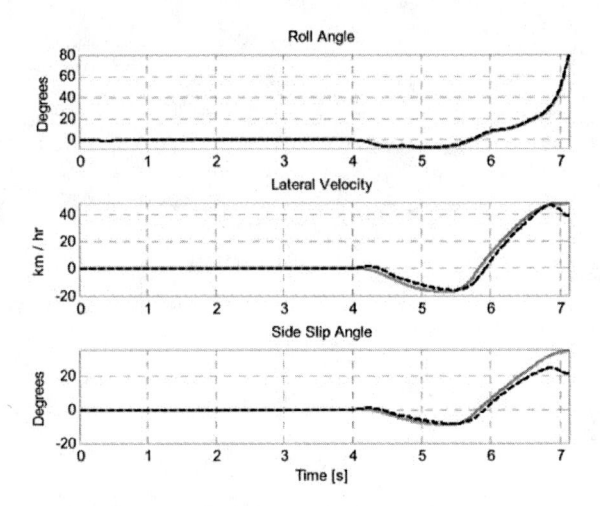

Figure 10. Estimated (dashed) and Measured (solid) Signals in a Simulated Fishhook Maneuver with Rollover

USE OF SIDE SLIP INFORMATION IN ROLLOVER DISCRIMINATION

The most common type of rollover is tripped rollover. In this event vehicle encounters a tripping mechanism (e.g. a curb, soft soil, or a crack in the pavement) while sliding in lateral direction. In order for a rollover to occur, the vehicle must possess momentum in the lateral direction that is sufficient to raise the vehicle center of mass above the tripping line. The minimum lateral velocity, which is sufficient to roll over a vehicle upon tripping, is referred to as the critical sliding velocity. If the lateral velocity is below the critical sliding velocity, the vehicle may experience a significant roll angle after tripping, but will not rollover. Unfortunately, the decision regarding the deployment of countermeasures must be made in the first phase of roll motion, when the traces of measured lateral acceleration and roll rate may be remarkably similar in rollover and non-rollover cases.

In Figure 11, the data collected from two soil trip tests are shown. Both tests were performed on the same vehicle, but at different speeds, one below and the other above the critical sliding velocity. It is seen that the traces of roll rate and lateral acceleration are almost identical prior to 0.194 sec., the time when a deployment decision must be made. The same phenomenon holds true for other trip type events such as curb trips.

Since it is impossible to discriminate between both events using information from these two sensors, it is impossible to make a reliable deployment decision using only that information. This difficulty has been reported by other researchers [11].

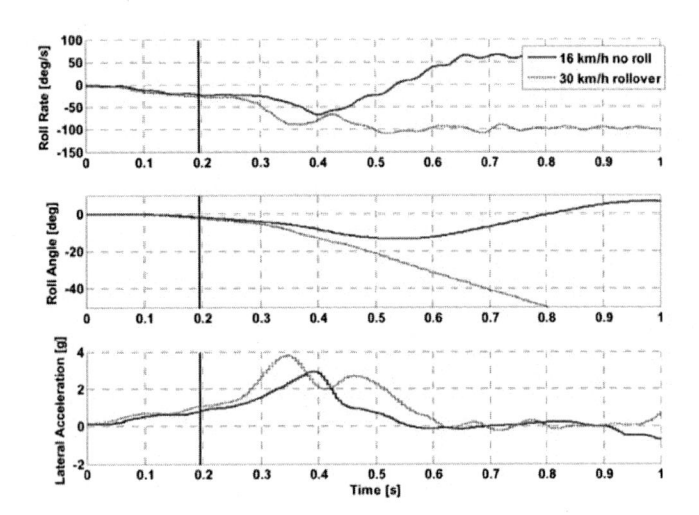

Figure 11. Comparison of Roll and No Roll Soil Trip Events

A rollover discrimination algorithm which uses the estimates of vehicle lateral velocity and side slip angle to help discriminate between rollover and near rollover events has been developed. The algorithm has been calibrated for different vehicle types to demonstrate its performance capabilities. The deployment time improvements for this algorithm compared to the previous algorithm without lateral velocity and side slip are shown in Table 1.

Table 1. Tripped Rollover Event Deployment Time Improvements

Test	Speed (km/h)	Improvement	Test	Speed (km/h)	Improvement
Curb	17	28%	Soil	25	31%
Curb	18	33%	Soil	25	39%
Curb	18	29%	Soil	25	45%
Curb	19	25%	Soil	25	41%
Curb	19	37%	Soil	25	35%
Curb	19	38%	Soil	25	41%
Curb	19	32%	Soil	25	29%
Curb	19	50%	Soil	28	25%
Curb	19	32%	Soil	29	26%
Curb	20	42%	Soil	29	28%
Curb	22	39%	Soil	29	29%
Curb	23	39%	Soil	30	25%
Curb	24	38%	Soil	30	29%
Curb	24	42%	Soil	30	37%
Curb	24	42%	Soil	30	38%
Curb	24	33%	Soil	30	30%
Curb	24	29%	Soil	30	41%
Curb	24	41%	Soil	30	35%
Curb	24	26%	Soil	30	26%
Curb	24	28%	Soil	31	29%
Curb	24	29%	Soil	31	28%
Curb	24	33%	Soil	31	31%
Curb	24	37%	Soil	32	49%
Curb	25	42%	Soil	32	29%
Curb	28	24%	Soil	33	28%
Curb	28	24%	Soil	36	33%
Curb	28	29%	Soil	36	72%

Significant improvements in the deployment times, typically in order of 30-40%, were achieved as a result of using side slip information. These results have been achieved while still maintaining high level of immunity to various abuse, misuse, and driving conditions.

SUMMARY/CONCLUSIONS

In this paper an algorithm for estimation of roll angle and vehicle side slip angle was described. Sensory information from ESC system and occupant protection system were combined to enhance both estimates. Specifically, yaw rate and estimated longitudinal velocity were used to provide an auxiliary estimate of roll angle, while roll angle estimate was used to remove the gravity component from measured lateral acceleration prior to estimating the lateral velocity. The proposed estimation algorithm is relatively simple and does not use vehicle or tire models or parameters.

The results of analysis, simulation and vehicle testing have demonstrated that

• Removal of sensor bias is important in achieving optimal performance.

• The proposed algorithm for sensor bias compensation worked well, but a short period of nearly straight driving is helpful in learning the bias.

• The estimates of roll angle are accurate in normal driving and quick transient maneuvers, driving on banked roads and during rollovers. Small errors may occur when vehicle is in a nearly steady-state condition and the side slip angle is large at the same time. These conditions occur very seldom on extremely slippery surfaces, e.g. on ice.

- Side slip velocity and side slip angle estimates are sufficiently accurate in all realistic driving conditions. Maneuvers involving slow drift of vehicle on slippery surfaces are the most challenging.

- Use of side slip information significantly improves deployment decisions in tripped rollover events; typical improvement in deployment time is in the order of 30-40%.

REFERENCES

1. van Zanten, A.T., "Bosch ESP Systems: 5 Years of Experience," SAE Technical Paper 2000-01-1633, 2000.

2. Hac, A. and Bedner, E., 2007, "Robustness of Side Slip Angle and Control Algorithms for Vehicle Chassis Control", ESV Conference, paper No. 07-0353.

3. Koibuchi, K., Yamamoto, M., Fukuda, Y., and Inagaki, S., "Vehicle Stability Control in Limit Cornering by Active Brake," SAE Technical Paper 960487, 1996.

4. Klier, W., Reim, A., and Stapel, D., "Robust Estimation of Vehicle Sideslip Angle - An Approach w/o Vehicle and Tire Models," SAE Technical Paper 2008-01-0582, 2008.

5. Hac, A., Brown, T., and Martens, J., "Detection of Vehicle Rollover," SAE Technical Paper 2004-01-1757, 2004.

6. Schiffman, J. K., 2000, Vehicle Rollover Sensing Using Short-term Integration", US patent No. 6,038,495.

7. Fennel, H. and Ding, E.L., "A Model-Based Failsafe System for the Continental TEVES Electronic-Stability-Program (ESP)," SAE Technical Paper 2000-01-1635, 2000.

8. D'Silva, S., Sundaram, P., and D'Ambrosio, J.G., "Co-Simulation Platform for Diagnostic Development of a Controlled Chassis System," SAE Technical Paper 2006-01-1058, 2006.

9. SAE International Surface Vehicle Recommended Practice, "Vehicle Dynamics Terminology," SAE Standard J670e, Rev. July 1976.

10. Hahn, H., 2001, "Rigid Body Dynamics of Mechanisms - Theoretical Basis", Springer Verlag.

11. Kröninger, M., Lahmann, R., Lich, T., Schmid, M. et al., "A New Sensing Concept for Tripped Rollovers," SAE Technical Paper 2004-01-0340, 2004.

12. Stengel, R. F., 1986, "Stochastic Optimal Control. Theory and Application", chapter 4, John Wiley & Sons, New York.

CONTACT INFORMATION

Aleksander Hac
Delphi Corporation
aleksander.b.hac@delphi.com

David Nichols
Delphi Corporation
david.j.nichols@delphi.com

Daniel Sygnarowicz
Delphi Corporation
daniel.sygnarowicz@delphi.com

APPENDIX

The process of estimating state variables of dynamic systems from partial measurements has been extensively investigated. Common approaches include a deterministic Luenberger observer, Kalman-Bucy filter and various frequency-domain approaches, such as used in H_∞ control theory. They all yield essentially the same structure of the observer equation. Here Kalman filter theory [12] is used to derive the basic structure of the estimator given by equation (7). First, the Kalman filter for the time-invariant system is summarized.

Consider a dynamic system described by the following state and output equations:

$$\dot{\mathbf{x}} = \mathbf{A}\mathbf{x} + \mathbf{B}\mathbf{u} + \boldsymbol{\xi}$$
$$\mathbf{y} = \mathbf{C}\mathbf{x} + \boldsymbol{\eta}$$

(A1)

Here $\mathbf{x}$ and $\mathbf{y}$ are the state and output vectors, respectively, $\mathbf{A}$, $\mathbf{B}$ and $\mathbf{C}$ are the system, input and output matrices with constant parameters. The vectors $\boldsymbol{\xi}$ and $\boldsymbol{\eta}$ are the system disturbance and measurement noise, respectively. They are assumed to be uncorrelated Gaussian white noise processes with the covariance matrices

$$E[\boldsymbol{\xi}(t)\boldsymbol{\xi}^T(\tau)] = \mathbf{Q}\delta(t-\tau)$$
$$E[\boldsymbol{\eta}(t)\boldsymbol{\eta}^T(\tau)] = \mathbf{R}\delta(t-\tau)$$

(A2)

Here E denotes the expected value, superscript T a transpose of a vector, $\mathbf{Q}$ and $\mathbf{R}$ are symmetric and positive definite matrices describing noise intensities.

If the system (A1) is completely observable (i.e. the observability matrix, $[\mathbf{C}^T, (\mathbf{CA})^T, (\mathbf{CA}^2)^T, \ldots]^T$ is of full rank), then the optimal estimate of the state vector $\mathbf{x}$, $\mathbf{x}_e$, is given by

$$\dot{\mathbf{x}}_e = \mathbf{A}\mathbf{x}_e + \mathbf{B}\mathbf{u} + \mathbf{K}(\mathbf{y} - \mathbf{C}\mathbf{x}_e)$$

(A3)

The observer gain matrix, $\mathbf{K}$, is given by

$$\mathbf{K} = \mathbf{P}\mathbf{C}^T\mathbf{R}^{-1}$$

(A4)

where $\mathbf{P}$ is a symmetric positive definite solution of the algebraic matrix Riccati equation:

$$\mathbf{A}\mathbf{P} + \mathbf{P}\mathbf{A}^T - \mathbf{P}\mathbf{C}^T\mathbf{R}^{-1}\mathbf{C}\mathbf{P} + \mathbf{Q} = \mathbf{0}$$

(A5)

The Kalman filter is an estimator with the gain matrix, that provides an optimal balance between our reliance on the state equation and the measured outputs. While strict mathematical formulation requires the state and output disturbances to be white noise processes, in engineering practice the levels of these noises (as indicated by intensity matrices $\mathbf{Q}$ and $\mathbf{R}$) are indicative of accuracy or the degree of reliance on state equations and measured outputs, respectively.

Within this framework, the problem of estimating the roll angle of vehicle body can be cast as follows. If the roll angle is the state ($x = \phi$) then the state equation is

$$\dot{\phi} = \omega_{xc} + \xi$$

(A6)

where ω_{xc} is the measured and centered roll rate and ξ is the disturbance. While there is no direct measurement relating the measured variable via a linear algebraic output equation to the roll angle, a pseudo-measurement obtained from the kinematic relationship (6) can be used instead. Hence, the output equation can be written as

$$y = \phi + \eta$$

(A7)

Here the output $y = \phi_{ek}$ (pseudo-measurement from equation 6) and η is the measurement error. Equations (A6) and (A7) are the state and output equations with $x = \phi$, $y = \phi_{ek}$, $u = \omega_{xc}$, $A = 0$, $B = 1$, $C = 1$, all scalar quantities. It is straightforward to show that the Kalman filter estimate of roll angle, ϕ_e, is given by

$$\dot{\phi}_e = -k\phi_e + \omega_{xc} + k\phi_{ek}$$

(A8)

Here k is the filter gain. It is seen that equations (A8) and (7) are identical when b_{bl} is substituted for k. Let the noise intensities be given by $Q = q^2$ and $R = r^2$, respectively. Then it follows from equations (A4) and (A5) that the filter gain is

$$k = \frac{q}{r}$$

(A9)

If our reliance on the dynamic equation (A6) is high as compared to the output equation, then it is reasonable to assume that q is much smaller than r and the gain k is small (much less than 1). This is the case during quick transient maneuvers when the estimate from the kinematic equation, used by the output equation, is inaccurate. Conversely, during nearly steady-state maneuvers the kinematic estimate is accurate and one can rely more heavily on the output equation. Therefore the value of r should be lower and the filter gain k higher. Thus the proposed logic of switching between the low and high values of the filter parameter b_{bl} (i.e., the Klaman filter gain) depending on conditions of motion can be explained and justified on the basis of Kalman filter theory and the proposed estimator can be viewed as a Kalman filter with gain adaptation.

Sensor Data Fusion for Active Safety Systems

2010-01-2332
Published
10/19/2010

Jorge Sans Sangorrin, Jan Sparbert, Ulrike Ahlrichs and Wolfgang Branz
Robert Bosch GmbH

Oliver Schwindt
Robert Bosch LLC

Copyright © 2010 SAE International

ABSTRACT

Active safety systems will have a great impact in the next generation of vehicles. This is partly originated by the increasing consumer's interest for safety and partly by new traffic safety laws. Control actions in the vehicle are based on an extensive environment model which contains information about relevant objects in vehicle surroundings. Sensor data fusion integrates measurements from different surround sensors into this environment model. In order to avoid system malfunctions, high reliability in the interpretation of the situation, and therefore in the environment model, is essential. Hence, the main idea of data fusion is to make use of the advantages of using multiple sensors and different technologies in order to fulfill these requirements, which are especially high due to autonomous interventions in vehicle dynamics (e. g. automatic emergency braking). The technical challenge in the development of a serial product relies in the implementation with given sensors, as well as in the risk assessment of the system.

INTRODUCTION

In front impacts, automatic emergency brake systems reduce vehicle velocity to minimize accident damages. In a critical situation, the use of a warning signal allows the driver to react before activation of the automatic brake. If the driver does not react after the warning signal, the automatic brake is activated [2, 3]. Due to this autonomous intervention in vehicle dynamics, requirements in terms of reliability and accuracy in the information about the vehicle surroundings are very high.

At low speed approaches, and depending whether the driver is attentive or not, the optimal warning time point may differ. In case of inattention, the warning signal has to come early enough so that sufficient reaction time is given to the driver. However, attentive drivers wait normally until latest possibility for convenient braking. The braking time point in this case is usually after the warning signal - provided at time point appropriated for inattentive drivers, and therefore attentive drivers might be disturbed with warning signals.

Another type of dilemma arises because the degree of danger can be deescalated either by braking or by evading the object. Especially in approaches at high speed, the object distance for collision avoidance by braking is longer than for object evasion. Thus, convenient evasion of the object is still possible at the time point for avoiding the collision by brake activation. Therefore, knowledge whether to evade the object is possible becomes crucial for solving this second dilemma. Hence, in addition to vehicles in the same lane, to take parallel traffic into account is of great advantage.

With relation to accident statistics, Figure 1 illustrates the main collision types within the field of effect of automatic brake systems for cars and similar vehicles, like pick-ups or vans. From this figure, a system that takes front traffic into account, may contribute to reduce damages in about 27% of all accidents.

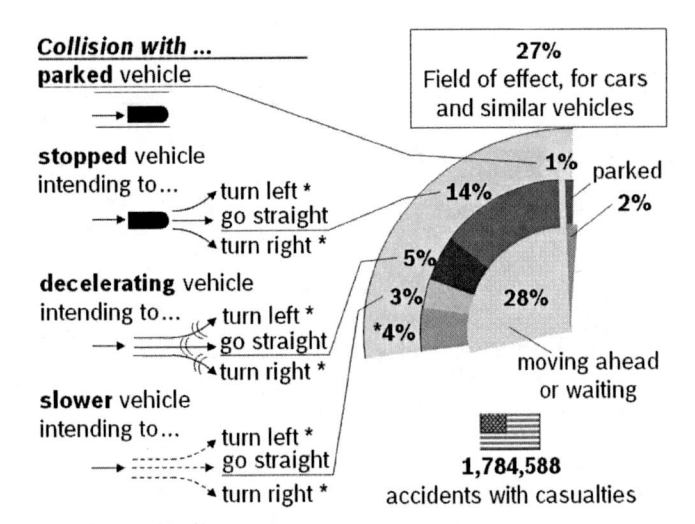

Figure 1. Accident statistics. Field of effect of collision avoidance braking systems for cars and similar vehicles, like pick-ups or vans. Sources: StBA, GIDAS, GES, NHTSA, Year 2006.

Requirements for the prototype system described in next sections have been derived systematically top-down from function to data fusion, and from data fusion to sensors. Required quality in detection and object data has been established for frontal collisions, in which the own, and right and left parallel lanes are occupied by stopped vehicles.

CONCEPTUAL ISSUES

Modeling of vehicle surroundings addresses the representation of relevant data (referred as *environment model* or *surround model* or *data model*) and processing of sensor measurements. For an automatic emergency brake system, the data model represents processed sensor information, which is provided to the next processing step, the *situation interpretation*, for risk assessment of the traffic situation (e.g. position of collision relevant objects relative to own vehicle and its uncertainty, calculated from distance measurements, and represented by a state vector with mean and variance values).

DATA MODELS

The environment model represents data about objects in vehicle surroundings expressed by probabilities and their density functions. The main factors to consider in the selection of the model are: the class of relevant objects, type of information about them, degree of detail of probability density functions and processing algorithms.

One possible choice is to model an object relying on a parameterized *single best estimate*, which is inferred from sensor data. This estimate is summarized in a state vector - expected to have a relatively low number of components per object. The components of the state vector describe, for example, position, dynamics and their uncertainties of objects in vehicle surroundings. Uncertainties are commonly approximated by Gaussian distributions due to the low number of necessary parameters as well as low required processing complexity.

However, in this case, model accuracy and certainty depend on the degree of matching of the real densities with Gaussian (see [5, 6]).

An alternative to the single best estimate approach is to approximate probabilities and density functions by a finite number of values, each roughly corresponding to a space in the data model. Some methods, like *grid-based*, decompose model space in cells, which contain the cumulative probability of density functions. Other methods, approximate density functions by finitely many samples called *particles*. No strong parametric assumptions are made, and therefore these methods are able to cope with varying and very different density functions. The number of parameters to approximate densities is variable, and the quality of approximations depends on it. As the number of parameters increase, these methods converge to the true density. Nevertheless, the computational complexity becomes usually, for that reason, a critical issue (see also [6]).

For reduction of the amount of data, grid cells or particles can be clustered into polygons. Polygon periphery is defined with vertices and edges, which results in a *polygon-based* data model. Thus, for example, one polygon would correspond to a certain space of the data model with cumulative probability higher than a certain threshold. However, in order to achieve a significant data reduction, the number of vertices must be lowest possible, while keeping the model accuracy sufficiently high. As for grid and particles, computational complexity can be a critical aspect, if no significant data reduction is achieved.

DATA PROCESSING

Processing of sensor data is necessary, since measurements are usually corrupted by noise and limited by factors like resolution. Furthermore, if all necessary object quantities are not provided, inference of non-measured quantities becomes necessary. *Object hypothesis* is a term that will be used to refer processed sets of sensing data produced by the same object and assumptions about its motion.

Selection of data processing algorithms depends on the data model used. Examples of references are: [5] with different methods for the single best estimate data model, and [6] for grid- and particle-based data models.

fusion levels

In the development of sensor data fusion, a major conceptual aspect is the selection of the level at which the information from the different sensors is processed. The main choices are *feature level* and *object level*. Figure 2 illustrates the main signal processing steps from sensors to situation interpretation, and fusion levels.

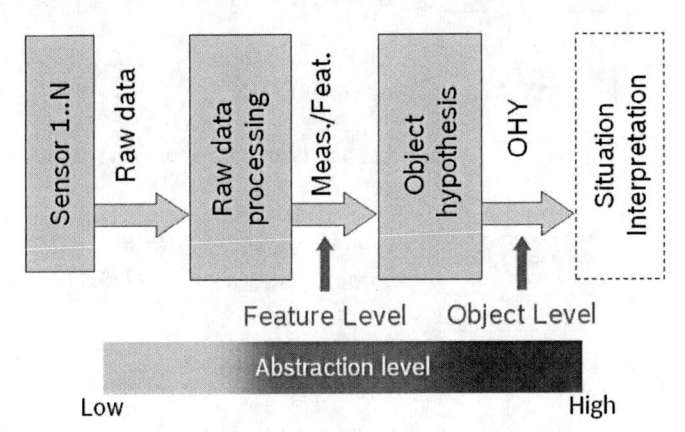

Figure 2. Main levels for sensor data fusion. OHY: Object hypotheses.

Sensor data fusion at feature level processes all available data at the same place. Object hypotheses are generated from measurements and features provided by all sensors. For example, radar locations (distance and angle measurements) are processed together with disparities from stereo-video. At this level, the difficulties that a certain sensor might have for initiating object hypotheses are supported by information from other sensors.

By the alternative sensor data fusion at object level, each single sensor generates object hypotheses based on its own measurements. For example, object position[1] is computed independently by radar and by stereo-video from locations and disparities respectively; data fusion integrates object position from each sensor into a fused object position. An advantage of this level is that algorithms can be optimized for a single sensor technology, and degradation of single sensor information does not affect to other sensors.

centralized and decentralized algorithm architectures

Depending on the fusion level, there are two basic algorithm architectures: centralized and decentralized for data fusion at feature level and object level respectively.

Figure 3 illustrates the centralized architecture. In this architecture, fused object hypotheses are computed by using early processed sensor raw data, i.e. measurements and/or features, from all sensors.

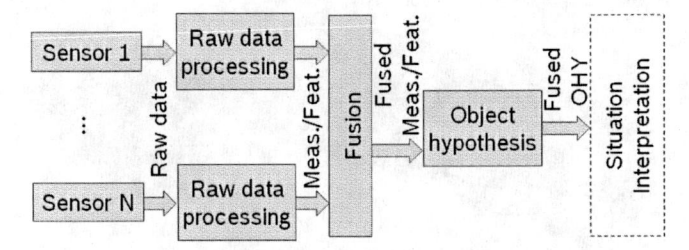

Figure 3. Centralized architecture. OHY: Object hypotheses.

Figure 4 illustrates the decentralized architecture. In this architecture, early processed sensor raw data is used to initiate and update single sensor object hypotheses. In the fusion step, fused object hypotheses are computed from single sensor object hypotheses.

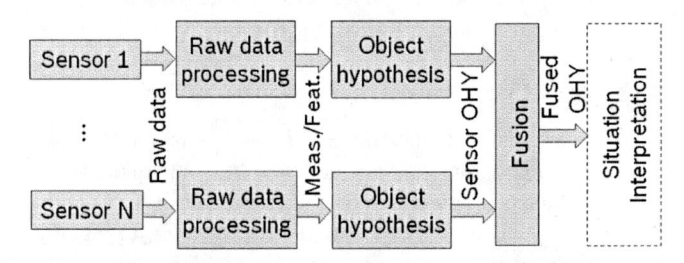

Figure 4. Decentralized architecture. OHY: Object hypotheses.

PROTOTYPE WITH AUTOMATIC EMERGENCY BRAKE SYSTEM

Within the framework of the public funded project AKTIV [7], a vehicle prototype has been equipped with environment sensors, electromechanical components for automatic brake activations and computer resources for data processing.

SENSING SYSTEM

After deriving requirements from data fusion to the sensing system, the combination of Long Range Radar and stereo-video was selected to deal with frontal collisions. Radar provided measurements about longitudinal position and dynamics, and position in lateral direction. Video provided data of position and dynamics in longitudinal and lateral directions, as well as object dimensions (width and length). Figure 5 illustrates these sensors.

---

[1] Note that this is a subspace of the object hypothesis

Figure 5. Sensing system. Stereo-video (left), Long Range Radar (right).

Multiple sensors provide redundant and additional data. This, together with sensor data fusion, allows computation of a data model with higher accuracy and reliability, than using one single sensor. Furthermore, if different technologies are used, higher robustness in object detection against single sensor false alarms is achieved, since each sensor is sensitive to different types of objects (see also [1]).

DATA FUSION

A single best estimate data model with Gaussian distributions for approximating uncertainties is used. In comparison to the other approaches, lower amount of data - and bus load, and lower computational needs for calculating and storing the relevant quantities supports the selection of this model. State vectors include the quantities: position and dynamics (velocity and acceleration) in Cartesian coordinates, object dimensions, uncertainties (variances) in these data, and probability for assessing the collision relevance of a certain object.

Sensor data is fused at object level, and processing algorithms are implemented with decentralized architecture. Important advantages of this approach are: First, parallel processing in the generation of single sensor object hypotheses is possible, which contributes to reduction of the overall cycle time. Second, in comparison to fusion at feature level with centralized architecture, easier adaption of algorithms to sensing variants allow cost reduction in development of different product lines.

Radar and stereo-video generate object hypotheses using only their own data. Fused object hypotheses are then computed at each cycle time in the processing steps illustrated in Figure 6.

Single sensor object hypotheses provide data relative to their sensor mounting position. Hence, the first step consists on spatial alignment of data to a common reference - block ALI. The selected point is the middle of the rear edge of the vehicle. This reference point is the origin of a Cartesian coordinate system, which is used every cycle time for data fusion.

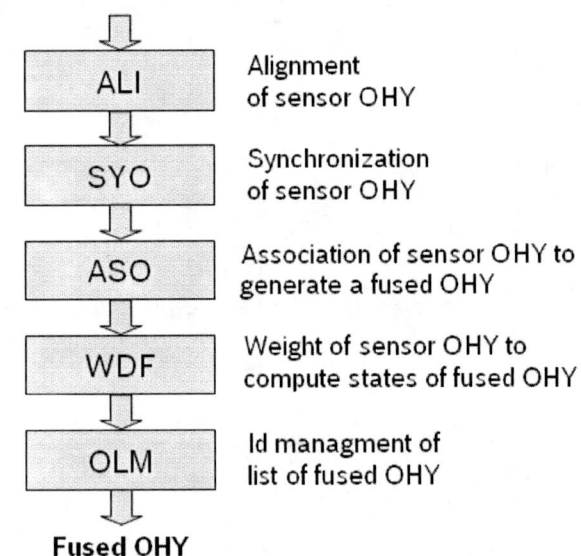

Figure 6. Main processing blocks for generation of fused object hypotheses. OHY: Object hypotheses.

Furthermore, sensors have different cycle times and latency, and therefore, data is synchronized to a "fusion time point" - block SYO. In this step, predictions are carried out for calculating the state vector at this time point. This prediction time must be as short as possible in order to minimize propagation of uncertainties.

Data association determines the correspondences between single sensor object hypotheses from all sensors, i.e. to determine which hypothesis from each sensor was generated by the same object - block ASO. To that end, a gating-based method is used (see [5]). Basically, the difference between each aligned and synchronized object hypothesis from stereo-video and radar is calculated. If the difference of a pair does not exceed a certain maximum, the gate is satisfied and the association succeeds. For calculating the difference of a pair, single differences of a subspace of the state vector, e.g. between position components in Cartesian coordinates, are calculated. The total difference is calculated as function of all single differences. Since assumed-zero-mean Gaussian statistics are used, the gate - or maximum difference - can be determined as the three-standard-deviation level. Association conflicts might occur, if one single sensor object hypothesis would satisfy the gate for forming more than one pair; for example, in a scenario in which objects would be very close to each other. In this case, information about past associations and about the object class becomes advantageous to solve these conflicts.

After having determined the pairs, the state vector of the fused object hypothesis is calculated as the weighted sum of state vectors of single sensor object hypotheses - block WDF.

The variance and the differences between components in state vectors are used to compute the weights.

Management of the list of fused object hypotheses is principally used to avoid generation of double hypotheses and to assign a unique id-number during the life of each hypothesis - block OLM.

It is worth mentioning that no tracking loop is performed in the data fusion. Association probability - computed in block ASO - is stored in a two-dimensional matrix. Row and column are indexed by the id-number of radar's and video's object hypotheses used to form the pair respectively. Thus, knowledge about which object hypothesis from radar and stereo-video formed the pair in pervious cycles is available for update of fused object hypotheses. This requires object hypotheses, both in sensors and in fusion, to have the same id-number over time.

RISK AND PERFORMANCE ASSESSMENT

Analysis of false interventions with real traffic data [3] concluded that, the higher the reduction of relative velocity, the higher probability of rear-end crash with following vehicles. This means that the potential damage grows with the deceleration caused by a false intervention of the automatic brake. Therefore, an extensive analysis of functional risks is essential.

Within this context, some of the aspects to consider in the development and test of the sensor data fusion for future serial products are:

• The surround sensing model must have sufficiently high potential to satisfy requirements for autonomous brake interventions.

• The rate of false brake interventions due to errors in the environment model has to be ensured by means of extensive continuous tests.

To that end, measurable parameters were used for performance evaluation of the sensor data fusion implemented in the vehicle prototype. Furthermore, these parameters allowed tracking of system and algorithm performance in the development phase. Following parameters were used:

• *Rate of Correct Association*: All sensors detect the same object part, and data association succeeds. This parameter allows measurement of the functional benefit of fusion algorithms.

• *Rate of False Association*: In this case, sensor data from different objects is wrongly associated. This type of error might result in a false intervention of an automatic brake. The risk of damage in a non-dangerous situation might be excessively high, and therefore this rate becomes the main parameter for ensuring correct functional performance of the fusion.

Figure 7 illustrates the own vehicle (E) approaching to stopped vehicles in congestion (A, B, C), and shows correct and false association cases, with information about object position provided by two different sensors.

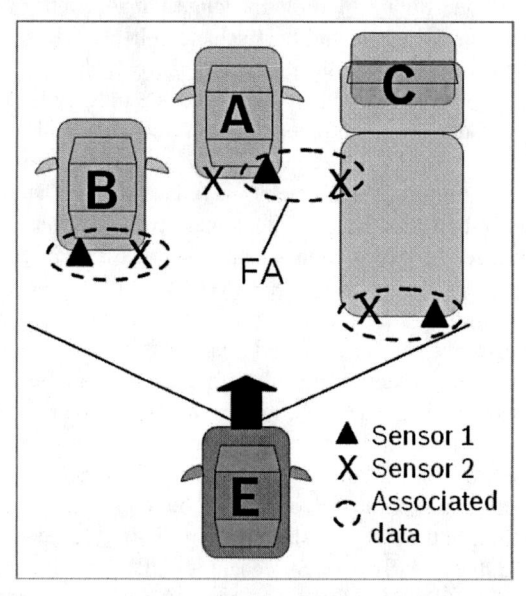

Figure 7. Approach to stopped vehicles in congestion. A, B, C are stopped vehicles. FA: False Association.

In addition to the parameters above, the ultimate system performance depends also on the accuracy of the quantities in the data model. The reason is that, position and dynamics of collision relevant objects, as well as their uncertainties, are the main information used in predictions for synchronization and in the situation interpretation for risk evaluation.

Intensive tests with data recorded from public roads showed the data fusion implemented to have potential to satisfy risk requirements for autonomous brake interventions. So far the Rate of Correct Associations observed was over 90%, and no False Associations were observed. The potential benefit was shown in prepared non-destructive scenarios. Here, frontal collisions up to about 50kph - host vehicle velocity - could be avoided in approaches to stopped vehicles with right and left lanes occupied.

Risk assessment must still be enhanced for serial production. The final benefit, i.e. the maximum reduction of host velocity, of an automatic emergency brake system will be limited by the maximum risk of false brake interventions allowed (see [4]).

EXEMPLARY DATA

Figure 8 illustrates stopped objects and object detection with stereo-video. The L-shapes in the picture show object position, width and length (the latter only for vehicles with lateral side visible). Figure 9 illustrates the position and dimensions (part of the information contained in the single sensor object hypothesis) of the objects in Figure 8.

Figure 10 and Figure 11 show the longitudinal distance to the object in the own lane and host vehicle velocity (vhost) over time for the scenario illustrated in Figure 8. In these two figures, the decision of "collision avoidable only by braking" is made considering the object only on own lane, and objects on own and parallel lanes respectively. At this decision time point, the automatic emergency brake is activated. The time difference between these points in these two situations is due to the different information about vehicle surroundings used for decision-making. Thus, if objects on parallel lanes and object width are used, earlier recognition of crash unavoidability can be achieved, than taking only objects on own lane into account, and therefore brakes can be earlier activated (compare deceleration in vhost-curves). Note that, in the two situations, host vehicle velocity is within the range in which evasion of the object on own lane is still possible after the time point for avoiding the collision by brake activation. In the situation illustrated in Figure 11, the crash is avoided due to early enough recognition that the collision can be avoided only by braking, i.e. evasion of front vehicle does not avoid an impact with another object on parallel lanes.

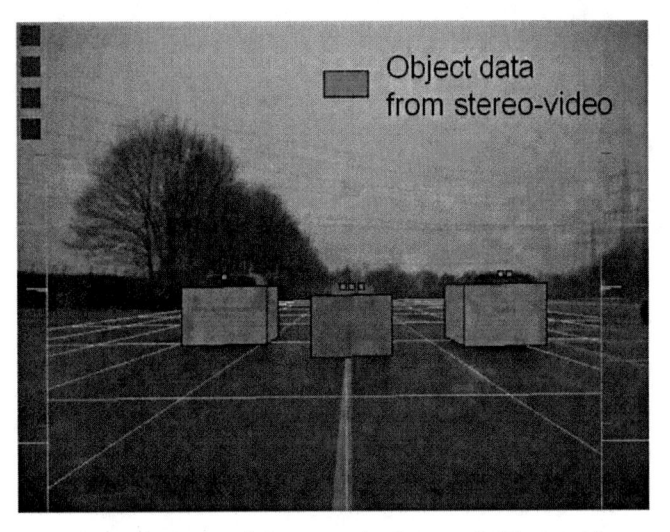

Figure 8. Image (left camera) of stopped objects with parallel lanes occupied, and single frame object detection with stereo-video (semi-transparent areas).

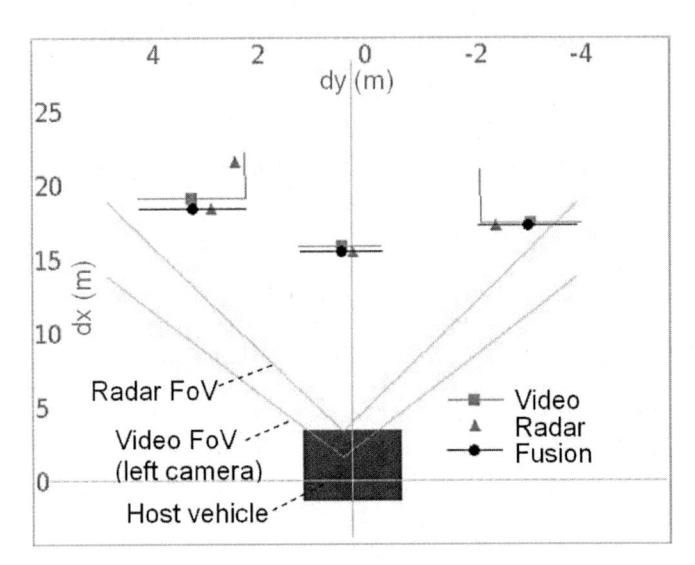

Figure 9. Representation of object position and dimensions (width and length) in Cartesian coordinates (dy, dx). Approximate radar's and stereo-video's field of view (FoV).

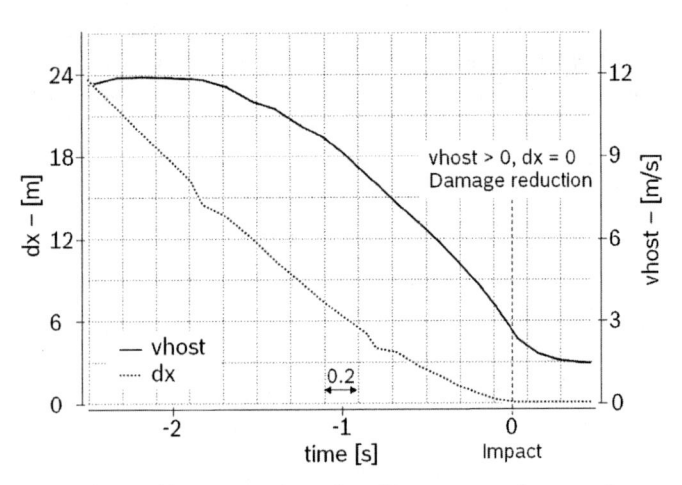

Figure 10. Approach to the object on own lane, and automatic brake activation considering only own lane for decision "collision avoidable only by braking". Relative longitudinal distance - dx (left axis) and host vehicle velocity - vhost (right axis) over time.

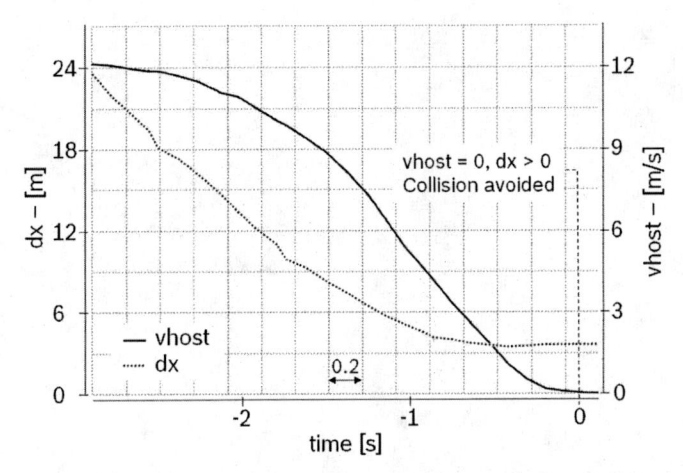

Figure 11. Approach to the object on own lane, and automatic brake activation considering own and parallel lanes for decision "collision avoidable only by braking". Relative longitudinal distance - dx (left axis) and host vehicle velocity - vhost (right axis) over time.

SUMMARY

The challenge in modeling of vehicle environment for active safety is to deal with uncertainties in a highly dynamic environment. Highways, open roads and cities are very different scenarios that may need particular attention. For risk evaluation of the situation, the probability of collision is calculated for relevant objects in vehicle surroundings. To that end, predictions are computed to estimate the situation after a certain time. Since uncertainties might grow drastically within the required prediction time, the accuracy of the information about the objects must be sufficiently high. Limitations of computational resources are a further factor to take into account. Therefore, in the development of serial products, a trade-off between performance and complexity has to be met.

From analysis of accident statistics, higher effectiveness in collision avoidance and damage reduction is achieved, if parallel traffic is taken in the situation assessment into account. The main reason is that possible evasion trajectories can be rejected due to objects in right and left lanes, and therefore an automatic brake can be activated earlier than if only the own lane is considered. The object width is also important information for recognition of evasion paths between objects.

A vehicle prototype was equipped with an automatic emergency brake system. A Long Range Radar and stereo-video provided information about objects in vehicle surroundings. Each sensor processed own measurements to compute object information. By means of sensor data fusion, sensor objects were integrated into fused objects. The use of multiple sensors and different technologies showed high robustness in object detection, because one sensor supported the other when difficulties and/or conflicts occurred.

Experiments with data recorded from public roads showed the environment model provided by the data fusion to have potential to satisfy safety requirements for autonomous emergency brake systems. The potential benefit of using information about objects on parallel lanes was shown in prepared non-destructive scenarios; higher reduction of host vehicle velocity was achieved, than when taking only the own lane into account.

Next steps will focus on further intensive tests in order to ensure sufficient safe performance for autonomous brake systems. Since the probability of rear-end collision with following traffic grows with the deceleration caused by autonomous brake activations, the highest possible reduction of velocity will be determined by the highest risk of false brake interventions allowed for a serial product.

REFERENCES

1. Hoetzer, D., Freundt, D., Lucas, B., "Automotive Radar and Vision Systems - Ready for the Mass Volume Market", presented at Vehicle Dynamics Expo 2008, USA, October 23, 2008.

2. Haering, J., Wilhelm, U., Branz, W., "Entwicklungsstrategie für Kollisionswarnsysteme im Niedrigpreissegment", VDI-FVT-Jahrbuch 2009: 60-64, December 2008.

3. Branz, W., Staempfle, M., "Kollisionsvermeidung in Längsverkehr - die Vision vom Unfallfreien Fahren rueckt naeher", presented at 3. Tagung Aktive Sicherheit durch Fahreassistenz, Germany, April 7-8, 2008.

4. Ebel, S., Wilhelm, U., Grimm, A., Sailer, U., "Wie sicher ist sicher genug?", presented at 6. Workshop Fahrerassistenzsysteme, Germany, September 28-30, 2009.

5. Blackman, S.S., Multiple-Target Tracking with Radar Applications, Artech House, Norwood, ISBN 0-89006-179-3, 1986.

6. Thrun, S., Burgard, W., Dieter, F., Probabilistic robotics, The MIT Press, Cambridge, ISBN 0-262-20162-3, 2005.

7. Project AKTIV, http://www.aktiv-online.org/english/projects.html, January 2010.

CONTACT INFORMATION

Jorge Sans Sangorrin
Robert Bosch GmbH
P.O. Box 300240, D-70442 Stuttgart, Germany

ACKNOWLEDGMENTS

This work has been partly funded by the German ministry of economics (BMWi) - Project Adaptive und kooperative Technologien für den intelligenten Verkehr (AKTIV).

SAE International

Frontal Crash Testing and Vehicle Safety Designs: A Historical Perspective Based on Crash Test Studies

2010-01-1024
Published
04/12/2010

Randa Radwan Samaha, Kennerly Digges and Thomas Fesich
George Washington Univ.

Michaela Authaler

Copyright © 2010 SAE International

ABSTRACT

This study tracks vehicle design changes and frontal crash test performance in NHTSA's NCAP and IIHS consumer information tests since the mid-90s for the Honda Accord and Toyota Camry. The objective was to provide insights into how passenger cars have changed in response to frontal consumer information tests. The history of major design changes for each model was researched and documented. The occupant injury measures from both NHTSA and IIHS were computed and the ratings compiled for several generations of both vehicles.

Changes in vehicle crash pulse and occupant injury measures from both NCAP and IIHS tests, and from Canadian low speed rigid barrier tests, when available, were used to assess driver frontal protection for various vehicle generations. Loading of the rigid barrier in NCAP tests was used to evaluate front end stiffness changes over the years. Overall, results indicate that the consumer testing programs have induced many improvements in both the Camry and Accord vehicle structures, and restraint systems technologies. Front end stiffness increased with the first Camry platform change after the introduction of the IIHS test as contrasted with a decrease in the front end stiffness for the Accord. Overall, advances in restraint technologies, specifically seat belt load limiters, seemed to permit the resulting higher acceleration to be accommodated at the NCAP and IIHS test conditions. The Camry, in contrast to the Accord, appeared to be more optimized for the NCAP test condition, where it recorded an increase in the risk of chest injury and possibly head injury in crashes with a smaller occupant size and at a lower speed than the current NCAP and IIHS test speeds.

INTRODUCTION

In the United States (U.S.), crash tests to provide consumer information about vehicle safety have been performed since 1978. The initial tests conducted by the National Highway Traffic Safety Administration (NHTSA) were frontal tests and crashed vehicles into a rigid frontal barrier at 35 mph (56 km/h). The Insurance Institute of Highway Safety (IIHS) began frontal testing into a deformable offset barrier at 40 mph (64.4 km/h) in 1995. This study tracks vehicle design changes and frontal crash test performance in NHTSA's New Car Assessment Program (NCAP) and IIHS frontal consumer information tests since the mid-90s for two popular passenger car models. The objective was to gain insights into how vehicles have changed in response to frontal consumer information tests. The study attempted to address the following questions:

- What is the nature of technologies in typical five star vehicles? Are such vehicles stiffer after the introduction of the IIHS frontal test?

- How have the manufacturers addressed the need to gain five stars and still provide protection at lower crash severities?

- What have been the major safety improvements over the years, and what technologies were involved?

FACTORS THAT INFLUENCE VEHICLE STIFFNESS AND RESTRAINT DESIGNS

Since its origin in 1978, the NCAP test condition has been a crash speed of 35 mph with full frontal vehicle engagement into a rigid barrier oriented perpendicular to the direction of travel. Several studies have been performed to determine the effect of this test condition on vehicle fleet stiffness. Park et al. (1999) studied NCAP tests of 175 light trucks, vans, and sport utility vehicles (LTVs), beginning with model year (MY) 1983, and concluded that LTVs have generally not become stiffer over fourteen years of NCAP testing, and that the less stiff LTVs have higher NCAP ratings [1]. However, Swanson et al. (2003) concluded that there has been a steady increase in the passenger car fleet stiffness over the period of 1982 to 2001, based on a study of load cell force, dynamic displacement, and measured crush of passenger car NCAP testing [2]. More recently, using energy equivalent static stiffness, Digges et al. (2009) showed that the average stiffness of cars and sport utility vehicles tested in the NCAP program has increased over the years [3].

Compared to the NCAP test condition, the IIHS offset deformable barrier crash test is particularly demanding of the vehicle structure, as it allows only 40 percent of the vehicle width to manage the crash energy. Nolan and Lund (2001) examined the relationship of vehicle mass, stiffness, and front-end length to the structural rating for IIHS offset tests of 1995 to 2001 model year passenger cars and LTVs [4]. They found that the majority of vehicles achieved improved structural performance without significant changes to front end stiffness for the first half-meter of crush. Such vehicles had essentially the same stiffness in the front crush zone but had rapidly increasing stiffness as the deformation approached the occupant compartment. This finding did not apply to vehicles which experienced catastrophic structural collapse when the IIHS offset test was initiated.

Since the early 1990s, the U.S. frontal crash protection standard (FMVSS 208) required vehicles be designed to meet a 30 mph (48 km/hr) full frontal, vehicle into a rigid barrier test with unbelted 50th %tile male dummies at the driver and right front passenger positions. FMVSS 208 has been amended several times in recent years to address injuries attributed air bags that were overly aggressive and to increase the speed of the crash test. The initial change, beginning in MY 1998, permitted air bags that satisfied a requirement to protect unbelted occupants by subjecting 50th %tile male dummies at the driver and right front passenger positions to a sled test that produced a standard 30 mph crash pulse. Subsequently, the standard was changed so that beginning in model year 2005 the unbelted 50th %tile male dummies were certified using a vehicle crash test into a rigid barrier at 25 mph (40 km/hr). Static tests of the vehicle's driver air bag with close-in 5th %tile female dummy were also required. The aggressiveness of the passenger air bag was controlled by static tests with child dummies or by disabling the air bag when a small occupant was present. Other requirements introduced at that time included a crash test with driver and right front passenger 5th %tile female belted dummies in an offset frontal crash at 25 mph into a deformable barrier. This test was to control the adequacy of the crash sensor and the aggressiveness of the air bag when subjected to this common crash occurrence. The standard also required 30 mph tests into a rigid barrier with two front seat belted dummies. Tests were required with both 50th %tile male and 5th %tile female dummies. In the most recent changes, the crash speed for both the belted 5 %tile and the 50th %tile dummies has been raised from 30 mph to 35 mph.

The changes in the U.S. standards have permitted the air bags to be less aggressive and better tailored to protect belted occupants in a 35 mph crash into a rigid barrier. However, based on lower severity tests conducted by the Canadian Government, the injury risk in lower severity crashes may be higher than in the 35 mph crash required by NCAP [5].

APPROACH

The Toyota Camry and Honda Accord were analyzed for this study. The two passenger cars were chosen according to the following guidelines: popular models with similar weight and generations spanning many years, for which both NHTSA and IIHS crash tests were available.

The history of major design changes for each model was researched, including changes in safety technologies, weight, and size over time. The occupant injury measures were computed and the frontal crash test ratings from both NHTSA and IIHS were compiled for several generations of both vehicles. Changes in vehicle crash pulse and occupant injury measures from both NCAP and IIHS tests, and from Canadian low speed (40km/h) rigid barrier tests, were used to assess driver frontal protection for various vehicle generations. The force versus displacement curves, derived from barrier load data and occupant compartment displacements, were used to assess changes in the vehicles' front end stiffness over the years.

DATA SOURCES

MODEL HISTORY

The history of model design changes was compiled from manufacturer press releases, as well as news websites; information was confirmed from the various test protocols for the model years tested by NHTSA and IIHS. The following was researched and documented:

- Times and dates of vehicle model changes
- Changes in safety technology
- Weight and size of the vehicle over time

TEST DATA

The data for the NCAP frontal crash tests and the Canadian low speed frontal crash tests were obtained from the online NHTSA Vehicle Crash Database [6]. In addition to the raw data, detailed reports, images, and videos for each crash test are available from this database. The data, reports, images, and videos for the IIHS frontal offset tests were obtained from the online IIHS TechData repository [7]. All available crash tests for each of the two vehicle models were analyzed.

U.S. NCAP Frontal Test: This test simulates a head on collision between two similar vehicles with a speed of 35 mph each. A vehicle is subject to a full frontal impact to a rigid barrier with load cells. The impact speed is 35 mph (56 km/h) with a 0° impact angle. Belted 50th %tile male HIII dummies in the driver and right front passenger seats are used to record data. The position of the dummies is specified by the test procedure of FMVSS 208. More than 100 data channels are collected and recorded, and multiple normal and high speed cameras are used for documentation.

Dummy Instrumentation

• Head tri-axial, chest tri-axial, pelvis tri-axial accelerometers

• Chest displacement potentiometer

• Upper neck transducers

• Right/left femur load cells

• Lower leg instrumentation

Vehicle Instrumentation

• Belts (lap/shoulder) load cells

• Floor pan rear left/right, engine top and bottom, brake caliper left/right accelerometers

• Dash panel accelerometer

• Sill accelerometer

• Rear seat cross member left/right accelerometers

Rating Criteria - The NCAP rates cars on a scale of one to five stars, based on the Head Injury Criterion (HIC) and the chest deceleration. In addition, the femur load of the left and right femur is reported, but not included in the star rating. It is important to note that NCAP ratings can be compared only among vehicles of similar weight. Given the same frontal ratings for two vehicles, one heavier than the other, the heavier vehicle will typically offer better protection in real-world frontal crashes. In a frontal test, the kinetic energy depends on both the speed and weight of the vehicle tested. The heavier vehicle has to manage more kinetic energy in the NCAP frontal test.

Table 1. NCAP Frontal Crash Test Ratings [8]

Stars	=	Chance of Serious Injury
★★★★★	=	10 percent or less chance of serious injury
★★★★	=	11 percent to 20 percent chance of serious injury
★★★	=	21 percent to 35 percent chance of serious injury
★★	=	36 percent to 45 percent chance of serious injury
★	=	46 percent or greater chance of serious injury

Canadian Low Speed Barrier Test: As part of a joint research program with NHTSA, Transport Canada has performed seventeen crash tests of 2005 and 2006 vehicles onto a full rigid barrier with a 0° impact angle. The Canadian tests, also available online from the NHTSA Vehicle Crash Database, differed from the NCAP test in impact speed (40 km/h) and dummy size (two belted 5th %tile female Hybrid III dummies were positioned in the driver and right front passenger seats). Therefore, these tests can be used to compare the performance of a vehicle under NCAP conditions with the performance of the vehicle under less severe conditions on a small sized dummy, as the instrumentation on the vehicle and dummy are similar.

IIHS Frontal Offset Test: This test simulates an offset crash between two similar vehicles travelling at 40 mph. The barrier used by IIHS is a deformable aluminum honeycomb structure. The test is conducted at an impact speed of 40 mph (64.4 km/h) with a 40% overlap and a 0° impact angle. As such, about ½ of the frontal structure is loaded and only a portion of the vehicle's front end has to manage the entire crash energy. The IIHS test conditions are more severe to the vehicle structure, but less severe to the restraint systems. One 50th %tile male HIII dummy is positioned in the driver seat using the UMTRI ATD Positioning seating procedure, which is based on a survey of preferred driving positions of volunteers [9, 10]. The IIHS seating is typically more rearward than the mid-track seat position of the NCAP test procedure.

Measurements - For the IIHS crash test evaluation, 28 measures are recorded. They include similar measurements on the dummy as stated for frontal NCAP. In addition, the intrusion into the safety cage is measured on nine points at the driver side in the front of the compartment. These measurements include the steering wheel movement, the instrument panel, and the foot well intrusions.

Rating Criteria - The IIHS frontal test evaluates the structural performance, the injury measures, and the restraints/dummy kinematics. As for the NCAP frontal test, the IIHS ratings can be compared only among vehicles of similar weight. Ratings are given for:

• Structure/safety cage: Amount and pattern of intrusion into occupant compartment

• Injuries measures: Head/neck injury, chest injury, leg/foot injury (left and right)

• Restraints/dummy kinematics

Each of these ratings is classified as Good (G), Acceptable (A), Marginal (M), or Poor (P). In addition, an overall evaluation with the same four different ratings is given. It should be noted that the IIHS rating is not only based on measured and calculated numbers, but also on factors like the dummy kinematics and some undesirable patterns of deformation that are not represented by the numerical values of the measurements. The IIHS evaluation guidelines for rating injury measures are provided in the Appendix. The ratings for compartment intrusion for the foot rest, toe pan, and brake pedal are as follows. The final Structure rating typically will not be more than one rating level better than the worst measurement.

• "Good" rating is less than 15 mm intrusion

• "Acceptable" rating is between 15 and 22.5 mm intrusion

• "Marginal" rating is between 22.5 and 30 mm intrusion

• "Poor" rating is over 30 mm intrusion.

METHODS

The NHTSA Signal Analysis software were used to process the data, compute injury measures, and perform analysis of the barrier load cell data. The raw data was filtered according to SAE J211-1[11]. The vehicles' performances under the three test conditions were compared. The following measures were tracked, when available, over the vehicle model years.

• **Head Injury Criteria, HIC36** (computed according to FMVSS 208 for both driver and passenger)

• **Neck Injury Criteria, Nij** (computed from upper neck forces and moments)

• **Maximum chest deflection**

• **Shoulder belt loads** (time histories and maximum values, unfortunately, not all were recorded)

• **Pelvis deceleration**

• **Femur loads**

• **Foot well intrusions**

• **Vehicle crash pulse** (Acceleration was taken from longitudinal accelerometers not placed near the crash zone, i.e., away from the localized deformations. If available, the acceleration of the rear seat, the rear floor pan, or the rear sill was used. If possible, two signals from comparable positions (e.g. rear seat left and right) were filtered separately and averaged.)

Force-displacement and force-time histories

To gain a better understanding of the actual crash event and front end structure stiffness, force-displacement and force-time curves were evaluated. The force-displacement characteristics for a given model were generated by the total force exerted on the barrier plotted against the dynamic compartment crush. The dynamic crush or displacement was calculated by twice integrating the vehicle crash pulse. This approach is typically used in modeling crush characteristics of vehicle front end structures. The underlying assumption is that the NCAP vehicle to rigid-full-barrier crash test can be modeled as a simple one mass-spring system with the vehicle front end structure represented by a nonlinear spring.

TOYOTA CAMRY - DATA, ANALYSIS, & DISCUSSION

MODEL HISTORY

The Toyota Camry has been the bestselling car in the United States in 9 out of the last 10 years [12]. In 2007, the sixth generation was introduced to the U.S. market, being the latest version of a car that went into production to replace the Corona in Toyota's model lineup in 1983 [13]. In this study, the investigated models are the 3rd, 4th, 5th, and 6th generations, spanning model years 1992 to 2007. Both the 3rd and 5th generations were based on entirely new platforms. The 2002 model year Camry was the first major redesign after the IIHS test program was initiated, the first all-new platform in 10 years. In the following, data has been collected from Toyota [10], as well as from websites, and were confirmed by the various test protocols for the model years tested by NHTSA and IIHS. The safety features for the investigated generations are highlighted below [12, 13, 14, 15, and 16].

3rd Generation (1992-1996)

• New platform

• Driver front airbag

• Passenger front airbag in 1994

4th Generation (1997-2001)

• "Revamped"- Additional two inches in wheelbase for a more spacious cabin, more legroom for rear-seat passengers, a lowered beltline, enhanced impact protection, and traction control

• Optional front side torso airbags and antilock brake standard in 1998

• Mid-model "facelift" in 2000 (exterior styling enhancements with a new front fascia that features a new grille and bumper design)

5th Generation (2002-2006)

• New platform

• Multi-stage advanced driver and passenger airbags in 2002

• Optional front seat-mounted torso airbags and front and rear head curtain airbags

• Anti-lock brakes standard 2005

• Belt force limiters introduced in 2005

• "Facelift" for the front for 2005 (freshened exterior styling with redesigned headlights, taillights, grille, and wheels)

6th Generation (2007- 2010)

• Redesigned for 2007 model year: longer wheelbase and wider track for a comfortable ride and roomy interior

• Equipped with belt force limiters and pretensioners, as well as a driver knee airbag, in addition to the front, side, and head curtain airbags already introduced in earlier model years

• First Camry available in a Hybrid version

• "Facelift for 2010 (exterior styling enhancements with a new front fascia that features a new grille and bumper design and larger projection headlamps)

Data Sources

The following test data were analyzed for the 3rd, 4th, 5th, and 6th generations (MY 1992-2007):

• 8 × NHTSA NCAP: 35 mph full frontal rigid barrier, 50th %tile male Hybrid III dummy

• 2 × Transport Canada Research: 35 mph full frontal rigid barrier, 5th %tile female dummy

• 4 × IIHS Consumer Program: 40 mph offset frontal deformable barrier, after 1995

• 1 × Transport Canada Research: 25 mph full frontal rigid barrier, 5th %tile female dummy

VEHICLE PERFORMANCE AND DATA ANALYSIS

As shown in Table 2, Toyota Camry frontal NCAP and IIHS ratings have improved over the last 15 years, and are currently the best rating from both programs.

Table 2. Overview of NCAP and IIHS Ratings for the Camry/

Model Year	NHTSA driver	NHTSA passenger	IIHS Overall Evaluation	Structure Safety Cage	Head/Neck	Chest	Leg/foot left	Leg/foot right	Restraints/dummy kinematics	IIHS MY Rating *Year Tested
1992	4	4								
1994	4	3	A	A	A	G	A	G	G	1995* (1994-1996)
1997	4	4	G	G	A	G	G	G	G	1997* (1997-2001)
2000	4	5								
2002	5	4	G	G	G	G	G	A	G	2002* (2002-2006)
2004	4	4								
2005	5	5								
2007	5	5	G	G	G	G	G	G	G	2007* (2007-2010)

Front End Structure Characteristics

Figures 1 and 2 show the force displacement characteristics of four generations of Camry based on the barrier force data measured in the 35 mph NCAP tests. It is evident that there was a structural change in 2002 with the introduction of the new platform resulting in an increase in frontal stiffness at 400 mm of displacement or crush, coupled with a decrease in stiffness at 200-300mm of displacement.

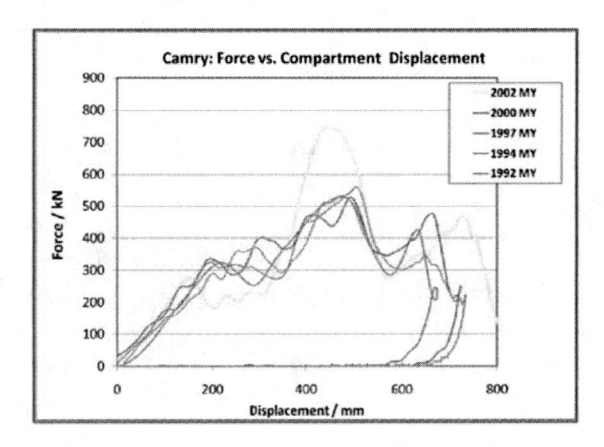

Figure 1. Camry Force-Displacement in NCAP Frontal Barrier Tests - MY 1992-2002

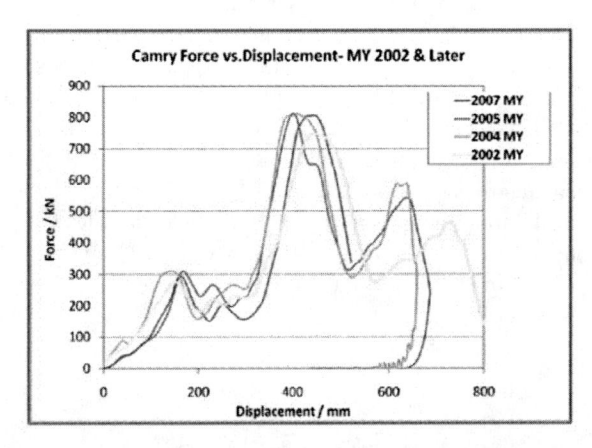

Figure 2. Camry Force-Displacement in NCAP Frontal Barrier Tests - MY 2002-2007

The most pronounced difference in the force-displacement characteristic introduced with the new platform in 2002 is a second force peak at approximately 400-425 mm of displacement with increased loads of around 800 kN compared to maximum loads of around 550 kN in the older models. This increased front end stiffness for the 5th generation is also recorded for the 6th generation tested for the model year 2007. The increase in front end stiffness characteristics of the Camry starting in MY 2002 coincides with an 8% increase in weight from 1386 kg to 1498 kg from the 3th to the 5th generation (Table 3); a 6% increase in the 2002 model year. Also, this change of stiffness characteristics is in line with the finding by Nolan (2001) that vehicles typically achieved improved structural performance in the IIHS offset test by essentially keeping the same stiffness in the front crush zone but rapidly increasing stiffness as the deformation approaches the occupant compartment [4].

Table 3. Dimensions of Toyota Camry (3rd through 6th generations)

Generation	Weight (kg)	Length/ Wheel base (mm)	Width (mm)
3rd (1992-1996)	1386	4770/ 2619	1770
4th (1997-2001)	1415	4789/ 2672	1780
5th (2002-2006)	1498	4805/ 2720	1795
6th (2007- 2010)	1498	4805/ 2776	1820

Some of the changes in force-displacement characteristics could be attributed to changes in the geometry of the car. With the introduction of the 5th generation MY 2002, the Camry was designed with a substantially shorter front end (shown in Table 4). While this allowed more occupant room without a major increase in the exterior dimensions, the decrease in front-end length implies a reduction in available crush distance, which typically leads to an increase in force to manage the crash energy. The total length of the front end, calculated by subtracting the distance "rear surface vehicle to firewall" from the total length of the vehicle indicates that the front end length decreased over 17% from 1240 mm in 1994 to 1029 mm in the 2007 model.

*Table 4. Length Dimensions of Camry Models**

Model Year	Total Length (mm)	Front End Length (mm)	Front to Engine Distance (mm)
1994	4761	1240	871
2000	4784	1174	829
2004	4813	1132	691
2007	4754	1029	730

*Based on measurements from NHTSA test reports [6]

The engine block is also closer to the front end. The distance from the front of the vehicle to the front surface of the engine block decreased 21% from 871 mm in the 1994 model to 691 mm for the 2004 model, but slightly increased to 730 mm in the 2007 model. It can be shown that as a result, the heavy and massive engine block becomes more involved in the crash. The displacement of the engine towards the passenger compartment increased from 173 mm in the 2000 model to 242 mm in the 2004 model. It should be noted that all vehicles have been tested on 4-cylinder engines with almost constant engine displacement of between 2.2 and 2.4l, so there should be no effect of engine size or the number of cylinders on the geometry of the front end.

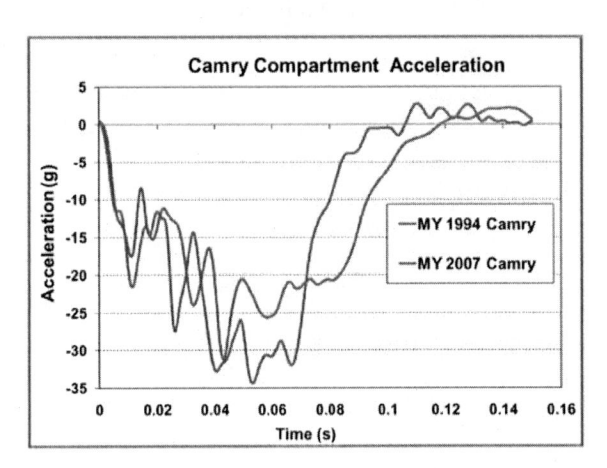

Figure 3. Camry Compartment Acceleration in NCAP Frontal Tests

The Camry front end changes have resulted in a higher acceleration and narrower pulse experienced by the occupant compartment, as shown in Figure 3.

Shoulder Belt Loads

The shoulder belt load of both the driver and passenger decreased for the NCAP tests over time. The introduction of shoulder belt force limiters in 2005 can be observed by plots of the recorded belt loads for the driver in Figure 4. The belt load for the lighter 5th %tile dummy in the available 2005

Canadian test at 25 mph is lower than the load recorded under NCAP conditions.

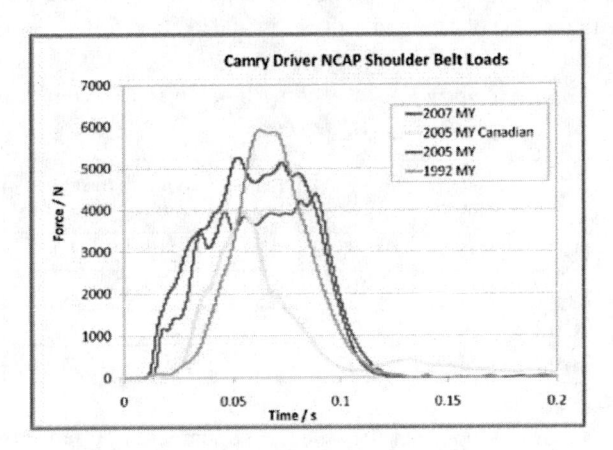

Figure 4. Camry Driver Shoulder Belt Force vs. Time in Frontal Barrier Tests

The substantial reduction in the shoulder belt loads achieved by the 2007 Camry is encouraging. Foret-Bruno et al. (2001) observed 50-60% reduction of moderate thoracic injuries (AIS2+) and 85% reduction of serious thoracic injuries (AIS3+) in cases with 4 kN load limitation, based on a study of 347 accidents involving belted occupants in vehicles equipped with 4 or 6 kN load limiters [17].

Head Injury Criterion

Figures 5 and 6 show no trend for the driver head injury criteria (HIC) in spite of structural changes; values were well under the threshold of 1000. Even with the increased stiffness characteristics in 2002 model years, the driver HIC drops in 2002 with the introduction of multi-stage advanced airbags.

IIHS Head/Neck ratings for MY 1997, 1995, and 2002: Although the IIHS test for the 1997 MY resulted in a HIC_{36} of 470 and a HIC_{15} of 422, the back of the dummy head impacted the B-pillar during rebound, producing a peak head resultant acceleration of 127G, which is considerably above an 80G injury threshold. This resulted in the "Acceptable" Head/Neck Rating for 2007. In comparison, the 1995 MY IIHS tests also had rebound with an impact to the B-pillar, producing a peak of 58G. This, in combination of what was assessed as uncontrolled torso kinematics, resulted in the "Acceptable" Head/Neck Rating for that year as well. Although the IIHS tests show a constant increase in HIC from 470 in 1997 to 651 in 2002, the Head/Neck Rating rises from "Acceptable" to "Good" that year, as the dummy kinematics were better controlled.

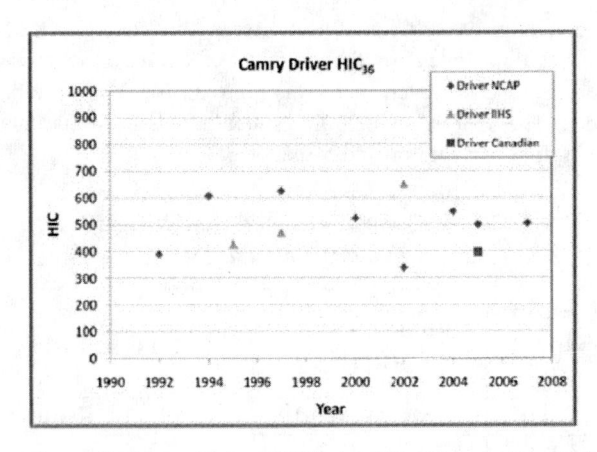

Figure 5. Camry Driver HIC vs. Model Year in Frontal Crash Tests

For NCAP, the elevated passenger HIC for MY 1994 is responsible for the three star rating. This high HIC value occurs when the passenger airbag was first introduced. Consequently, the passenger HIC decreased over the years with the notable observation that HIC is 34% higher for the 5th %tile female dummy in the 25mph Canadian test than the 50th %tile male HIII dummy in the 35mph NCAP test.

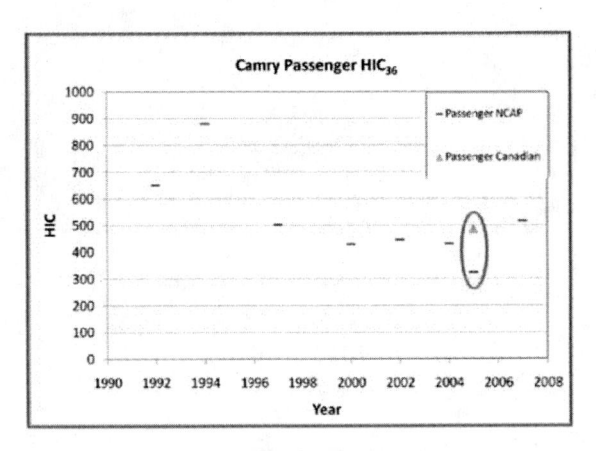

Figure 6. Camry Passenger HIC vs. Model Year in Frontal Crash Tests

Femur Loads, Tibia Indices, and Foot well Intrusions

The femur loads for the IIHS driver, shown in Table 5, were well below the 7.3 kN cut-off for a "Good" rating over the years. However, they were further reduced beginning in 2002, which coincides with the increased front end stiffness of the Camry as discussed above.

The NCAP driver femur loads were well below the threshold of 10 kN; however, there is no trend over model years (shown in Figures 7 and 8).

With a value of 0.79, initially an "Acceptable" rating in 1995, the left tibia index for the IIHS driver decreased over the model years; however, there was no trend for the right tibia index. The 0.99 value in 2002 resulted in an almost marginal rating for the right tibia index that year.

Table 5. Camry Femur Forces and Tibia Indices in IIHS Frontal Offset Tests [7]

Camry Model Year	Femur Force (kN) Left	Femur Force (kN) Right	Max Tibia Index Left	Max Tibia Index Right
1995	3.0	3.8	0.79	0.37
1997	3.9	2.4	0.57	0.68
2002	1.4	1.8	0.39	0.99
2007	1.3	1.7	0.25	0.49

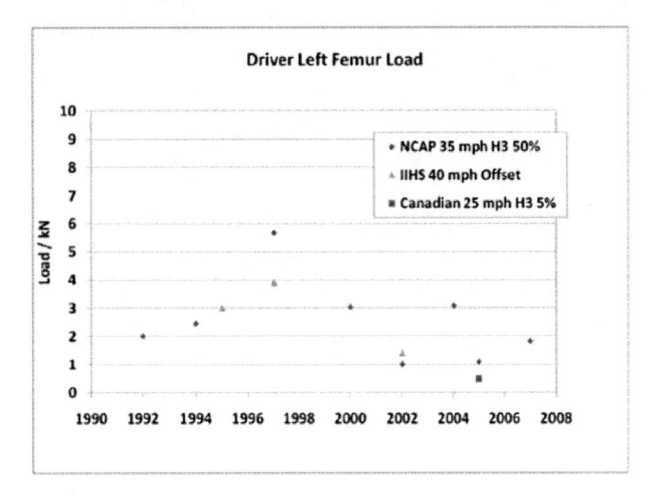

Figure 7. Driver Left Femur Loads

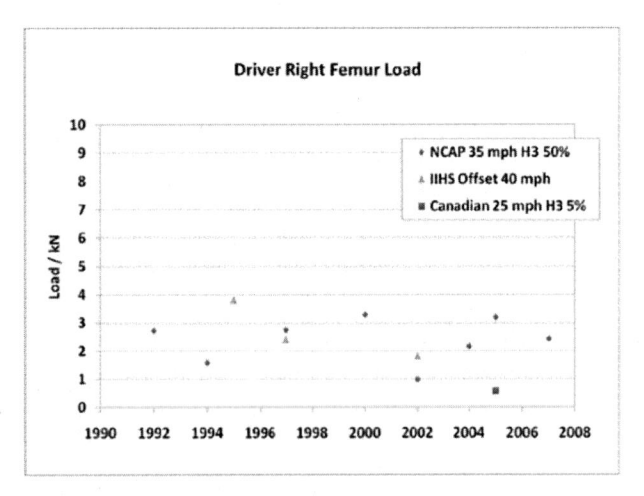

Figure 8. Driver Right Femur Loads

The IIHS driver foot well intrusions decreased substantially in 1997, resulting in a improved rating of "Good" in the Structural/Safety Rating, and again decreased considerably with the new platform introduced in 2002 (Table 6).

Table 6. Camry Foot Well Intrusions in IIHS Frontal Offset Tests

Camry Model Year	Footrest (cm)	Left (cm)	Center (cm)	Right (cm)	Brake Pedal (cm)
1995	14	25	23	21	14
1997	4	11	12	11	7
2002	4	6	7	5	7
2007	4	6	5	4	4

It is worth noting that the wheel base increased 10 cm from the 3rd to the 5th generations (from MY 1995 to MY 1997), as shown in Table 3, while the overall length increased only 1.5 cm. This increase in cabin room, along with the decrease in stiffness after 200-225 mm of displacement, shown in Figure 9, could have contributed to the decrease in foot well intrusions in 1997.

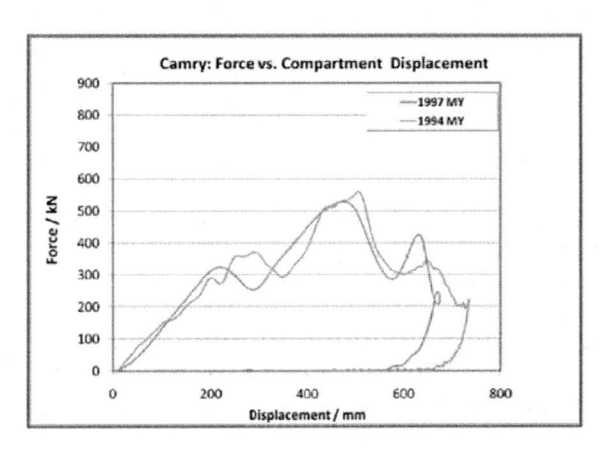

Figure 9. NCAP Force-Displacement Characteristics, MY 1994 and 1997

Chest Deflection

Figures 10 and 11 show that the NCAP chest compressions did not have a clear trend over the years. The driver chest compressions in the IIHS tests decreased over the years, but were higher than NCAP. It is notable that the chest deflection is 24% higher for the driver dummy (and 29% higher for the passenger dummy) for the 2005 model in the 25 mph speed Canadian test than in the 35 mph NCAP test.

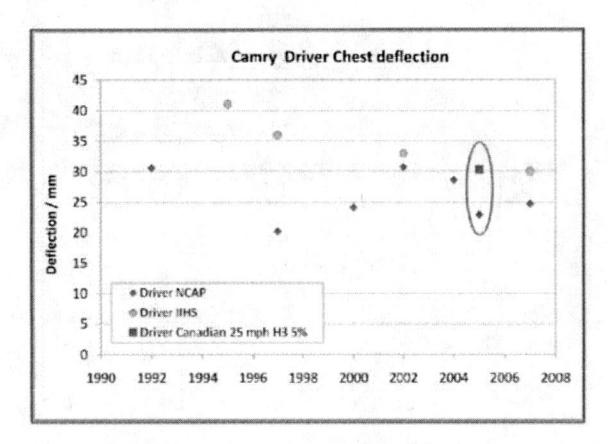

Figure 10. Camry Driver Chest Deflection vs. Model Year in Frontal Crash Tests

For the tested 2005 model, the NCAP driver chest deflection is 23 mm, whereas the Transport Canada test shows a deflection of 30 mm. If the Injury Risk Curve presented by Laituri et al. (2003) is used to compare these deflections, it can be seen that the NCAP test yields an injury risk of AIS 3+ injuries of almost zero; whereas the Canadian result yields an injury risk of approximately 5-10% [18]. Although both values are small, the difference is obvious.

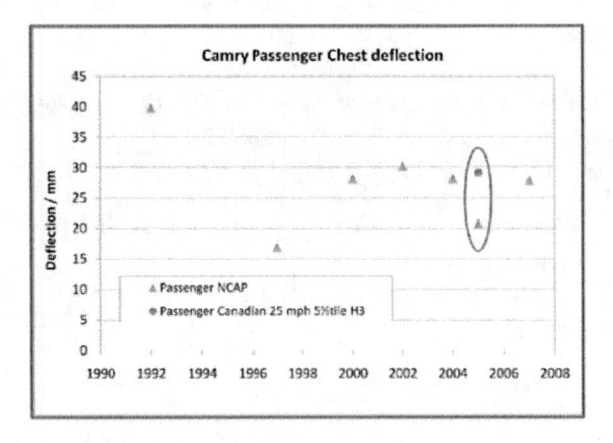

Figure 11. Camry Passenger Chest Deflection vs. Model Year in Frontal Crash Tests

Additional Injury Measures

• The NCAP driver pelvis deceleration decreased from 1995 to 2007 from around 70 G to 45 Gs; whereas they remained at a low level around 40-50G for the passenger, as compared with the 100 G threshold.

There was no clear trend in both the IIHS and NCAP driver N_{ij} values, which typically ranged from 0.2 to 0.4, well below the threshold of 0.8 for a "Good" rating. The $N_{com-ext}$ (0.44 vs. 0.22) and $N_{ten-ext}$ (0.31 vs. 0.23) for the driver dummy were slightly higher in the Canadian tests with the 5th %tile female HIII dummy.

HONDA ACCORD - DATA, ANALYSIS, & DISCUSSION

MODEL HISTORY

The Honda Accord has been manufactured since 1976 and has always been one of the top five selling vehicles in the U.S. market. In this study, the investigated models are the 5th, 6th, and 7th generations spanning model years 1994 to 2007. The safety features for these generations are highlighted below [19, 20, and 21].

5th generation (1994-1997)

• Driver and passenger airbag

• Three point lap and shoulder belt

• Dual locking shoulder belts

• Anchorage height adjustable belt

6th generation (1998-2002) new features

• Unit body, high strength cabin, and ladder sub frame

• Right front and all rear shoulder belt retractors are convertible from emergency to locking for ease of children restriction use

• New features in MY 2000

• Multi-stage deployment system in airbags

• Knee bolster

• New features in MY 2001

• Stabilizer front bar (26.5mm)

• Driver and passenger advanced airbag system

• Belt pretensioner

• Energy management feature (including belt force limiter)

7th generation (2003-2007)

• Redesign of the Stabilizer front bar (25.4mm)

• Claimed that the new model is 27% stiffer

• The engine slips under the occupant compartment in a crash

Data Sources

The following tests data were analyzed for the 5th, 6th, and 7th generations (MY 1994-2007):

• 5 × NHTSA NCAP: 35 mph full frontal rigid barrier, 50th %tile male Hybrid III dummy

- 3 × IIHS Consumer Program: 40 mph offset frontal deformable barrier, after 1995
- 1 × Transport Canada Research: 25 mph full frontal rigid barrier, 5th %tile female HIII dummy

VEHICLE PERFORMANCE AND DATA ANALYSIS

As shown in Table 7, the Honda Accord frontal NCAP and IIHS ratings have improved over the last 15 years, and are currently the best rating from both programs.

Table 7. Overview of NCAP and IIHS Ratings for the Accord

Model Year	NHTSA driver	NHTSA passenger	IIHS Overall Evaluation	Structure Safety Cage	Head/Neck	Chest	Leg/foot left	Leg/foot right	Restraints/dummy kinematics	IIHS MY Rating *Year Tested
1994	4	3	A	A	G	G	P	A	G	1995* (1994-1997)
1997	4	4								
2000	4	4	A	A	G	G	G	P	G	1998* (1998-2002)
2001	5	5								
2003	5	5	G	G	G	G	G	G	G	2003* (2003-2007)

Front End Structure Characteristics

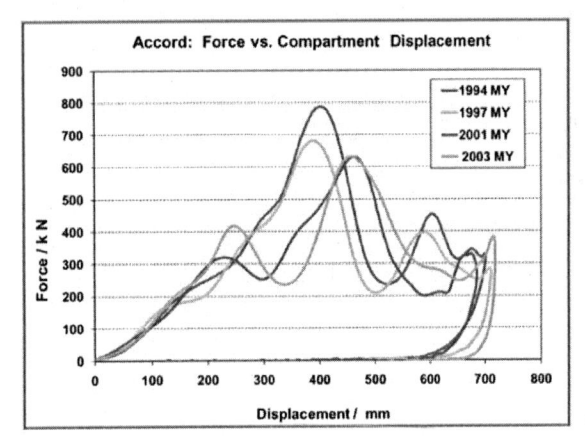

Figure 12. Accord Force-Displacement in NCAP Frontal Barrier Tests

Figure 12 shows the force-displacement characteristics of four generations of the Accord based on the barrier force data measured in the 35 mph NCAP tests. It is evident that there was a substantial front end structural change in 2001 that was further refined in 2003 model year. There was a decrease in stiffness after 250-300 mm of displacement, which matches up with the introduction and further redesign of the Stabilizer front bar.

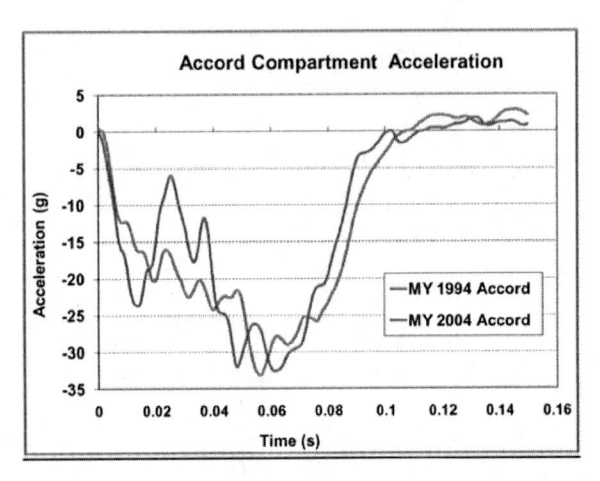

Figure 13. Accord Compartment Acceleration in NCAP Frontal Tests

The reduction in front end stiffness in the 6th and 7th generation Accords occurs even with the weight increase of 7% from 1315 kg to 1410 kg from the 5th to the 7th generation (shown in Table 8), and the small 3.5% decrease in front end length from 1130 mm in MY 1992 to 1090 mm MY 2004 (shown in Table 9). These changes are reflected in the crash pulse with reduced accelerations before 40ms but similar maximum accelerations between the 5th and 7th generation Accord, as shown in Figure 13.

Table 8. Dimensions of the Accord (5th through 7th generations)

Generation	Weight (kg)	Length (m)	Width (mm)
5th (1994-1997)	1315	4675	1580
6th (1998-2002)	1390	4603-4800	1780
7th (2003-2007)	1410	4815	1820

*Table 9. Length Dimensions of Accord Models**

Model Year	Total Length (mm)	Front End Length (mm)	Front to Engine Distance (mm)
1994	4657	1130	685
2000	4787	1092	580
2004	4800	1090	580

*Based on measurements from NHTSA test reports [6]

*Based on measurements from NHTSA test reports [6]

Shoulder Belt Loads

The introduction of belt pretensioners and load limiters in the 2001 model year can be observed in Figures 14 and 15. The Accord achieved a substantial reduction the shoulder belt loads beginning in 2001, which is expected to correspond to a substantial reduction of thoracic injuries in real world accident as shown by Foret-Bruno et al. (2001) and described earlier in the section for the Camry [17].

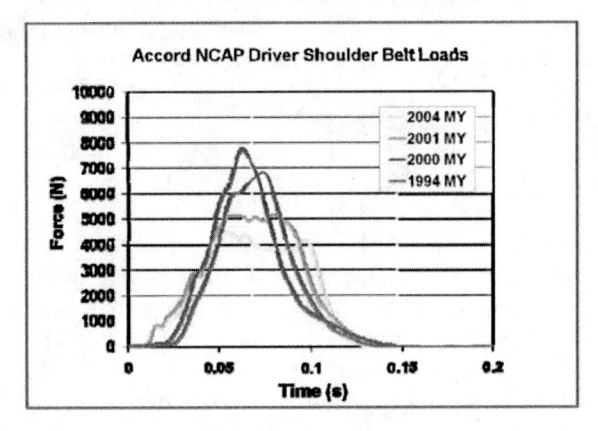

Figure 14. Accord Driver Shoulder Belt Force vs. Time in NCAP Frontal Barrier Tests

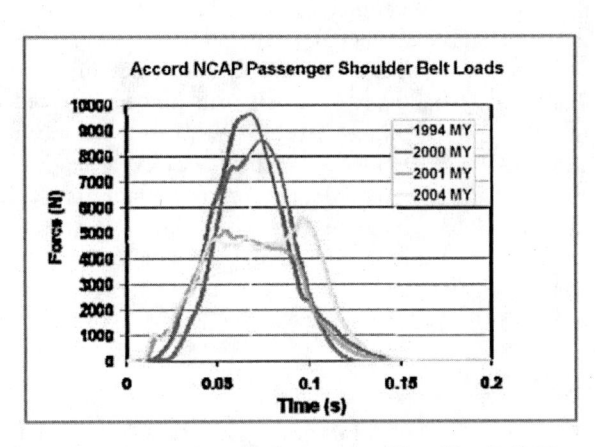

Figure 15. Accord Front Passenger Shoulder Belt Force vs. Time in NCAP Frontal Barrier Tests

Head Injury Criterion

Figure 16 shows a substantial drop in the NCAP driver HIC and a more modest drop in the NCAP passenger HIC for the 2001 vehicle, the model year in which advanced airbags systems, belt pretensioners, and force limiters were introduced for both the driver and passenger.

In a study of NCAP data of 1998 to 2001 model year cars and light trucks, Walz (2004) found statistically significant reductions of HIC by 232 for both the drivers and passengers of vehicles equipped with pretensioners and force limiters when compared to vehicles without such improvements [22].

Belt pretensioners retract the safety belt almost instantly in a crash to remove excess slack, while force limiters allow the belts to yield in a crash, preventing the shoulder belt from applying a high load on the occupant chest.

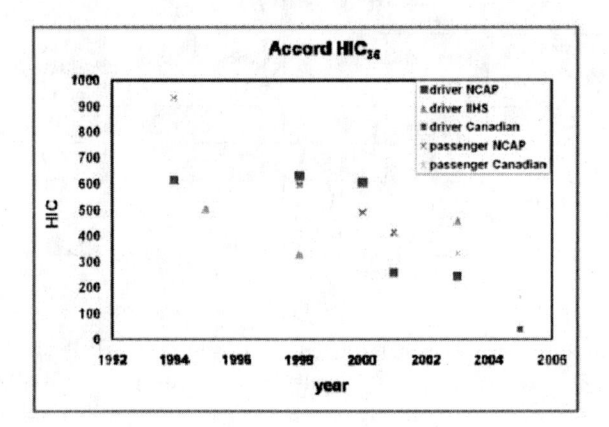

Figure 16. Accord Driver HIC vs. Model Year in Frontal Crash Tests

Femur Loads, Tibia Indices, and Foot Well Intrusions

Even with the decrease in front end stiffness in the 7th generation, there is a decrease in the IIHS femur forces and a substantial reduction in tibia indices and foot well intrusions in the IIHS frontal offset test as shown in Tables 10 and 11. The ratings for the leg/foot improved from "Poor" to "Good" in the 7th generation (2007 MY), and the rating for the Structure/Safety cage improved from "Acceptable" to "Good".

The left and right femur loads of the NCAP driver and passenger ranged between 5718 N and 1307 N, at least 40% under the threshold of 10 kN; however, a trend could not be determined over the years.

Table 10. Accord Femur Forces and Tibia Indices in IIHS Frontal Offset Tests [7]

Accord Model Year	Femur Force (kN) Left	Femur Force (kN) Right	Max Tibia Index Left	Max Tibia Index Right
1995	3.3	1.0	1.7	0.91
1998	3.6	2.0	0.47	1.39
2003	0.4	0.6	0.29	0.49
2008	1.0	0.3	0.41	0.4

The introduction of a unit body and high strength cabin in MY 1998 (6th generation) Accord seems to result in considerable reductions in compartment intrusion at the foot well area in the IIHS test as shown in Table 11.

Table 11. Accord Foot Well Intrusions in IIHS Frontal Offset Tests [10]

Accord Model Year	Footrest (cm)	Left (cm)	Center (cm)	Right (cm)	Brake Pedal (cm)
1995	16	25	27	23	22
1998	9	18	21	17	12
2003	6	10	13	9	12
2008	7	9	8	6	8

Chest Deflection

Figure 16 shows that the decrease in shoulder belt loads in MY 2001 translated to considerable reduction in chest compression for both the driver and passenger. In her study of NCAP data of 1998 to 2001 model year vehicles, Walz (2004) found statistically significant reductions of chest compression by 10.6 mm for both the drivers and passengers of vehicles equipped with pretensioners and force limiters, as compared to the drivers and passengers of vehicles without these safety improvements [22].

It is worth noting that the chest deflection in the 2005 model lower speed Canadian test for the 5th %tile dummy is at the same level as for the 50th %tile male HIII dummy driver and passenger in the 35 mph NCAP test.

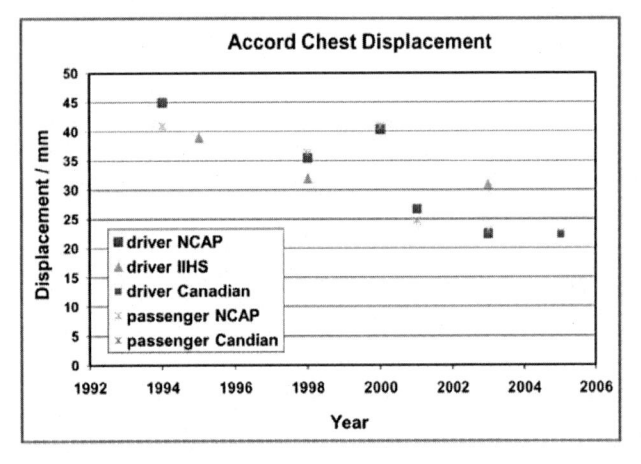

Figure 16. Accord Driver Chest Deflection vs. Model Year in Frontal Crash Tests

Additional Injury Measures

• The NCAP pelvis Gs were basically level over the years, at least 40% below the 100G threshold for pelvis accelerations, with a range of 53-56Gs for the driver (with a decrease to 47Gs with the introduction of the 7th generation Accord) and a range 50-59Gs for the passenger.

• The NCAP N_{ij}'s decreased constantly over the years, but always under the threshold of 1; driver N_{ij}'s were reduced more from 0.79 to 0.12 and the passenger N_{ij}'s decreased from 0.49 to 0.28. The IIHS driver N_{ij}'s were very low, with a range of 0.23-0.33 and no exhibited trend.

SUMMARY/CONCLUSIONS

Results indicate that consumer testing programs have induced many improvements in both the Camry and Accord vehicle structures and restraint systems technologies over the years. Both models have improved over the last 15 years, and currently have the best rating from both frontal NCAP and IIHS testing programs.

Front end stiffness increased with the first Camry platform change, along with a reduction in front end length, after the introduction of the IIHS test. This is contrasted with a decrease in the front end stiffness for the Accord, which had a slight reduction of front end length. The Camry remained at a similar overall length with increasing cabin room, while the Accord increased overall length and cabin room, and maintained a similar front end length. However, both models underwent an increase of 7-8% in weight over the generations studied, spanning the mid 1990s to the current model. The changes in the Camry front structure resulted in higher compartment accelerations as compared with the Accord, which experienced similar compartment peak accelerations as the earlier model years. Overall, the advances in restraint technologies, specifically seat belt load limiters, seemed to permit the resulting higher acceleration to be accommodated at the NCAP and IIHS test conditions. However, the Camry, in contrast with the Accord, appeared to be more optimized for the NCAP test condition, where it recorded an increase in the risk of chest injury and possibly head injury in crashes with a smaller occupant size and at a lower speed than the current NCAP and IIHS test speeds.

REFERENCES

1. Park, B.T., Hackney, J.R., Morgan, R.M., Chan, H. et al., "The New Car Assessment Program: Has It Led to Stiffer Light Trucks and Vans Over the Years?" SAE Technical Paper 1999-01-0064, 1999.

2. Swanson, J., Rockwell T., Beuse, N., Park, B., Summers S., Summers, L., "Evaluation of stiffness measures from the US New Car Assessment Program," Proceedings of the 18th International Technical Conference on the Enhanced Safety of Vehicles (ESV), Paper No. 527-O, 2003.

3. Digges, K., Sahraei, E., Samaha, R., "Opportunities for Improved Rear Seat Child Safety," presented at Protection of Children in Cars, 7th International Conference, December 2009, Munich, Germany.

4. Nolan, J., Lund, A., "Frontal Offset Deformable Barrier Crash Testing and Its Effect On Vehicle Stiffness," Proceedings of the 17th ESV, Paper No. 487, 2001.

5. Digges, K., Dalmotas, D., "Benefits of a Low Severity Frontal Crash Test," 51st Annual Proceedings of AAAM, 2007; 51:299-317.

6. NHTSA Vehicle Crash Test Database, http://www.nhtsa.gov/, September 2009.

7. IIHS TechData, http://techdata.iihs.org/, September 2009.

8. www.safercar.gov for the NCAP Star Rating, September 2009.

9. IIHS, Offset Barrier Test Protocol, http://www.iihs.org/ratings/protocols/pdf/test_protocol_high.pdf, September 2009.

10. Manary, M.A., Reed, M.P., Flannagan, C.A.C., and Schneider, L.W., "ATD Positioning Based on Driver Posture and Position," SAE Technical Paper 983163, 1998.

11. SAE International Surface Vehicle Recommended Practice, "Instrumentation for Impact Test - Part 1 - Electronic Instrumentation," SAE Standard J211-1, Rev. July 2007.

12. Toyota, "50th Anniversary of Toyota, Product History - Camry," http://www.toyota.com, September 2009.

13. Toyota Vehicles, "Toyota Passenger Car Chronology," http://pressroom.toyota.com, September 2009.

14. Automotive News, "3rd Generation Camry Took Toyota to a New Level in '92", Maskerey, M. A., October, 2007.

15. http://en.wikipedia.org/wiki/Camry, December, 2007.

16. Edmunds, INSIDE LINE, "Toyota, Generations", DiPietro, J, http://www.edmunds.com/insideline/do/Features/articleId=46002, September 2009.

17. Foret-Bruno, J.-Y., Trosseille, X., Page, Y., Huère. J.-H. et al., "Comparison of Thoracic Injury Risk in Frontal Car Crashes for Occupant Restrained without Belt Load Limiters and Those Restrained with 6 kN and 4 kN Belt Load Limiters," SAE Technical Paper 2001-22-0009, 2001.

18. Laituri, T.R., Prasad, P., Kachnowski, B.P., Sullivan, K. et al., "Predictions of AIS3+ Thoracic Risks for Belted Occupants and Full-Engagement, Real-World Frontal Impacts: Sensitivity to Various Theoretical Risk Curves," SAE Technical Paper 2003-01-1355, 2003.

19. www.candiandriver.com, November 2007.

20. http://www.tc.gc.ca/en/menu.htm, November 2007.

21. Honda, Press and Media Center, http://corporate.honda.com/press/list.aspx, November 2007.

22. Walz, M., "NCAP Test Improvements with Pretensioners and Load Limiters, "Traffic Injury Prevention," VOL 5; PART 1, Pages 18-25, 2004.

CONTACT INFORMATION

Randa Radwan Samaha
rrsamaha@ncac.gwu.edu
+1 703 201 5277

APPENDIX

Frontal Offset Crashworthiness Evaluation Guidelines for Rating Injury Measures, IIHS June 2009, http://www.iihs.org/ratings/protocols/pdf/measures_frontal.pdf

Body Region	Parameter	IARV	Good – Acceptable	Acceptable – Marginal	Marginal – Poor
Head and neck	HIC-15	700	560	700	840
	N_{ij}	1.00	0.80	1.00	1.20
	Neck axial tension (kN)*	3.3	2.6	3.3	4.0
	Neck compression (kN)*	4.0	3.2	4.0	4.8
Chest	Thoracic spine acceleration (3 ms clip, g)	60	60	75	90
	Sternum deflection (mm)	−50	−50	−60	−75
	Sternum deflection rate (m/s)	−8.2	−6.6	−8.2	−9.8
	Viscous criterion (m/s)	1.0	0.8	1.0	1.2
Leg and foot, left and right	Femur axial force (kN)**	−9.1	−7.3	−9.1	−10.9
	Tibia-femur displacement (mm)	−15	−12	−15	−18
	Tibia index (upper, lower)	1.00	0.80	1.00	1.20
	Tibia axial force (kN)	−8.0	−4.0	−6.0	−8.0
	Foot acceleration (g)	150	150	200	260

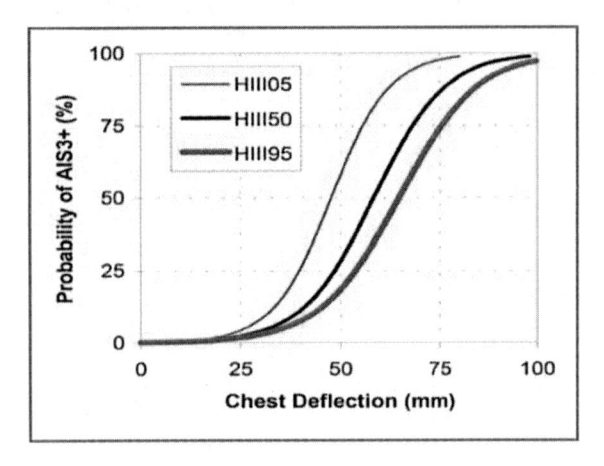

Figure A-1. Risk-curve for various Hybrid III-dummies (source: Laituri chest risk 2003-01-1355)

ROAD SAFETY

2009-01-0592

Synthesizing a System for Improving Road Safety in China

Hongtao Yu, Jiahe Zhang, Yuankui Meng and Shili Ni
CATARC Shanghai Operation

Weijian Han
Ford Motor Company

Peng Ren
Clemson University

Copyright © 2009 SAE International

ABSTRACT

Improvement of road safety is becoming a key task for sustainable mobility in China. A joint study was conducted by China Automotive Technology and Research Center (CATARC) and Ford Motor Company to identify global experience and adopt best practices with Chinese local conditions. The project carried out: 1, Collecting road safety data in China and compare the data systems with US/UK systems; 2, Developing a method to identify most critical factors that cause accidents and fatality; 3, Reviewing experience and lessons learned in US and UK to reduce accidents and fatality; 4, Establish a strategy and recommendations for effectively improving road safety in China within a shorter time period.

INTRODUCTION

China has maintained a fast economic growth since entered the WTO family. As a result of the market demands, the rapid increase of car sales caused the explosion of vehicle in use in China. Cars are no longer the unreachable luxury in daily life for normal people. It is urgent to address road safety issues before getting worse. A most effective way is to identify global experience and lessons learned and adopt best practices with Chinese local conditions. This will allow China to develop a systems approach that includes human behaviors, vehicle safety technologies, traffic management, and road infrastructures to reduce road accidents and fatalities in a shorter time. The objective of the project is to investigate road safety data and the data system in China; identify the most critical areas/factors that cause accidents and fatality in terms of human behaviors, vehicle safety technologies, and road infrastructure; and assemble global best practices to target those critical areas/factors.

THE ROAD SAFETY DATA SYSTEM IN CHINA

REPORT & RECORD SYSTEM

Reliable data is essential to this study. It is important to understand how the road traffic data in China are recorded and reported. Therefore, compatibility can be appreciated by comparing the China's traffic data system with the systems in UK and US.

China road accident recording system

Record is the base of a road accident data system. An accident record contains the first-hand information from the scene of accident. It is the source of all follow-up data and the foundation of further analysis.

1. Workflow of road accident recording

The traffic police from Public Security Bureau in China are responsible for investigating all road accidents. The traffic police record an accident and serves as the origin road safety data. The Public Security Bureau has a network to administrate information flow of road accidents from local to municipal level, and from province to central government.

In the latest "Regulation of Traffic Accident Processing Procedure", the procedure is divided into two types: simplified procedure and normal procedure. At the accident scene, the traffic police will identify the actual accident severity based on the condition of accidents; and choose simplified or normal procedure to handle the accident. Simplified procedure can be used when there is no injury and little property damage. Traffic police will register the accident, identify responsibility of all parties, and write a report that states details of the accident. After signing the report, the involved parties can leave the scene and wait for following up procedures, like compensation etc. Traffic police will use normal procedure when accidents have injuries or severe damage to properties. Traffic police will arrange emergency medical service and protect the accident scene while resume the traffic. After that, the traffic police attended the scene will conduct the accident investigation to record all original information as soon as possible.

After finishing all investigations, the traffic police will put all information together and filing a complete document for the accident. Any change later in the investigation will be updated timely after filing the document. When the accident is finally settled, this document will become a first and complete record for the accident, which is the foundation of the road accident recording system.

2. Road accident recording forms

The traffic police have a range of methods to record all information with technology advancement: paper document, photo and video etc. Accidents can be recorded from many aspects with specific details that will help accident reconstruction, analysis, and research.

Traffic police will usually use investigation notes, scene sketch and scene photography. The investigation notes contains all parties' basic personal information; vehicle safety specification and actual loading; accident basic information; violation behaviors of all parties; road environment; other related facts for the accident. The scene sketch includes the dimension, shape, location and relationship of all elements of the scene. Both the scene sketch and the scene photos will be kept in accordance with the investigation notes.

3. Road accident recording content and definitions

In "Regulation of Traffic Accident Processing Procedure", accident information recorded is defined as the following six categories and each category follows standards issued by the Ministry of Public Safety:

- Basic personal information;
- Vehicle safety specification and actual loading;
- Basic accident information;
- Violation behaviors and injudicious action of all parties;
- Road environment;
- Other related facts for the accident.

China road accident reporting system

In order to achieve a full understanding of road accident reporting system in China, we looked into various aspects of the road traffic accident reporting system, including statistical methods, procedures, data report and release.

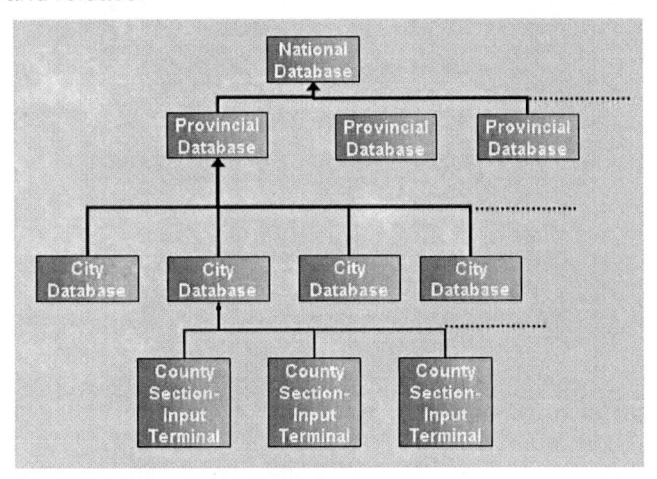

Figure 1

1. Road accident information management system

The road accident information management system (Figure 1) is a working process that traffic policemen input all data into the system, and then data are compiled layer by layer at the information center administrated by the Ministry of Public Security. The information center has a national database that has categorized and summarized data for the nation and each province.

2. Process of reporting system

The road traffic accident data system is a computerized information-based management system. Traffic police use computer to input the recorded data into the system. The process is from bottom to top. Data are generated at district/county, and aggregated to municipal and provincial level. Finally, each province reports it's summarized data to the information center the Ministry of Public Security.

3. Content and scope of reporting system

Scope of the China road traffic accident data system: fatalities, injuries, and property loss. Accidents occurred on all public roads, roads on campus of schools, mines, construction site, railway crossing or ferry, etc.

The system provides daily report and monthly report. The daily report has data gathered within 24 hours, including parties and vehicle involved, time, accident scene, fatality and injury, accident form, scene form; dangerous goods, initial accident contributory investigation.

The monthly report is covers from the 21st day of previous month to the 20th day of the following month.

The monthly report includes more information than items in the daily report system: property loss, weather, visibility, road condition, illumination condition, traffic control and sign, and accident cause.

4. Release of the traffic accident data

There are two ways of the accident data releasing: regular news conference and the annual road accident report.

ACCIDENT DATA SYSTEM COMPARISON BETWEEN CHINA, US AND UK

Both the US and UK have a well-developed highway accident data system. A comparison of systems is made between China, US and UK. The first step is to compare data system structures, then identify their similarities and differences on statistical categories.

Comparison on basic structures

1. China road safety data system

The China system has been equipped with complete computer network, which is named "National Road Traffic Accident Data System". With this data system, local policemen can easily input all accident information into the national data system at their terminal device after their conducted the investigation activities. The information entered into the data system are encoded from the scene investigation notes, scene sketch and other materials. The encoding procedure is complied with the "Road Accident Information Collection Table" for the input procedure. After input at each local terminal, all data will be pooled together automatically and uploaded real-timely to the Traffic Administration Bureau of the Ministry of Public Security.

Data from the "National Road Traffic Accident Data System" is the only source for "People's Republic of China Annual Report on Road Traffic Accident" as well as the National Bureau of Statistics of China. And the "Road Accident Information Collection Table" is the cornerstone of the whole system.

2. The US road safety data system

In the US, there are several road accident statistics systems for different research purposes and government agencies. The most important system is Fatal Analysis Reporting System (FARS). The mission of FARS is to make vehicle crash information accessible and useful so that traffic safety can be improved. In FARS, there are more than one hundred data elements to represent each road accident. FARS issues an annual report on road safety in the US around July.

FARS covers all fifty stats in the US, the District of Columbia, and Puerto Rico. All crash involving a motor vehicle traveling on the traffic way customarily open to the public and resulting in the death of a person (occupant of a vehicle or a non-motorist) within 30 days will be included in FARS.

As an independent system, the information in the FARS database are gathered from the state's own source documents, "Police Accident Reports (PARS)", "State vehicle registration files", "State driver licensing files", "State Highway Department data", "Vital Statistics", "Death certificates", "Coroner/Medical examiner reports", "Hospital medical records" and "Emergency medical service reports". Some or all information from these databases are coded on standard FARS forms and input into the FARS database.

For the whole data collection procedure, the FARS has strict control criteria for Personally Identifiable Information (PII) and Quality Control. First, FARS controls all access of PII information for the reason of illegal individual contact or track. On the other side, Quality Control is considered as a vital feature of FARS system. All information entered the database should follow statistical control chart, through consistency check and other checks to ensure timeliness, completeness, and accuracy.

3. The UK road safety data system

The road accident statistics system was found by Standing Committee on Road Accident Statistics (SCRAS). It was started in 1977, and then implemented in 1979. SCRAS is responsible for the supervision and administration of the STATS 19. The members of the SCRAS are from central government, local government authorities, Department for Transport (DfT) and police force, while SCRAS itself is administrated by DfT, environment authority and DETR. Annual road accident report in UK is released separately in DfT, the Scottish Executive (SE) and the National Assembly for Wales (NAfW). However, all these report are based on the STATS 19 data.

The form STATS 19 is filled by the police officer attending the accident scene. There is an obvious difference between UK and the US, but similar to China road accident recording procedure. The police officer at the accident scene first will investigate the accident scene and record all information carefully. Information collected during this procedure contains basic information, driver and vehicle details, all parties' information, additional notes, sketch plan and record of interview at scene. All these information later will be encoded according to the STATS 19 format and input into the database.

For the quality and accuracy of all data, STATS also have a series standards and handbooks to ensure that, like the STATS 20 "Instruction for the Completion of Road Accident Reports" etc.

Comparison of the data systems

1. System comparison

Since the most basic function of all these systems is same, the collecting procedure of these three systems, "National Road Traffic Accident Data System (NRTADS)", "FARS" and "Road Accident Collection System of UK", is following the same flow. First, an accident occurred somewhere on a public road, then the police officer will be at the scene to investigation and record all accident information. All the data will then be complied through a statistics system annually for the annual report and further research works. The similarity of these three systems is that they all collect following information: basic information of the road accident; involved parties' information and road and vehicle information.

There are differences among the three systems. First of all, the whole statistics procedure is conducted by police in China and UK. However it is the FARS Analysts who conducts all statistics and encodes of road accident information in the US; Secondly, the data source of FARS are much wider than NRTADS and STATS 19. FARS are generated from ten databases of the US, while in China and UK, the police road accident report is the only source for road accident database; Finally, NRTADS doesn't provide detailed data to the public, only the government authorities can access the database, while database of FARS and STATS 19 are accessible for the public in the US and UK.

2. Statistical data comparison

Table 1 shows the number of dominating distinctions among three systems. There is a unique category in this table for UK, "Contributory Factors". The purpose of establishing this separate category is to make it easier for users of this database to find the information about contributory factors in UK. Another obvious point in Table 1 is that FARS is the most detailed system, referring to 124 data elements of basic information, vehicle information and personal information.

Country Category	China	US	UK
Basic Info.	25	39	26
Vehicle Info.	14	36	25
Personal Info.	17	22 (Driver)/27 (Other)	18 (Casualty Only)
Contributory factors	N/A	N/A	9
Total	56	124	78

Table 1 Data categories in the three systems

Category	Common Elements				Unique		
	All	A-B	A-C	B-C	A	B	C
Basic Info	8	3	1	3	7	10	7
Vehicle Info	1	2	1	0	10	14	5
Personal Info	5	1	1	5	11	32	9
Road Info	3	3	0	1	4	5	1
Accident Type	0	1	0	2	0	11	7

* A=China, B=US, C=UK

Table 2 Comparison of data categories

Table 1 is based on each country's original statistical forms. We reorganized all the elements as five categories in Table 2. It shows the similarity and the distinction of three statistical systems.

3. Differences of three systems

China's road accident statistics system is mainly aimed at users of the government and traffic management authorities. This targeted user group determines its features: a. Emphasis on identifying the responsibility of the accident; b. Emphasis on follow-up treatment; c. Emphasis on the casualty in the accident.

The United State's FARS integrates a wide range of resources from several databases, which makes it most comprehensive among all three systems. FARS provides: a. Comprehensive information on vehicles in the accident; b. Detailed track info on Emergency Medical Services; c. Direct info on violation history of the parties in the accident; d. Accident event process info. This detailed information can be used by automotive companies, medical services, insurance companies, besides government agencies.

The United Kingdom's system STATS 19 focuses more on the detailed info of police agencies responsible for the accident as well as the pedestrians and students information than the other two systems. But the most obvious feature of STATS 19 is its separated statistic table on "Contributory Factors", which includes information on vehicles, roads etc. This separate table can be easily used, which is very convenient and humanized.

4. Findings of the comparison

Based on the characteristics of all the systems, apparent suggestion for improving China system would be:

First, open the data system to more users. It can be a valuable source of information for auto industry, consumers, medical services, insurance companies and many other industries. Second, make the data system support a real-time accessibility. The end users can use the data system effectively to improve road safety in current fast motorization process.

Injuries / Accidents * 10,000 = Injuries for Every 10000 Accidents (Injuries / 10,000 Accidents)
Direct Property Loss / Accidents = Direct Property Loss for Every Accident (Loss/Accident)
We examined each factor from the above three aspects to generate a list of factor severity to fatality, injury and property loss.

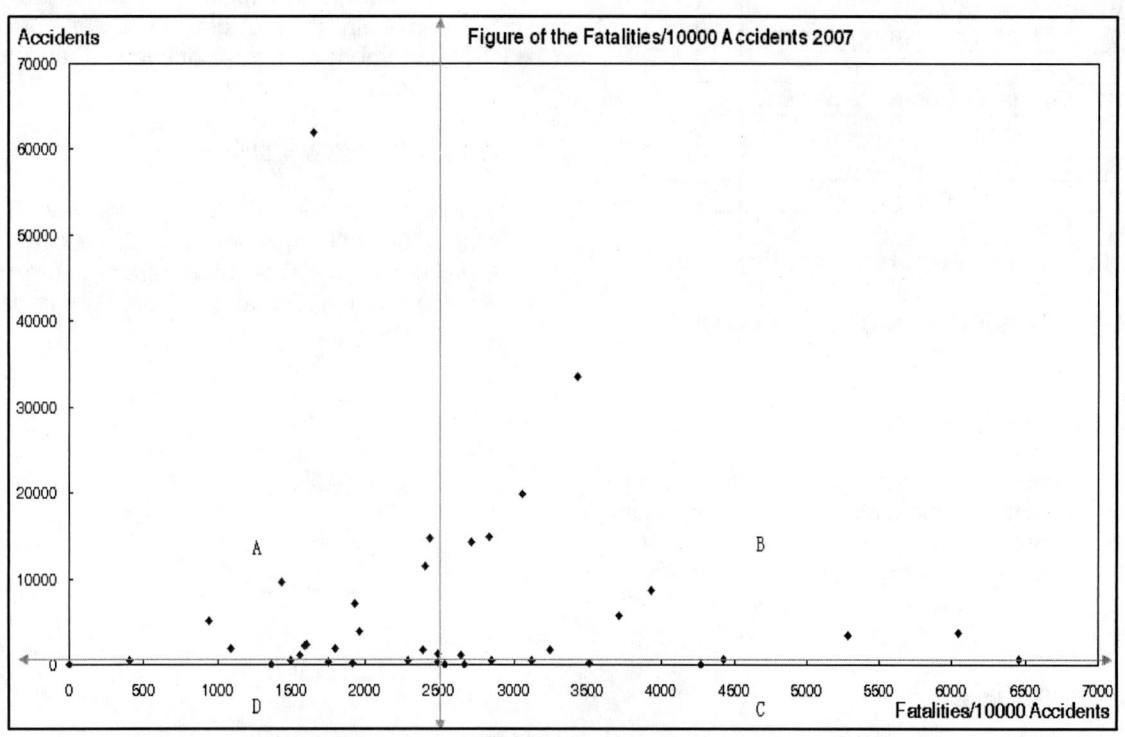

Figure 2: Fatality number and rate in 2007

DATA ANALYZING

THE CHINA ROAD SAFETY DATA

The Traffic Management Research Institute of the Ministry of Public Security publishes an annual road accident statistic report every year. This report provides statistic data of the accident, fatality, injury and direct property loss in the perspective of the factors that cause accidents, time, location, road condition, etc. In this project, we used the reports from 2000 to 2007. We looked into all factors and prioritized them in terms of the number of accidents associated with each factor and percentage of the fatality, injury and loss of each factor.

THE MOST CRITICAL TOP 10 FACTORS OF ROAD ACCIDENT

Data processing

The rate of fatality and injury were calculated on a ten-thousand-accident base, as shown in the following formulas:
Fatalities / Accidents * 10,000 = Fatalities for Every 10000 Accidents (Fatalities / 10,000 Accidents)

Critical factor determination

Figure 2 demonstrates an approach to filtrating critical factors. The figure shows the perspective of fatality in 2007. The X axis is the rate of fatalities/10,000 accidents of the factors, and the Y axis is the total number of accidents of the factors. Each blue diamond dot represents a factor. The vertical red line is the average fatality rate of all factors. The factors on the right side of this line have a greater rate than the average. The red (movable) horizontal line can be used to determine the most critical factors that caused highest number of accidents and fatality rates. These two red lines divide the figure into four areas: A, B, C and D. The factors in area B are the factors with the highest number of accidents and fatality rates. The top 10 most critical factors in 2007 were determined by moving the horizontal red line vertically up until ten factors appeared in the area B.

We used the same method to examine a total number of 120 factors from 2000 to 2007, and generated the top 10 deadly factors in China.

103

<u>The top 10 deadly factors</u>

The top 10 most deadly factors in 2007 are listed in Table 3.

Rank	Factors	Score
1	Speeding	10
2	No License	9
3	Wrong-way	8
4	Illegal Driveway Occupant	7
5	Drunk Driving	6
6	Illegal driving onto roadway	5
7	Illegal Loading	4
8	Fatigue	3
9	Illegal Parking	2
10	Illegal Usage of Lamplight	1

Table 3: Top 10 Factors of Fatality for 2007

SYSTEMS APPROACH: MOST EFFECTIVE STRATEGY FOR CHINA

A SYSTEM DEFINITION

Although tremendous efforts have been applied to improve road safety in China, a clearly defined system is needed to address road safety issues from human behavior change, vehicle safety technology advancement, and road infrastructure improvement at the same time.

<u>Human behavior change</u>

Human behavior is the essential to road safety. Advanced vehicle safety technologies and road infrastructure can provide a safe environment to road users. People may not be aware they are engaged in dangerous behaviors on road, especially at the early stage of the motorization process. On the other hand, there are always some rule breakers. More than 90% percent of the accidents in China can be traced back to human mistakes. It is extremely important to promote safety consciousness of road users and change their behavior through enforcement, propaganda and education.

<u>Vehicle safety technology advancement</u>

Vehicle safety technologies have greatly increased the safety of vehicle, especially the invention of the safety belt and air bag, provided a great protection for the driver and passenger. For example, among the top 10 deadly factors, five of them could be improved by advanced vehicle safety technologies: Speeding, illegal driveway occupant, drunk driving, no license, and fatigue.

- Speed-limit equipment was invented for speed control, which has only been used in some luxury vehicles.
- A lane detection technique can help vehicles get rid of illegal driveway occupant. The technique will identify and help to adjust the driving way if the vehicle goes outside the roadway.
- Drunk driving and no license can be controlled by an intelligent lock, which can only been started up by breathing test and license identified.
- Fatigue drive detector is a new kind of instrument can help drivers get rid of fatigue driving. The instrument can warn drivers if their pupil is deviant roving.

<u>Road infrastructure improvement</u>

Figure 3 presents a trend of highway development in China. Overall traffic safety in China could be improved significantly by the road infrastructure development, particularly the fast growth of high-grade roads (e.g. express way).

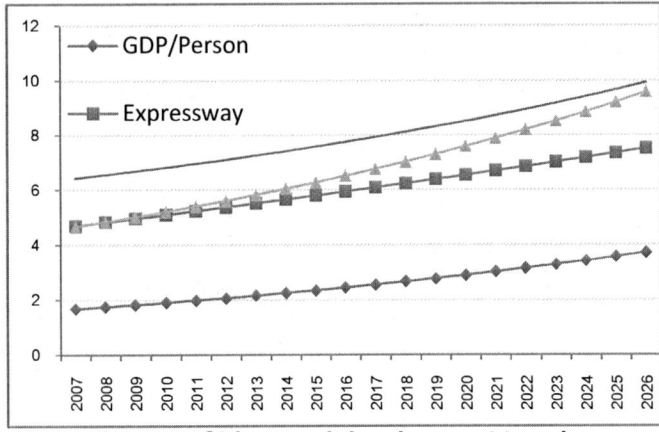

Figure 3. China road development trend

ADOPT BEST PRACTICES FOR THE SYSTEM

Western countries have proceed to the high level of the road safety, such as the pedestrian safety, seatbelt usage, baby and child passenger safety, first aid after accident, safe communities and etc. It is important for China to learn the lessons from western countries and adopt best practices. Here are some examples:

1. Speeding. Many western countries developed effective way to control speed. The measures include speed limit, speed limit enforcement; roadside and traffic light cameras; intelligent speed adaptation-information.
Chinese government put a huge amount of efforts on the holistic countermeasures in the past years. The speed limit was set in different levels and sections of roadway or highway like 40, 80, and 120km etc. Cameras were set up in the urban areas to supervise speeding and the other illegal activities on roads.

2. Drunk driving is a global problem of road safety.
There is no dry law and zero-tolerance in China, children can buy alcohol easily. In China, there are two penalty levels of drunk driving according to the new statute, fine and detention are the basic punishment, and the driving

license would be withdrawn immediately when detected 0.08% BAC; the less charge is between 0.02% and 0.08% BAC, the license would be withdrawn by the second time within a year.

3. Fatigue driving is a serious problem, especially for professional drivers and the people with bad lifestyle. Lack of sleeping, long-distance driving and some medicine will cause fatigue. A new high technological instrument is developed to help drivers get rid of fatigue in UK. The instrument will alarm when the pupil of the driver is detected as abnormal. Drivers in US was limited their driving time less than 10 hours a day. In China, the fatigue is defined as following: the driving time exceeds 8 hours every day; continual driving more than 4 hours while the break is less than 20 minutes.

4. No license is a lethal factor of road safety. Uncertified driver will greatly increase the risk of accidents. There are four categories of measure which can solve the problem in the western countries: pre-driver education, driver training and testing, graduated licensing, and penalties.

CONCLUSION

RECOMMENDATION ON IMPROVING THE ROAD SAFETY IN CHINA

For some factors, with the severe enforcement during these years, the road accident and casualty have been improved. Some serious factors have been well controlled, such as speeding, drunk driving and fatigue etc. However, some new problems appeared, such as no license and disregarding traffic light. It is strategically important for China to develop a national framework to reduce road accidents and fatality as shown in Figure 4. This will provide a platform to address to a long-term solution to road safety problem during the rapid motorization process.

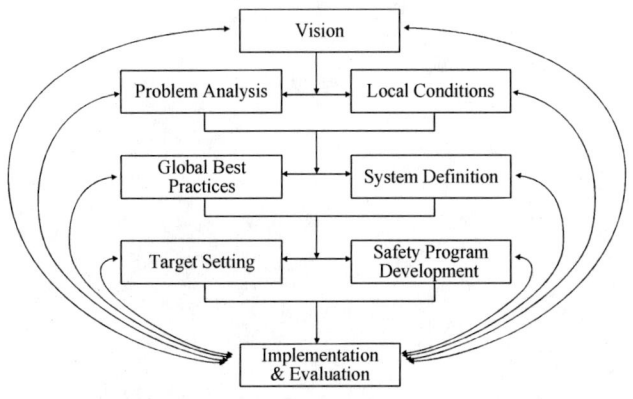

Figure 4. Framework for Road

REFERENCES

1. Road Accident Statistical Annual Report, 2004~07;
2. Law of the People's Republic of China on Road Traffic Safety, 28th October, 2003 issued & 1st May, 2004 in force;
3. International Blood Alcohol Limits: http://www.driveandstayalive.com/articles%20and%20topics/drunk%20driving/artcl--drunk-driving-0005--global-BAC-limits.htm;
4. Codes for road traffic accident scene, GA 17.1-17.11-2003 (22nd October, 2003 issued & in force);
5. Codes for traffic accident information, GA 16.1-16.11-2003 (22nd October, 2003 issued & in force);
6. Department for Transport of UK, Instructions for the Completion of Road Accident Reports. STATS 20. www.collisionreporting.gov.uk;
7. Department for Transport of UK, Accident Statistics. STATS 19. www.collisionreporting.gov.uk;
8. Sketch drafting for road traffic accident, GA 49-93 (22nd March, 1993 issued; 1st April, 1993 in force);
9. The filing document of traffic accident, GA 40-2004 (9th August, 2004 issued; 1st October, 2004 in force);
10. The photography of examination in road traffic accident, GA 50-2005 (7th September, 2005 issued; 1st November, 2005 in force);
11. Traffic Administration Bureau of the Ministry of Public Security, Collection of traffic safety laws and regulations (2005 edition). Beijing: Publish House of the Chinese People's Public Security University;
12. U.S. Department of Transportation, FARS Coding and Validation Manual. www.fars.nhtsa.dot.gov;

CONTACT

CATARC Shanghai Operation:
Hongtao Yu: yht@shcatarc.com.cn
Jiahe Zhang: zjh@shcatarc.com.cn
Yuankui Meng: myk@shcatarc.com.cn
Shili Ni: nsl@shcatarc.com.cn
Ford Motor Company:
Weijian Han

DEFINITIONS, ACRONYMS & ABBREVIATIONS

Definitions
Wrong-way: Driving on the wrong side of driveway.
Speeding: Driving with the exceeded maximum speed per hour as indicated on the speed limitation mark.
Illegal driveway occupant: Motor vehicle drives on improper driveway.
Drunk driving: Driving after drunk.
Illegal overtaking: Overtaking by disobeying the law and regulation.
No license: Driving without driving license, no matter licensed or not.

Vehicles illegally encounter: Motor vehicle driving on roads without separator or median is not obey the statute taking proper action.
Disregarding traffic lights: Ignoring the signal of traffic light.
Fatigue: Driving while tired or sleepy.
Illegal driving onto roadway: Driving on the roadway without permission.

Acronyms & Abbreviations

CATARC: China Automotive Technology and Research Center;
US: United States of America;
UK: United Kingdom;
WTO: World Trade Organization;
FARS: Fatal Analysis Reporting System;
NCSA: National Center for Statistics and Analysis;
NHTSA: National Highway Traffic Safety Administration;
DOT: Department of Transportation;
PARS: Police Accident Reports;
PII: Personally Identifiable Information;
DfT: Department for Transport;
SE: Scottish Executive;
NAfW: National Assembly for Wales;
NRTADS: National Road Traffic Accident Data System;
EMS: Emergency Medical Services;
Loss: direct property loss;
BAC: blood alcohol concentrations

2009-01-2917

Structural Improvement for the Crash Safety of Commercial Vehicle

Libo Cao, Zhonghao Bai, Jun Wu, Chongzhen Cui and Zhenfeng Niu
The State Key Laboratory of Advanced Design and Manufacturing for Vehicle Body

Copyright © 2009 SAE International

ABSTRACT

Statistic analysis on commercial vehicle crash accidents in China were done by using the annual traffic accident reports from Ministry of Public Security. The Chinese crash safety rules on commercial vehicle were introduced. The main reasons which cause severe injury to the passenger in the cab in frontal crash accidents were studied. HYPERMESH software was used to do the finite element modelling of the frontal structure and cab of a production truck. The swing hammer impact simulation was conducted by using LS-DYNA software and the results were compared with the test results to validate the model. A new supporting structure for the cab to improve the safety of the passenger in cab was proposed. Meanwhile, an extendable and retractable longitudinal beam energy absorbing structure was also studied by using the finite element model. The simulation results show that these structures can obviously improve the frontal crash safety of the commercial vehicle.

INTRODUCTION

In China, more and more large and heavy commercial vehicles are used in freeway and highway transportation, the statistical data is shown in Table 1. It can be seen that, the population of commercial vehicle has an increasing tendency, and the growth rates are more than 10% in 2007 and 2008. Their crash safety problem is becoming increasingly serious. According to the statistics of traffic accidents between 2003 and 2007, there were more than 2.2 million traffic accidents causing about 458,000 fatalities totally in China. Among them, 635,000 cases and 175,887 fatalities include commercial vehicles, which account for about 38.4% of total accidents and 27.9% of total fatalities. The direct economic loss exceeds 4 billion Yuan [1]. Thus, commercial vehicle safety has brought more and more attention to the government and researcher.

Among various of reasons of the high fatalities and mortality in commercial vehicle included accidents, four reasons may be the most important:

1. In China, most of commercial vehicles are non-nose structure. The energy-absorption length before cab is very short. The passenger will be injured or loss of life by squeezing in severe crash [2]

Table 1 Comparison of heavy commercial vehicle population

Year	Heavy commercial vehicle population (million)	Average annual growth rate (%)
2005	1.68	-
2006	1.74	3.57%
2007	2.05	17.82%
2008	2.35	14.6%

2. Due to large mass and strong stiffness of commercial vehicle, the passengers in passenger car are very dangerous when it crash with commercial vehicle [3].
3. Large blind area exists because of the bulky body of commercial vehicle.
4. Narrow body width and high gravity center lead commercial vehicle easy to roll over.

Several criterions relate to the safety of commercial vehicles have been enacted in China, such as GB 11567.2-2001, Motor Vehicles and Trailers-Rear underrun protection requirement; GB 20182-2006, The Cab of Commercial Vehicles-External Projection; GB 11567.1-2001, Motor Vehicles and Trailers-Lateral Protection Requirement. Two other draft criterions have also been proposed. They are The Protection of the Occupant of the Cab of a Commercial Vehicle [4] and Front Underrun Protective Requirements for Commercial Vehicle. They will be enacted in the near future.

In order to improve the crash safety performance of non-nose commercial vehicle, the cab must be designed in such a way that guaranteed a sufficient survival space in accident [5]. So, the energy absorption length in front of the cab should be elongated. Two methods can be used for this purpose. One is extending the energy absorption element out of the frontal structure. Another one is allowing the cab move back relative to the longitudinal beam. So, an extendable and retractable energy absorption system was proposed and a new structure for the suspension structure of the cab was also studied in this paper.

METHODOLOGY

A production non-nose commercial vehicle was chosen to be studied. The frontal pendulum impact test was conducted according to the draft criterion The Protection of the Occupant of the Cab of a Commercial Vehicle, which is similar to ECE R29. The finite element model of the commercial vehicle was developed and validated by compared with the test results. Then, an extendable and retractable energy absorption structure which can be installed in longitudinal beams was designed. The suspension structure of the cab was modified. Their effectiveness for the improvement of the crash safety performance was simulated and analyzed.

FRONTAL PENDULUM IMPACT TEST

The pendulum which was used in frontal pendulum impact test is made of steel. The mass is 1,500±250kg and evenly distributed. The pendulum is 2,500mm wide and 800mm high with flat and rectangular striking surface. It is freely suspended by two beams which are rigidly connected to it. The distance between two beams is not less than 1000mm. Two beams are at least 3,500mm long from the axis of suspension to the geometric centre of pendulum. The gravity center of the pendulum is 50 +5/0 mm below the R-Point of the driver's seat, and positioned in the median longitudinal plane of the vehicle. Test setting and the test site were shown in Figure1 and Figure 2.

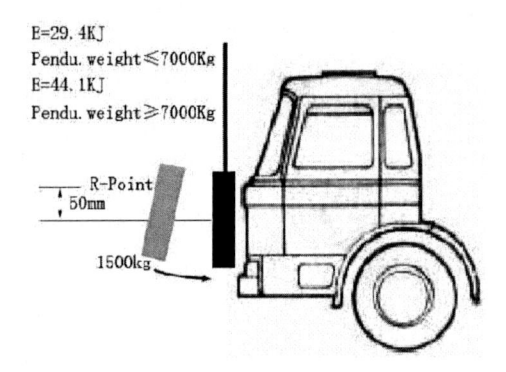

Figure1. Schematic diagram of the test

Figure 2 Test site

To avoid the vehicle moving during the test, some anchoring chains were used to fix vehicle in both transverse and longitudinal orientation before test. The chains were connected to the frame. An accelerometer was mounted on the center of gravity of pendulum. The impact energy of the pendulum was set to 44.1KJ. The pendulum stroked the cab in the horizontal direction towards the rear of the cab and parallel to the median longitudinal plane of the vehicle.

DEVELOPMENT OF THE FINITE ELEMENT MODEL

HYPERMESH software was used to do the modeling. The whole model consists of cab body, front and rear suspension of cab, front part of frame, engine and its accessories. The cab body was mainly meshed by Belytchko Tasy shell elements (Type 2). Solid elements were used to simulate engine and bracket. Cab body was jointed by spot-weld element. Other parts, such as suspension, were jointed together by bolts, rigid beams and joints. The overall model consists of 323268 nodes and 315762 elements, as shown in Figure 3.

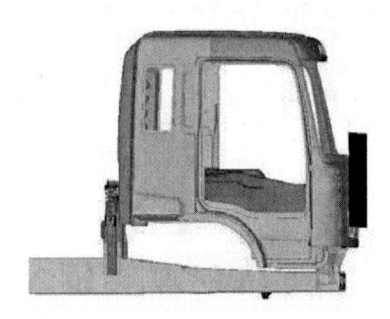

Figure 3 Finite element model

The pendulum was simulated by rigid plate. Automatic single surface contact was adopted to simulate the contact between the components of the cab. Automatic surf-surf contact was defined between the rigid pendulum and the frontal components of the cab.

VERIFICATION OF THE FINITE ELEMENT MODEL

The front and rear end of the frame were restrained in the simulation to restrict the transverse and longitudinal displacement. The rigid pendulum impacted the front face of the cab with an initial velocity of 2.19 rad/s. The impact energy was 44.1 kJ. The acceleration of the pendulum's gravity center was indicated by the dashed line in Figure 4. The test result was also shown in Figure 4 by continue line. It can be seen from the contrast curves that the peak values are approximately 22g and 19g respectively and the instantaneous of the peak value are in accordance with each other. So, the model was used in further study.

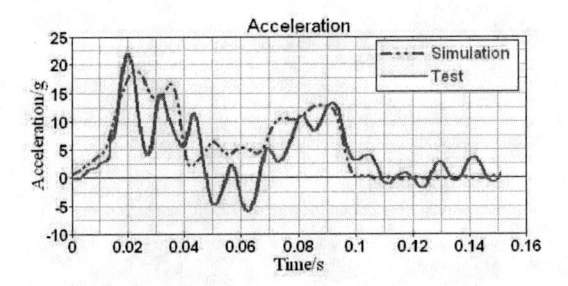

Figure 4 Comparison of the acceleration of pendulum's gravity center

EXTENDABLE AND RETRACTABLE ENERGY ABSORPTION SYSTEM

An extendable and retractable energy absorption system was developed. It consists of two auxiliary energy absorption equipments, an electromagnetic valve, a control unit, a gasholder and some connection pipeline, as shown in Figure 5. The auxiliary energy absorption equipments can be installed in the longitudinal beams of the frame. Their front ends connect to the bumper. The control unit consists of an accelerometer and some hardware and software. When the vehicle take emergency brake and the brake acceleration satisfy the system trigger condition [6, 7], the electromagnetic valve will turn on and the compressed air in the gasholder will rush into the auxiliary energy absorption equipments, which located in the equipments, and the bumper forward and locked in less than 100 ms. Then, the energy absorption length of the vehicle is prolonged. If crash does not happen, the beams can be retracted into the equipments.

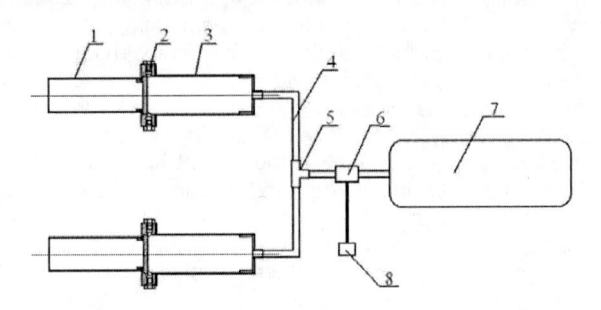

Figure 5 Structure principle of the system
(1 Energy absorption beam 2 Self-lock mechanism
3 Auxiliary energy absorption equipment 4 Pipeline
5 Junction 6 Electromagnetic valve 7 Gasholder
8 Control unit)

TRANSLOCATIVE STRUCTURE

Usually, the cab of non-nose commercial vehicle connects with frame by brackets. These brackets must have enough strength and stiffness to support the cab. However, if the strength and stiffness are too large, the cab can't move backward during crash and the deformation of the cab will be large. The brackets may also be damaged and can't provide the cab with reliable support. For example, in the frontal pendulum impact test, as shown in Figure 2, the brackets of the tested vehicle were broken, as shown in the circle in Figure 6. So, a translocative structure which can be used in the front brackets was designed, as shown in Figure 7. If the impact force is large enough, the bottom rail of the cab can move backward relative to the suspension supporting plate.

Figure 6 Broken bracket

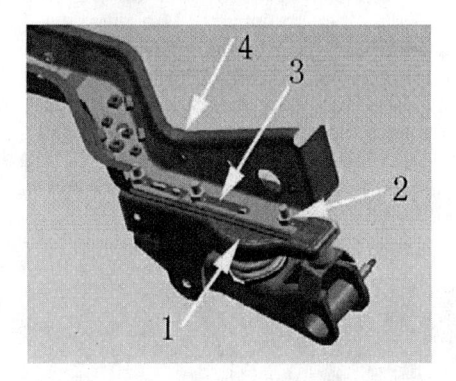

Figure 7 Translocative structure
(1 Suspension supporting plate 2 Bolt
3 Guiding grooves 4 Bottom rail)

RESULTS AND ANALYSIS

The extendable and retractable energy absorption system was manufactured and tested in Hunan University, as shown in Figure 8. Two high speed cameras were used to record the test process. When the air pressure in the gasholder reach to 0.6 MPa, the beams can be push out in 71 ms, as shown in Figure 9.

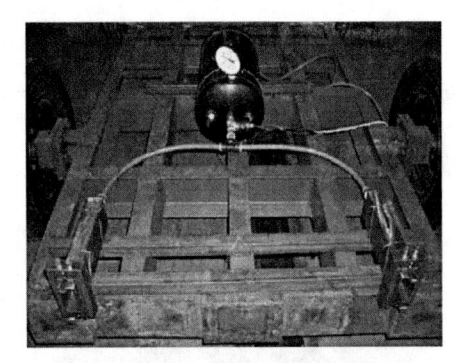

Figure 8 extendable and retractable energy absorption system

Two simulation models of the commercial vehicle installed with or without the extendable and retractable crash energy absorption equipment were used to do the same frontal crash simulation. They crashed with a rigid wall at the speed of 36km/h, as shown in Figure 10. The maximum intrusion of the cab and the energy absorbed by the vehicle were compared with each other, as shown in Table 2. It can be seen from Table 1 that the maximum intrusion of the cab is 83 mm or 35 mm respectively without or with this equipment. The difference between them is 48mm, or 57.8% reduced. That is to say the passenger's survival space can be increased notably. On the other hand, the total impact energy is 101.5kJ. The energy absorbed by the vehicle is 82.1kJ or 67kJ respectively without or with this equipment. It means that the crash energy transferred to the vehicle and passenger can be reduced.

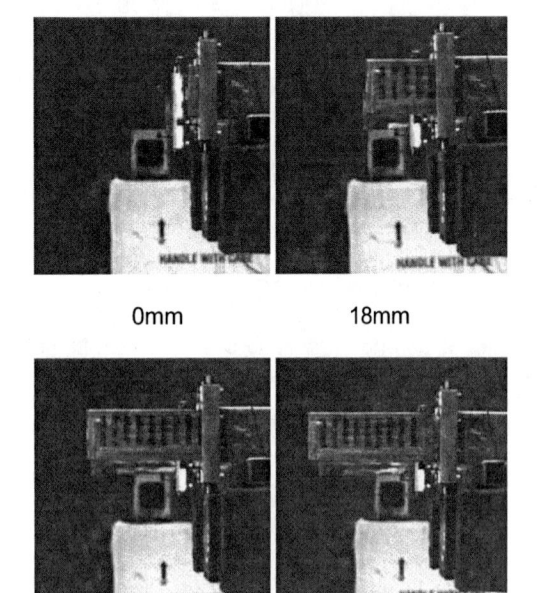

Figure 9 The pushed out process of the beam

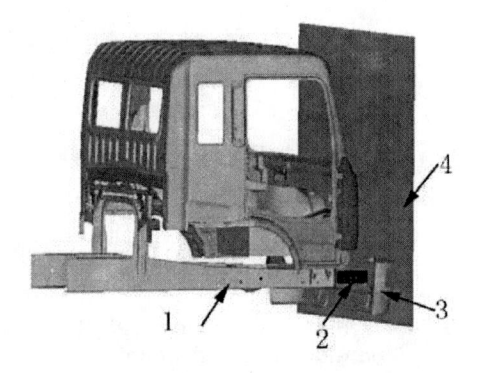

Figure 10 Simulation model with energy absorption equipment
(1 Frame 2 energy absorption beam 3 Frontal bumper 4 Rigid wall)

Table 2 Comparison of the simulation results

	Intrusion(mm)	Absorbed energy (KJ)
Without the equipment	83	67
With the equipment	35	82.1
Differences	48	15.1

In order to study the effect of the translocative structure to the safety performance of non-nose commercial vehicle, Two models with or without the translocative structure were used to do the frontal pendulum impact test simulation. Table 3 indicated the comparison of the simulation results. It can be seen from Table 2 that the backward displacement of the cab is 193 mm or 230 mm respectively without or with the translocative structure. That is to say, the relative movement between cab and frame has been increased and the deformation of the cab can be decreased. Therefore, the translocative structure can provide a better protection for occupants in the front impact of the non-nose commercial vehicle.

Table 3 Comparison of simulation results

	Backward displacement(mm)
Without translocative structure	193
With translocative structure	230
Differences	37

The comparison of acceleration curves was shown in Figure 11. The figure indicates that the acceleration curve of vehicle with translocative structure was much smoother than the vehicle without this structure. So, with the structure, the energy absorbed by the vehicle was larger and steadier.

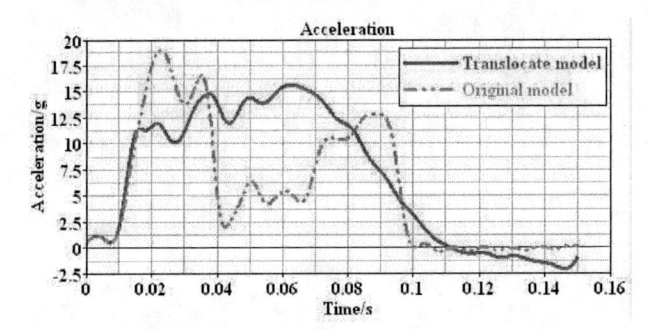

Figure 11 Acceleration comparison

CONCLUSIONS

The statistic data indicate that commercial vehicle accident is one of the major types of traffic accident in China. It has caused high fatalities and mortality. The main reason for it is that the non-nose commercial vehicles haven't enough energy absorption space. Test results show that the original brackets of the cab can be damaged and cause dangerous to the passengers. So, the extendable and retractable energy absorption system and the translocative structure for the bracket of cab were developed. Test results show that the extendable and retractable energy absorption beam can be pushed out in time. Simulation results show that the deformation of the cab of the commercial vehicle with extendable and retractable energy absorption equipment and translocative structure can be significantly reduced in front crash. So, the crash safety performance of the commercial vehicle can be improved. In order to prevent the extending beams from injuring pedestrian on city road, the extendable and retractable energy absorption equipment can be set to work over certain vehicle speed, which corresponds with the speed on the freeway.

REFERENCES

1. The Ministry of Public Security Traffic Administration. (2007) The 2006 Annals of Statistics from Road Traffic Accidents in People's Republic of China. Traffic Management Research Institute of the Ministry of Public Security, Wu Xi, (in Chinese)
2. Li Sanhong, Guo Konghui, et al. Frontal Pendulum Impact Test and Computer Simulation of Commercial Vehicles. [J]. China Mechanical Engineering. 2005, 16(23) (in Chinese)
3. Zhao Youping. Study and Application of Computer Simulation Technology of Commercial Vehicle Crash. [D]. HuaZhong University of science and technology. 2005(in Chinese)
4. The protection of the occupants of the cab of a commercial vehicle (drafted) (in Chinese)
5. Horst Raich, et al. Safety Analysis of the New Actros Megaspace Cabin According to ECE-R29/02. [C]4th European LS-DYNA Users Conference.
6. Tang Mingfu, Cao Libo, et al. Crash Accident Discriminant System Based on the Brake Acceleration Integral. [C]. The Sixth International Forum of Automotive Traffic Safety. Xiamen. (in Chinese)
7. Libo Cao, Zhonghao Bai, et al. Crash energy absorption equipment combines active and passive safety. ICRASH 2006, July 4-6, 2006, Athens, Greece

CONTACT

Name, Libo Cao

E-mail, hdclb@163.com

Cell Phone: +86 130 5517 6227 (China)

2010-36-0034

Innovative Concepts for Smart Road Restraint Systems to Provide Greater Safety for Vulnerable Users - Smart RRS

Mario Nombela and Arturo Dávila
IDIADA Automotive Technology

Juan José Alba
Universidad de Zaragoza

Juan Luis de Miguel
Centro Zaragoza

Copyright © 2010 SAE International

ABSTRACT

Worldwide, 1.2 million people die in road crashes yearly; 43,000 just in Europe. This implies a cost to the European society of approximately 160 billion euro, making use of 10% of all health care resources. Sharp objects like crash barriers may lead vulnerable road users into serious injuries. Different road restraint system designs have been developed in recent years to improve vulnerable road users' safety.

SMART RRS is an FP7 SST 2007 RTD1 European collaborative project funded by the EC with the participation of 10 institutions from 5 countries. The project aims to develop a new smart road restraint system that will reduce the number of deaths and injuries caused in road traffic accidents by integrating primary and tertiary sensor systems in it, providing greater protection to all road users, warning motorists and emergency services of danger for prevention purposes and alerting emergency teams of accidents as they happen to minimize response time to the exact location of the incident. This new smart restraint system will:

• Reduce the number of accidents through better information on the actual state of the road and traffic flow (climatic conditions, traffic flow, obstructions, hazards, accidents).

• Eliminate dangerous profiles from road restraint systems (crash barriers) that currently endanger vulnerable road users.

• Optimise road safety by providing exact information of where and when accidents happen in real-time.

The project obtained interesting results from an in-depth review of motorcycle accidents, showing that the most aggressive elements for riders are protection systems installed on roadsides (continuous, punctual, rigid, wire rope). Also the accidents involving roadside protective systems include high speeds and the rider commonly impacts the barrier in an upward position, with severe outcome. Some of the most important injuries received by riders are blunt impacts to the head, member amputation and severe thoracic intrusion.

INTRODUCTION

The objective of the "Innovative concepts for smart road restraint systems to provide greater safety for vulnerable road users" (Smart RRS) project is to reduce the number of injuries and deaths caused by road traffic accidents to vulnerable road users such as motorcyclists, cyclists and passengers through the development of a smart road restraint system.

The goal will be obtained through the development of a new road restraint system that will include primary and tertiary sensor systems in it, providing greater protection to all road users, warning motorists and emergency services of danger for prevention purposes and alerting emergency teams of accidents as they happen to minimize response time to the exact location of the incident. This new smart restraint system will:

• Reduce the number of accidents through better information on the actual state of the road and traffic flow (climatic conditions, traffic flow, obstructions, hazards, accidents).

• Eliminate dangerous profiles from road restraint systems (crash barriers) that currently endanger vulnerable road users.

• Optimise road safety by providing exact information of where and when accidents happen in real-time.

THE "SMART RRS" PROJECT

Within the actions that have already taken place in the development of the project, the results obtained from Work Package 1 clearly show a tendency on the accidents that occur with motorcycles. The literature review that took place threw some interesting results concerning the characteristics of motorcycle and other vulnerable user accidents, having a general and a particular view in which the description of the injury mechanisms. The most common injuries are produced from contact with fixed objects, objects located on the side of the road or road restraint systems.

The main task tackled in this part of the project was the characterization of the main parameters of the selected accident set of cases: range of speeds at the point of impact, angles of impact, frequency of injuries by body region, etc. It is also important to know the physiological tolerance thresholds of the human body on each of the most commonly injured regions. In addition, a selection of the most common causes of death or severe injury related to the road restraint system related accidents was made.

The analysis showed that the impact of motorcyclists against a fixed object occurred in 4% of the cases in urban areas while it varies between 10 and 20% in rural areas. This is due to the fact that the velocity of the motorcycle in rural areas tends to be higher and also because of the design of the road, more prone to a rider exiting the road than on a city street. The most common objects impacted in this type of accident are trees/poles, roadside barriers and general road/housing infrastructure. In many of these cases, the rider would hit the obstacle in an upright position.

Having these results in consideration, special attention was put on roadside barriers and the effects on accidents with motorcycles. The most dangerous aspects of the guardrails with respect to a motorcyclist are the exposed guardrail posts and the sharp edges. Some guardrails have been equipped with special protective devices, from which we can find continuous and discontinuous or punctual systems.

In the case of a sliding motorcyclist, the discontinuous systems are worse than the continuous devices. Anyhow, post modifications together with post envelopes show a positive approach in decreasing risks for motorcyclists.

Unfortunately, this system has only proven effective at low speeds and specific angles. There is a high risk for a rider to directly hit one of the barrier posts while approaching a guardrail in a sliding position. For a distance of 2.5 m between the posts, the probability is more than 35% for an impact angle of 30º, increasing to more than 70% for a 15º angle of impact.

A more appropriate solution seems to be the addition of a lower extra guardrail. This system provides a better energy absorption system than concrete or wire barriers and helps in redirecting the rider along the original path. Nevertheless, these systems have their downside because sometimes the rider impacts the barrier in a very high angle and allocates too much energy and pressure on the neck and head, producing severe or fatal injuries. In some other cases, a limb (hand, arm, leg) might get caught in between the barriers or amputated with the sharp edges of the system.

The case of a rider hitting the barrier in an upright position is equally important and the associated risks of being thrown on or over the barrier are significant. In this case, most of the injuries occur when, after a shallow impact, the rider slides and tumbles in the top of the supporting posts. Also, if the height of the barrier is too low, the rider might be thrown over and hit another obstacle of the road (many times the obstacle intended to be avoided with the barrier).

Several testing procedures have been developed and applied in different countries to evaluate the risk of a rider becoming injured when sliding into a motorcyclist protection system. They all have an impact angle of 30º. One of the protocols includes two different orientations of the rider's longitudinal axis and the speeds range in between 55 and 70 kph. Only one protocol selects different impact points for the system, being it a punctual or continuous system.

Another procedure includes the motorcycle. In this case, the angles are 12º and 25º with a speed of 60 km/h. This simulates a rider going over the barrier. All these procedures require a Hybrid II/III dummy, in some case with replacement parts. The biomechanical limits applied in the tests are mostly a HIC value of 1000 and a neck extension moment of 57Nm.

In response to this information and attempts to decrease the severity of these collisions, protecting devices have been launched by various manufacturers. Numerous of these systems are being used at statistically documented, potentially dangerous curves. However, its extensive use has been limited for economical reasons.

EN 1317 is the official regulation in use across Europe covering road restraint systems. Its protocols address issues regarding safety barriers, crash cushions and terminal and transition of barriers, amongst other things. Specific to the safety barrier, EN 1317 defines criteria and process for impact testing for vehicle barriers.

Motorcyclist safety during roadside obstacle collisions is not precedence for all European countries and so due to the minority they represent; specific regulations have not always seemed a requirement. However, some countries have discovered this field problematic and have consequentially introduced additional protocols. The UNE-135900 (regulation used in Spain as an assessment of effectiveness for protective roadside devices for motorcyclists), and the French LIER test protocol (EQUS9910208C) (research into head and neck injuries during motorcycle-barrier collisions) are two major examples.

A comprehensive revision of the available regulations was conducted with the aspiration to determine the strengths and weaknesses of the protocols, exploring the relationship between injury severity and impact configurations. Later on, a new protocol was designed so that the future system evaluations have more strict requirements, yet are easier, cheaper and faster to design and produce.

UNE 135900. - The Spanish norm is one of the first to be applied in a legislative basis. This test procedure includes three possibilities for impact points according to the type of system to be tested. The approval of such test will then enable the system to be marketed and installed on Spanish roads. The characteristics of the evaluation procedure are:

Trajectory 1: post centred impact, applicable to punctual and continuous motorcyclist protective systems; 30° angle.

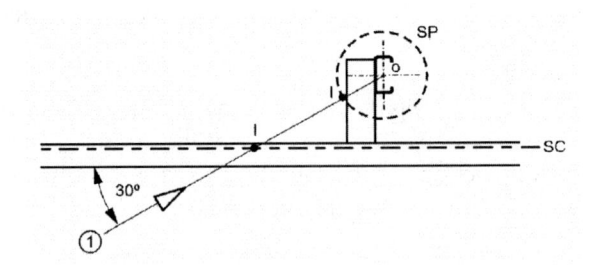

Post Centred Impact Trajectory

Trajectory 2: Post off-centred impact, applicable only to punctual systems; 30° angle.

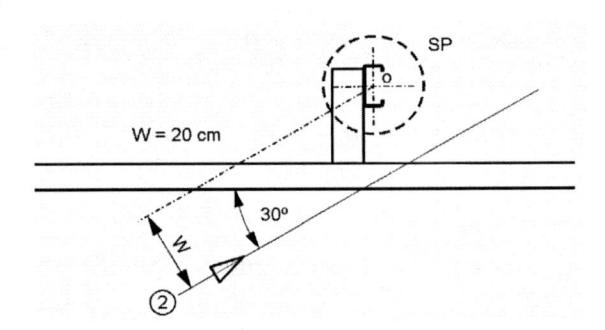

Post Off-centred Impact Trajectory

Trajectory 3: Mid span centred impact, applicable only to continuous systems; 30° angle.

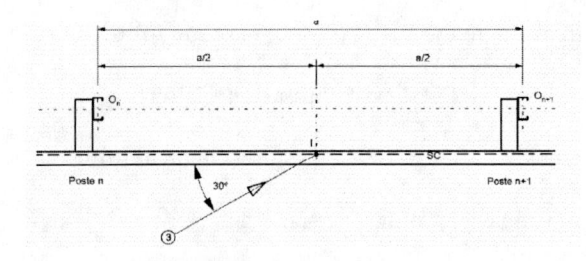

Mid Span Centred Impact Trajectory

For each of the trajectories, the system manufacturer can now choose to have the test either at 60 km/h or 70 km/h. The dummy must be a Hybrid III 50th percentile male, with a modified pelvis for laying position and a fuse clavicle. The launching system of the dummy must have no contact with it at least 2 m prior to impact.

The assessment of the system is based on the bio mechanic measurements of the HIC 36 and of the neck forces and moments. Limit values are defined and measured signals have to be contained into template curves defined by the norm. The legislation finally assesses the system as being of level I (very good protection of the motorcyclist) or level II (homologated but protection could be better).

UNE 135900 Setup

LIER EQUS9910208C. - French LIER protocol is very similar to the Spanish UNE protocol. In this case, a set of two tests need to be performed, with the only variation of the dummy positioning with respect to the barrier. The aim of having these two positions is to discriminate the results obtained in the case where the driver hits the barrier directly with the head against the results obtained from a crash with contact of the shoulder and head. Another difference found between protocols is that the LIER protocol is used only for continuous protective devices, having the impact located in a mid location along the barrier (not against a support). Velocity for these tests is 60 km/h.

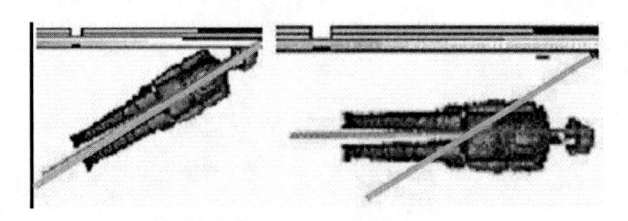

LIER Test Dummy Position

The biomechanical limits used to evaluate the LIER protocol are the following:

LIER Test Evaluation Criteria

Measurement	Biomechanical limit
Resultant head acceleration	220 g
HIC	1000
Neck flexional moment	190 Nm
Neck extension moment	57 Nm
Neck lateral flexion	-
Neck Fx	330 daN
Neck Fz traction	330 daN
Neck Fz compression	400 daN

A review of other available protocols and evaluation activities was made. This included some of the most important developments carried out in the field of simulation, crash testing, product design and protocol evaluation.

Simulation can be considered one of the most important tendencies, as most of the work analyzed focuses on the development and correlation of programs that allow virtual simulation of the systems (with all material characteristics), motorcycle and dummy models (for dynamic analysis and HIC measurements), and replication of real life accidents in order to associate a certain dynamic and mechanical performance to be later evaluated in bias of developing better protective devices.

The analysed projects consider different types of accident, use different simulation software and give several approaches to what a safe road restraint system should do and how to perform. Simulations were carried out to design continuous and punctual systems, using different dummy models. It is very important to consider what other institutions have been doing and what their approaches are, according to the specific needs and goals they are pursuing.

Simulated Barrier Test

Continuing the project, an in-depth state of the art analysis was performed. The main objective of this chore was to know which systems exist, where are they used, how they work and what are the main obstacles that such systems face, whereas on the functionality or on other accident related aspects. To establish a trend, two main systems were classified: punctual and continuous road restraint systems.

Punctual energy absorption systems are generally located in the guardrail posts, are made out of a polymer based material and are good for speeds up to 60 km/h. They protect the riders against sharp edges, yet allow them to go underneath the beams. The risk of going underneath the beams is that another hazardous object or cliff may be present. Also, the possibility of still suffering an amputation of a limb is present, mainly on the beam.

Punctual Protection system

Continuous systems that redirect the rider are generally metallic or plastic elements located underneath the guardrail and all along the barrier, serving as a limiting wall for the rider to lose energy and stay on the road. They protect the rider from contacting objects behind the guardrail. Some systems are looking to make a combination of energy absorption and redirection, providing the benefits of both systems. The risk they show is that sometimes the rider gets tangled between the beam and the restraint system, inducing severe injuries to limbs. Also, high angles represent a problem since the neck forces could be very high.

Continuous Protection System

A reference on active and tertiary safety systems for the road has also been performed, with useful information on the future of the road safety technologies. This is of special interest to our project due to the nature of our future design, which will include a tertiary safety system to communicate hazards and accident situations both to drivers and emergency services. A wide variety of systems are available that cover different sensing and communication functions.

After reviewing what exists and is being made in the road restraint system area, a new protocol was designed. This protocol is specifically designed to make use of the best characteristics of the evaluation methods analysed to be able to cover up for the deficiencies in them. In this way, a deficiency of a system is covered by another method, obtaining a stringent and efficient evaluation protocol.

The basic principles of this evaluation protocol are: subsystem simulation and testing, full system simulation and testing with the new requirements and a required UNE 135900 test. These elements are arranged in such a way that a new system can be developed in less time, more economically and with greater improvements in rider protection. The general flow chart of this entirely new process is presented next:

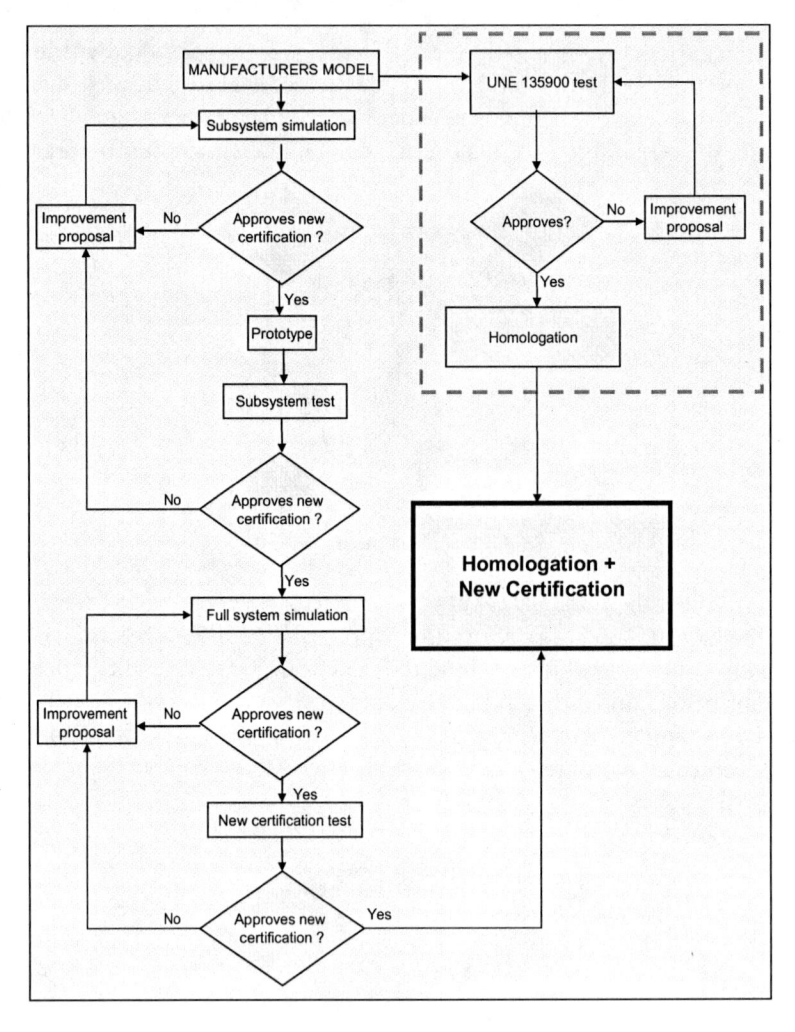

New Test Protocol Flowchart

The selection of tests and simulations to be made was decided according to the comparison of strengths and weaknesses of each of them. As the systems are designed apart and pursue a different goal, they become compatible between each other and provide in the end a better designed and tested product.

As a first step, the strengths and weaknesses of the UNE 135900 norm were defined. The result was that this evaluation protocol is visually very explicit, manufacturers can choose the test velocity and the evaluation impedes the system to allow any limb of the dummy to go through or get caught in between the system. The downsides of the protocol rely on the low repeatability and none specified atmospheric conditions for the test. Also, the fact that the dummy must be released of any guiding mechanism 2 m prior to impact induces slight variations that may give different results from test to test.

Later on, an evaluation on the Simulation of the UNE 135900 was made. In this case, the visual approach is still very important and the requirements are the same. As an extra point, this method allows for a quick change in configuration and has very good repeatability. On the weakness points, we see that there are no quantitative results coming from this system, only qualitative results. Also, the welding lines and other software related issues become an obstacle in getting faithful information from the simulation.

A new idea was then to add subsystem simulation and testing. The use of subsystems allows for an evaluation of the selected part of the body or infrastructure. All of this according to the preliminary studies done where the most aggressive systems and the most commonly injured body parts were specified. The objective is to simulate the subsystem tests according to maximum representativity and repetitiveness criteria, and at the same time simplifying impact absorber evaluation.

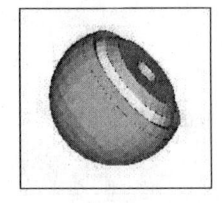

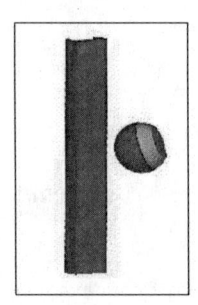

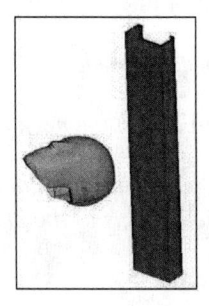

Simulated Subsystem Tests

Similar to their entire test counterparts, the simulation and real life testing of subsystems share almost the same strengths and weaknesses. Real life testing is visually explicit, provides quantitative data, yet takes longer to reproduce. Simulation is also visually explicit, repeatable, and quick for configuration changing but lacks quantitative values.

In this special case, a selection of two real life subsystem tests were compared, free flight test and guided flight test. As a special case, guided flight test is highly repeatable, trustworthy and allows for quick configuration changes. Adding all this up, we now have a protocol that discriminates faults in the systems before they are even manufactured, saving time in the design and cost in the entire life cycle of the product while enhancing safety on the roads. With this new evaluation protocol, road restraint systems will satisfy and overpass required homologation criteria.

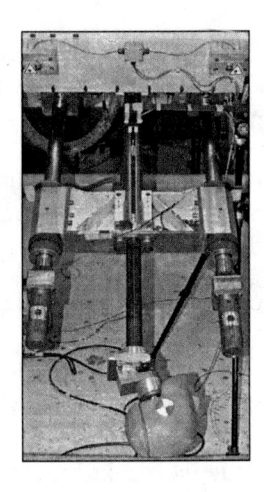

Guided Test Setup

The new test protocol includes subsystem simulation and testing, full system simulation and testing with a new specified configuration and a required UNE 135900 test. The new test configuration is the following:

- Impactor: Hybrid III 50th percentile dummy head with a total mass of 12,6 kg, including guiding system and dummy head.

- Impactor instrumentation:

    o *3 uniaxial accelerometers or 1 triaxial accelerometer inside the head.*

    o *1 6-axle load cell in the neck to obtain data from force and moments in directions X, Y, Z.*

    o *Although these values are not used as acceptance criteria, they will be useful in the future to design improvements to the protocol.*

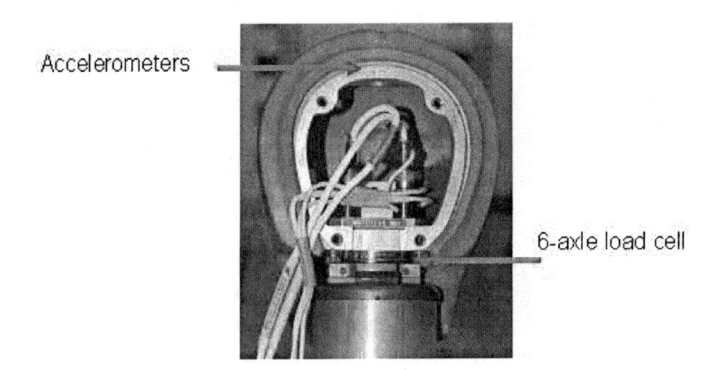

Dummy Head Instrumentation

- Protection system: a sample of the protection system will be mounted in order to include:

    o *A section with a post (with or without support, according to design)*

- *A section with a support*

- *A mid-beam section*

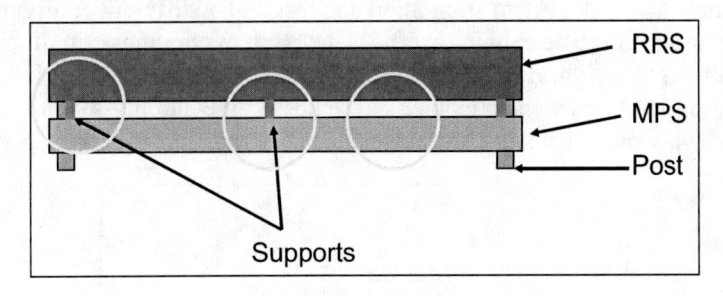

Selected Impact Points.

- Helmet: Commercially available integral helmet, with a 1,300 kg ± 0,050 kg mass and with a polycarbonate carcass. Another helmet may be used as long as it is demonstrated that it has the equivalent mass, geometrical characteristics and materials and that it complies with ruling 22 from E/ECE/TRANS/505.

The test will be carried out for three different angles on each of the three locations in the protection system. The evaluation will be designed during WP6, subtask 6.1 Definition of a new evaluation protocol for motorist protection systems. Nevertheless, we show a previous idea of how the evaluation will be performed.

The total HIC_{15} value for each of the three mentioned positions will be obtained from the subsystem simulations and the real tests, providing a coherent approach. The chart, when completed, will use a similar system to the one used in EuroNCAP, (by colour grading), and a final figure will add up. If this value is over the specified approving value, then the system must be re developed. Once the tests are satisfactory and a UNE 135900 obtained, the New Certification will be extended.

SRRS SUBSYSTEM EVALUATION CHART			
TEST EVALUATION	30°	45°	60°
POST			
MID-SPAN			
MID-SPAN W/ SUPPORT			

HIC > 1000	0 POINTS
650 < HIC < 1000	1 POINT
HIC < 650	2 POINTS

Maximum Possible Score	18
Minumum Score Required	12

Proposed Evaluation Chart

It is important to mention that at this stage of the development, the evaluation system or procedure is still a proposal and may be modified, according to the necessities and to the reliability of the proposed system. Nevertheless, the procedure will try to mimic other evaluation protocols in order to simplify the information given to the user.

The new protocol also includes a full system simulation and test with a different configuration, which will try to provide a more stringent test setup to be certain that the tested system complies with higher protection levels. The dummy used for the tests is a Hybrid III with modified back and pelvis (the one defined in UNE 135900). It is equipped with a homologated helmet and a leather suit to protect the dummy's skin and its components. The dummy will be launched lying on its back over a special sled. The position and trajectory are shown next:

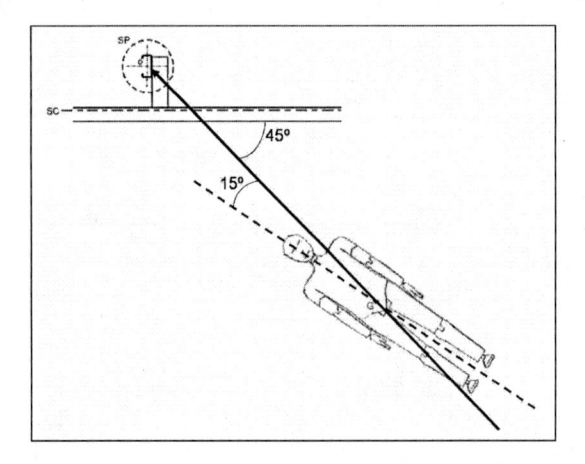

New Impact Trajectory

In this launching position, the vertebral axis of the dummy forms a 15° angle with the launching trajectory. In the same way, the approaching trajectory is aligned with the centre of gravity of the post (similar to UNE 135900, but with a 45° angle). This will generate a higher energy during impact that would need to be absorbed by the system yet the dummy will impact with the same parts of the body as in the UNE test, providing more stringent test but with the same principle applied in UNE 135900.

With this approach, we are trying to evaluate the performance of the system in a more aggressive angle for the user, where more energy needs to be absorbed by the restraint system because of the lower redirection capability. Nevertheless, the impact of the dummy will be with the same angle in order to provide correlation data for future actions.

The biomechanical indexes that are used to evaluate the severity of the motorist impacts are:

Biomechanical Indexes

	Head	Neck		
Level	HIC 36	Mcox[Nm]	Mcoy ext[Nm]	Mcoy flex [Nm]
Level 1	650	134	42	190
Level 2	1000	134	57	190

It is important to mention that HIC_{15} values are used for the evaluation in the subsystem tests, due to the fact that the time frame for these procedure is under 36 milliseconds and only HIC_{15} can be obtained. Nevertheless, both criteria are (HIC_{15} and HIC_{36}) are correspondent and the index values can be maintained.

The next step of the project is to design a new road restraint system, following the new protocol and the evidence found throughout the entire literature reviews and accident analysis. The system will be able to communicate with traffic, emergency services and to discriminate between accident types. The main goal of this system is to provide all users with the exact information required at the moment, to avoid accidents in a precautionary phase or, if an accident does happen, to mark the site, contact the emergency services and provide them with the most available information possible (location, type of accident, vehicles involved, road and weather conditions, etc).

The project continues its development phase and will soon be able to present initial drafts of the proposed new road restraint system. The system is being developed with the simulation methods presented earlier and will eventually be tested after manufacture for the final evaluation and certification.

SUMMARY/CONCLUSIONS

After a thorough analysis of the characteristics of motorcycle impacts in real life, the type of injuries suffered by the motorcyclists and their outcome, the design characteristics of actual road restraint systems for vehicles and vulnerable users, the available legislation and our future needs for designing a Smart Road Restraint System, a proposal of a new evaluation system and a set or requirements was written.

The main point in the development of this part of the project was to analyse what are the weaknesses and strengths of the actual legislative procedures and to include or propose improvements, in order to have a more accurate system validation and development. What was found throughout the research was that accidents tend to happen in several angles and several speeds. Up to date, the only difference considered in the UNE normative is the speed. We believe that this can further be improved if also a second angle of trajectory was included.

Our proposal, includes a new flowchart for activities, which in turn make a standard UNE 135900 test as a minimum requirement, and later on, begins with the development of subsystem test simulations where the amount of energy absorbed by the system will be measured. Once this simulation has been approved, then real tests will take place. If the barrier behaves as expected and approves, then the next stage is the full system simulation.

In the full system simulation, the analysed parameters are the degree of redirection provided to the rider and the biomechanical indexes obtained. Also, this simulation will include the new certification protocol that increases the angle of trajectory but maintains the same angle of impact of the dummy. With the same procedure as in subsystems, the simulation needs to be approved prior to conducting real tests. Once all of this has been approved, a new certification can be obtained with the combined results of the UNE 135900 (considered as the minimum requirement) and the new test, increasing the degree of safety provided by the motorcyclist protection system.

REFERENCES

1. Fonts Mestres P. (2007): *Estudi d'un Procediment d'Assaig d'Absorbidors de Xoc per a Motoristes, que s'instal·len a les Barreres Metàl·liques de les Carreteres I Autopistes* {(Study of a Test protocol for motorcyclists friendly crash absorbers installed on roads and highways' metallic barriers. In Catalán.)}, 2007, Department of Passive Safety, Idiada, Santa Oliva, Spain.
2. Andreone L., Galliano F. (2007): *Cooperative systems applications to improve Road Safety: the WATCH OVER project*, ITS Aalborg 2007.
3. MAIDS (2004): *In depth investigations of accidents involving powered two wheelers* {final report, ACEM}.
4. Hampton C. Gabler: *The risk of fatality in motorcycle crashes with roadside barriers*, Virginia Tech {Paper number 07-0474}.
5. Sala, & Astori (1998). *New concepts and materials for passive safety of motorcyclists.*
6. Proceedings of the IRCOBI Conference, September 1998, Göteborg, Sweden.
7. Federation of European Motorcyclists Associations (2000): *Final Report of the Motorcyclist & Crash Barriers Project*, 2000. http://www.fema.ridersrights.org/crashbarrier/index.html
8. Mulvihill C., Corben B. (September 2004): *Motorcyclist injury risk with flexible barriers and potential mitigating measures*, Monash University Accident Research Centre, September 2004. http://www.mraa.org.au/downloads/BarrierPostReport-FinalReport.pdf
9. VicRoads: *Victorian Motorcycle Road Safety Strategy* {www.vicroads.vic.gov.au}VicRoads (2000): *Road Design Guidelines: Safety Barriers,* March 2000.
10. Duncan C., Corben B., Truedsson N., Tingvall C. (December 2000): *Motorcycle and safety barrier crash testing: feasibility study*, Monash University Accident Reasearch Centre, Report No CR 201, December 2000. http://www.monash.edu.au/muarc/reports/atsb201.pdf
11. Czaika M. (September 2006): *Wire Rope Un-safety Barriers*, September 2006.
12. Bloch J.A. (November 2005): *Evaluation of Road Restraint Systems by an Accredited Test House in the Context of European Standardisation*, Building tomorrow's Transport Infrastructure in South East Europe conference, 16 – 17 November 2005, Belgrade.
13. Equipamiento Para la Señalización Vial AEN/CTN 135 (2005): *The evaluation of the behaviour of the systems for protection of motorcyclists in the barriers of security and railings (UNE 135900)*, Madrid, Spain, 2005.
14. Laboratoire d'essai INRETS Équipements de la Route (1998) : *Protocole d'Évaluation de Dispositifs de Protection pour Motoclistes* {(Motorcyclist Friendly Devices Assessment Protocol. In French.)}, 1998.
15. Aprosys SP4 FP6-PLT-506503 (2007): *Integrated Project on Advanced Protection Systems,* deliverable D421A, July 2007.
16. Quincy R. (1998): *Protocole d'essais de dispositif de retenue assurant le sécurité des motocyclistes*, Laboratoire d'essais Inrets Equipements de la Route (LIER), 1998.
17. Rogers N., Zellner Anoop Chawla J., Nakatani T. (September 2004): *Methodologies for motorcyclist injury prediction by means of computer simulation*, IRCOBI conference, Graz, Austria, September 2004.

CONTACT INFORMATION

Mario Nombela: mnombela@idiada.com / telephone number: +34 977 166 021

Arturo Dávila: adavila@idiada.com / telephone number: +34 977 166 021

Juan José Alba: jjalba@unizar.es

Juan Luis de Miguel, jl.demiguel@centro-zaragoza.com

DRIVER ASSISTANCE AND MODELING

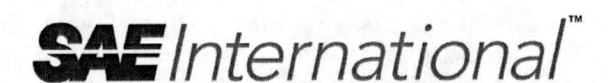

Calibration and Verification of Driver Assistance and Vehicle Safety Communications Systems

2010-01-0664
Published
04/12/2010

Mohammad Naserian
Vector North America

Kurt Krueger
Vector North America

Copyright © 2010 SAE International

ABSTRACT

The development of complex control systems requires appropriate tool that assists engineers in the development, test, and verification processes. In this paper we use vehicle safety as an example of a control system and demonstrate a test-bed that can be used from the beginning of development to end of line production. Safety applications such as Cooperative Forward Collision Warning (CFCW), Emergency Electronic Brake Lights, Lange Change Warning and Pre-Crash Sensing are given the highest development priority for future vehicles. There are several proposals for the safety application systems based on Dedicated Short Range Communications (DSRC). The control system for safety application can be prototyped, calibrated and tested by utilizing an off-the-shelf tool called CANape that has appropriate interfaces to the vehicle bus systems, Electronic Control Unit (ECU) and multimedia recording. The tests and data collection process can also be automated in the lab or in the field tests. CANape has assisted automotive manufacturers and suppliers in developing driver assistance, vehicle and traffic safety, comfort and infotainment systems in additional to power train and transmission control applications. This paper introduces a test-bed for development, tuning, testing and verification of vehicle-to-vehicle communications systems. The concept can be applied to development of any control system in vehicle environment.

INTRODUCTION

Statistics gathered by National Highway Traffic Safety Administration (NHTSA) indicated that about 35,000 people are killed in fatal vehicle accidents each year in the United States. In addition to fatal accidents, statistics show that around 2.9 million people are injured each year [1] and [2]. The driver's reaction time to critical events occurring on the road plays an important role in fatal accident statistics and can reduce the possibility or severeness of the crash [3]. Around 60 percent of roadway collisions could be avoided if the driver's reaction time is reduced by at least one-half second [4]. In Japan, Intelligent Transportation System and Safety is a joint national project where government and private sectors have been working together. It was estimated that the number of fatalities could be reduced by around 5,000 per year by using Vehicle Safety Communications (VSC) in Japan [5].

Recently, a collaborative effort between five OEMs (Daimler, Ford, GM, Honda and Toyota) and US Department of Transportation (DOT) has been formed to investigate improving the vehicle safety by utilizing new emerging technologies such as Dedicated Short Range Communications (DSRC) and global positioning (GPS) [6]. A DSRC-based vehicle safety communications that is equipped with GPS can provide significant additional information about the driving situation. This is beyond the capabilities of existing object sensing systems that are utilized in autonomous safety systems [7]. Safety applications such as Cooperative Forward Collision Warning (CFCW), Emergency Electronic Brake Lights, Lange Change Warning and Pre-Crash Sensing are highlighted with the highest development priority [6] and [7].

If vehicles are equipped with wireless devices, the ones that are in wireless transmission range of each other can establish a wireless ad hoc network. Mobile Ad Hoc Network (MANET) routing protocols such as Ad Hoc Distance Vector (AODV) [8] are not appropriate for safety applications because of the initial route discovery phase of the protocol that needs to be performed before data transmission. The latency requirement for the Intelligent Transmission Systems (ITS) safety applications is suggested to be less than 200 ms [9]. Short delivery latencies are hard to be guaranteed in MANET protocols due to route discovery and high mobility of nodes. Furthermore, in an ITS safety applications, the destination of a data packet (sink) is not a specific node and any nearby node is a potential destination. These facts have shifted the vehicle safety applications towards a broadcast-based approach rather than a forwarding-based approach. Most of the research is focused on the medium access protocol. To avoid the random and unpredictable nature of 802.11 [10], a Time Division Multiple Access (TDMA)-based slotted MAC protocol was proposed in [11]. A contention-free medium access was proposed in [12] through transmitting one-way tokens.

A Vehicular Collision Warning Communications (VCWC) protocol was introduced in [13] that identifies application requirements for vehicular cooperative collision warning considering congestion control for emergency warning messages. IEEE 1609 Standards for Wireless Access in Vehicular Environments (WAVE) [14] have been developed by the Dedicated Short Range Communications Working Group in IEEE. The standard is constantly being updated and has had 4 revisions since its first release. Authors of [15] reviewed the existing DSRC-based vehicle to vehicle communications protocols and compared their performance.

A simulation environment to simulate wireless communication systems based on IEEE802.11p was introduced for NS-2 Network Simulator [16]. The popular Network Simulator (NS-2) [17] has been used by many researchers for simulation and evaluation of new communications protocols. NS-2 is a great tool for wireless communications simulations but is not geared toward distributed networking applications, which employ Controller Area Network (CAN) [18] or FlexRay [19]. Once the wireless technology challenges are overcome via NS-2 simulations, the desired application must be tested in a vehicle environment. This paper provides a test-bed for rapid-prototyping, test and verification of safety communications systems proposed for vehicular applications.

RAPID PROTOTYPING, TEST AND TUNINGN

Simulation environments such as NS-2 are appropriate for network simulations and creating mobility models for the wireless nodes. NS-2 is not created for vehicle bus communications and is not capable of simulating vehicle bus conditions. After proof of concept step with extensive simulations (e.g. with NS-2), the application needs to be tested in the vehicle environment. Application may be implemented with off-the-shelf software using MATLAB and Simulink from The MathWorks [20]. CANape from Vector [21] supports model based development as well as different vehicle bus protocols and standard measurement and calibration protocols such as CCP (CAN Calibration protocol) and eXtended Calibration protocol (XCP) [22]. Vector provides free driver for setting up a CCP or XCP Slave that can be downloaded from its web site [23].

<figure 1 here>

The MathWorks and Vector have partnered in developing their software environments. CANape is a measurement and calibration tool that can also be used as an instrumentation interface to Simulink to allow calibration and data stimulation to the application model in the early stages of the development process. This way the transition from lab development to field testing will also be easier keeping the same database and project files. To connect to Simulink model a XCP over Ethernet link is utilized which provides flexibility of running these two applications (Simulink and CANape) on two different computers to access more processing power. At this stage, calibration of the application model is possible with stimulating recorded data from the vehicle communication bus.

More down into the development process, where the design structure may not be changed as frequently, dll file of the application model can be used in CANape test set ups. The compile of the developed algorithm in Simulink generates dll and A2L database files that can be imported in CANape where it can be tested and run against recorded data from vehicles. In CANape, parameters of the algorithm are tuned to improve the results at this step. Bugs and errors in the algorithm may be realized when experiment data is analyzed in CANape. The test results can be recorded in CANape and replayed for further analysis. This test set-up is depicted in Figure 1.

<figure 2 here>

PROTOTYPE VEHICLE TEST SET UP

After the first step of algorithm tuning introduced in the previous section, it is time for some field testing. For the beginning of the field test, at least two prototype instrumented vehicles are required so that the safety communications messages that are exchanged can be recorded, analyzed and verified. Such a test is required before the mass production of the safety communications ECU.

The suggested prototype vehicle equipment set up is shown in Figure 2. In this test set up, a fast and reliable data acquiring system is crucial for verifying and analyzing the test results. XCP calibration protocol allows communication to the electronic control units over any medium, including Ethernet. Using XCP over Ethernet allows high bandwidth up to 5 Mbps, where hundreds of signals can be acquired every

10 msec from the ECU. Each prototype test vehicle is equipped with a laptop that has CANape and hardware interface to vehicle bus (CAN or FlexRay). The video can be recorded and overlaid with detected objects and safety warnings issued to the driver of the host vehicle (Figure 3). Scripts can be written inside CANape for task automation and analyzing data. Data from GPS and video cameras are time-aligned with the acquired data from different sources including vehicle CAN bus and Electronic Control Units (ECUs) of the vehicle.

<figure 3 here>

CONCLUSIONS

This paper provided a conceptual test-bed for verification, tuning and rapid-prototyping of vehicle safety communications applications. The proposed rapid-prototyping model helps to catch the bugs or errors in early stages of the development. The algorithm can be tested against real data collected from vehicles where any required modification and tuning is possible before the field test. With the proposed prototype test vehicle model, the final verification and tuning to the model can be performed. In every test, data from DSRC receivers and transmitter, ECU, vehicle bus (CAN or FlexRay), GPS and video camera can be recorded, replayed and analyzed in CANape for further tuning and development of the application.

REFERENCES

1. "Early Estimate of Motor Vehicle Traffic Fatalities From January to October 2008", NHTSA, DOT HS 811 054, December 2008, http://nhtsa.gov/staticfiles/DOT/NHTSA/NCSA/Content/RNotes/2008/811054.pdf.

2. "Traffic Safety Facts 2003 to 2007", NHTSA, http://www-nrd.nhtsa.dot.gov/departments/nrd-30/ncsa/STSI/19\IA/2007/19\IA\2007.htm.

3. U.S. Department of Transportation, "A Methodology for Estimating Potential Safety Benefits for Pre-Production Driver Assistance Systems", U.S. Department of Transportation, National Highway Traffic Administration, DOT HS 810 945, May 2008.

4. Wang C. D. and Thompson J. P., "Apparatus and Method for Motion Detection and Tracking of Objects in a Region for Collision Avoidance Utilizing a Real-Time adaptive Probabilistic Neural Network", 1997 US. Patent No. 5,613,039.

5. Oyama S., "Vehicle Safety Communications: Progresses in Japan", IEEE International Conference on Vehicular Electronics and Safety, ICVES 2008, September 2008, pp. 241-241.

6. Krishnan, H., "Vehicle Safety Communications Project," Presentation at CAMP Vehicle Safety Communications Consortium, February 15, 2006, http://www.sae.org/events/ads/krishnan.pdf.

7. Crash Avoidance Metrics Partnership (CAMP), "Vehicle Safety Communications - Applications", VSC-A, First Annual Report, Submitted to the Intelligent Transportation Systems (ITS) Joint Program Office (JPO) of the Research and Innovative Technology Administration (RITA) and the National Highway Traffic Safety Administration (NHTSA), September 4, 2008.

8. Perkins C. E., Belding-Royer Elizabeth M., "Ad Hoc On Demand Distance Vector (AODV) Routing", IETF Internet-Draft, http://moment.cs.ucsb.edu/AODV/ID/draft-ietf-manet-aodv-13.txt, February 2003.

9. Xu Q., Sengupta R., and Jiang D., "Design and Analysis of Highway Safety Communication Protocol in 5.9 GHz Dedicated Short-Range Communications Spectrum", In the Proceeding of IEEE Vehiculat Technology Conference (VTC2003), Vol. 57, No. 4, 2003, pp. 2451-2455.

10. Bharghavan V., Demers A., Shenker S., and Zhang L. "MACAW: A Media Access Protocol for Wireless LANs", In the Proceedings of ACM SIGCOMM 94, London, UK, August 1994, pp. 212-225.

11. Tickoo A. O. and Sikdar B., "Queuing Analysis and Delay Mitigation in IEEE 802.11 Random Access MAC-Based Wireless Networks", In the Proceeding of IEEE INFOCOM, 2004.

12. Masamura M. and Okada H. "A Novel Contention-Free Medium Access Control Protocol for Inter-Vehicle Communications Systems", Mobile Adhoc and Sensor Systems, 2007. MASS 2007. IEEE Internatonal Conference, October 2007, pp. 1-6.

13. Yang X., Liu J., Zhao F. and Vaidya N. H., "A Vehicle-to-Vehicle Communication Protocol for Cooperative Collision Warning", In the Proceeding of the 1st IEEE International Conference on Mobile and Ubiquitous System Networking and Services, pp. 114-123, Boston, USA, August 2004.

14. U.S. Department of Transportation, "IEEE 1609 - Family of Standards for Wireless Access in Vehicular Environments (WAVE)", Intelligent Transportation Systems Standards Fact Sheet, January 2006.

15. Yin J., ElBatt T., Yeung G., Ryu B., Habermas S.,Krishnan H., Talty T., "Performance Evaluation of Safety Applications over DSRC Vehicular Ad Hoc Networks", Proceedings of the 1st ACM International Workshop on Vehicular Ad Hoc Networks, Philadelphia, pp. 1-9, 2004.

16. Chen Q., Jiang D., Taliwal V., and Delgrossi L., "IEEE 802.11 based Vehicular Communication Simulation Design for ns-2", In the Proceedings of the 3rd International Workshop on Vehicular Ad Hoc Networks (VANET), September 2006, pp. 50-56.

17. Fall K. and Varadhan K., "Ns Notes and Documentation Technical Report", University of California Berkeley, LBL, USC/ISI, and Xeron PARC, 2003.

18. www.can-cia.de

19. http://www.flexray.com

20. The Mathworks, "MATLAB and Simulink for Technical Computing'", http://www.mathworks.com

21. Vector, "CANape, The All-Around Tool for Measurement, Calibration and Diagnostics of ECUs'", http://www.vector.com/vi_canape_en.html

22. http://www.asam.net

23. http://www.vector.com/downloads/drivers/xcp.exe

CONTACT INFORMATION

Mohammad Naserian and Kurt Krueger
Address: 39500 Orchard Hill Pl. Suite 550, Michigan, 48375, USA
naserian@ieee.org
kskrueger@ieee.org
Phone: +1- 248-449-9290

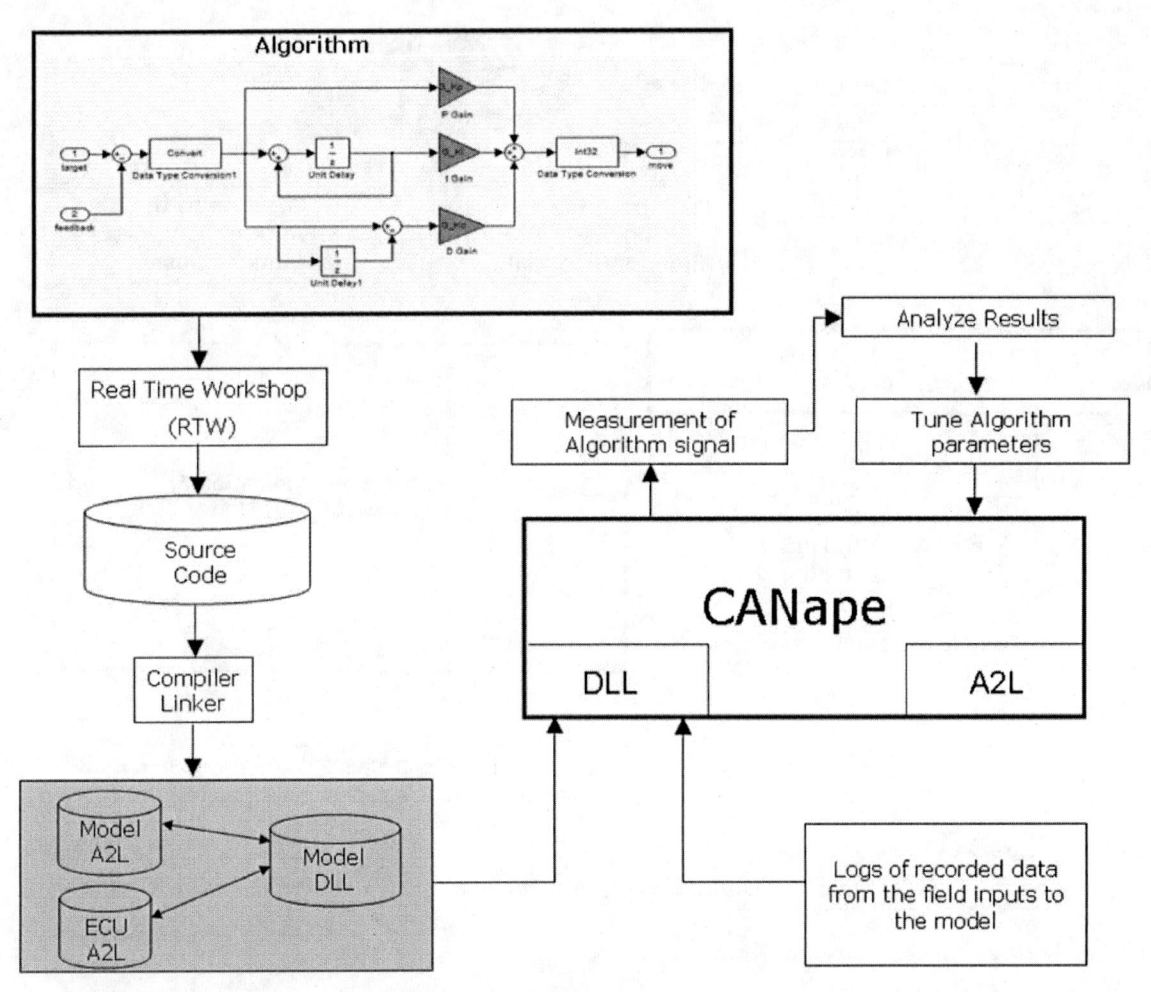

Figure 1. Application development using off-the-shelf products (CANape from Vector and MATLAB/Simulink from The MathWorks

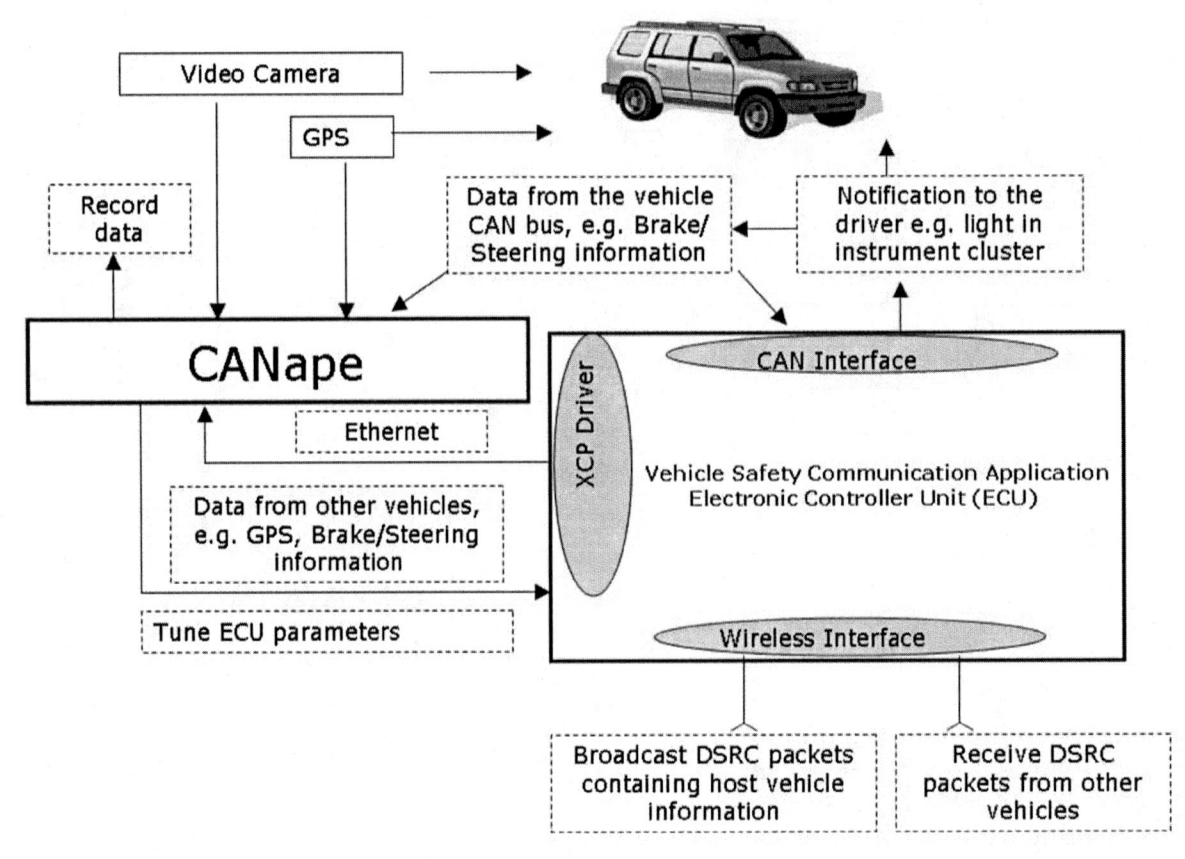

Figure 2. Vehicle Safety Communications application: prototype vehicle testing and verification

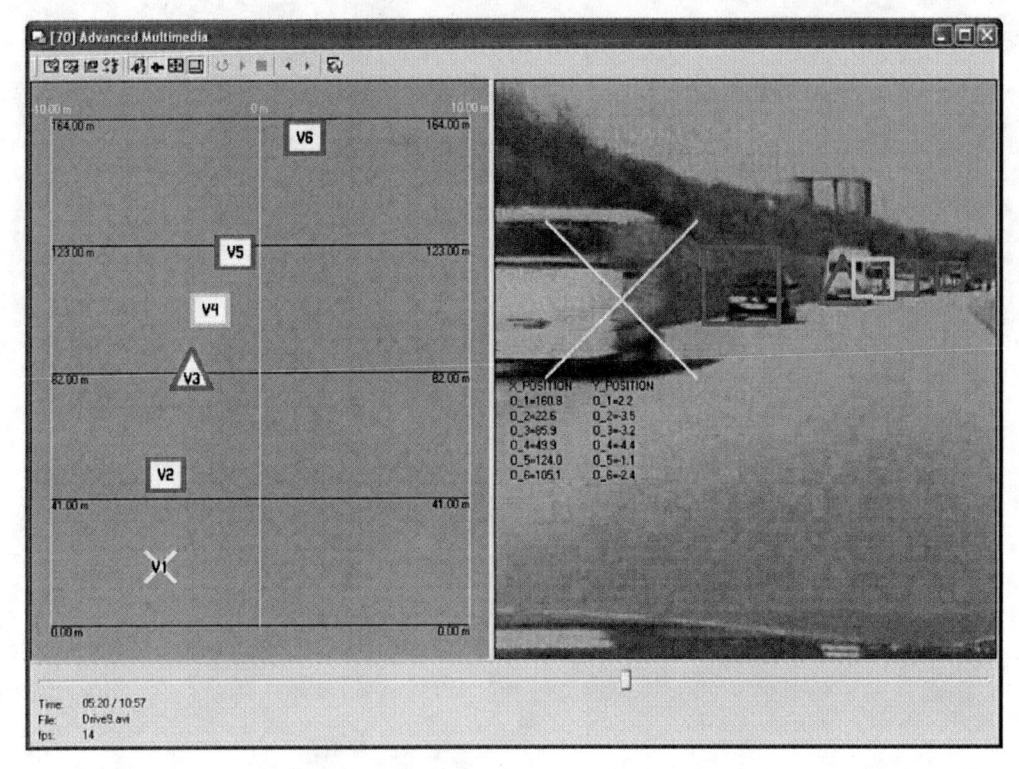

Figure 3. The location of neighboring vehicles relative to the host test vehicle and any issued warning signals can be viewed and verified in CANape's multimedia window. The overlaid graphic objects on the video and the top view is drawn based on the information of the received DSRC packets.

Driver Alcohol Detection System for Safety (DADSS). Background and Rationale for Technology Approaches

2010-01-1580
Published
01/01/2010

Susan A. Ferguson
Ferguson International LLC

Abdullatif (Bud) Zaouk and Clair Strohl
QinetiQ North America

Copyright © 2010 SAE International

ABSTRACT

The Automotive Coalition for Traffic Safety (ACTS) and the National Highway Traffic Safety Administration (NHTSA) have commenced a five-year cooperative agreement exploring the feasibility of, and the public policy challenges associated with, widespread use of in-vehicle alcohol detection technology to prevent alcohol-impaired driving. This effort, known as the Driver Alcohol Detection System for Safety (DADSS) program, aims to develop technologies that could be a component of a system to prevent the vehicle from being driven when the device registers that the driver's blood alcohol concentration (BAC) exceeds the legal limit (currently 0.08 g/dL throughout the United States). For DADSS installation as original equipment in new vehicles there are critical requirements to be met. Alcohol detection technology must be seamless to the driver and be able to quickly and accurately measure the driver's BAC non-invasively. DADSS devices must be compatible for mass-production at a moderate price, be durable, meet high levels of reliability, and require little or no maintenance. Potential technological approaches have been identified and thorough analyses undertaken to determine candidates for further development, utilizing a clear understanding of the processes by which alcohol is absorbed into the blood stream, distributed within the body, and eliminated from it. This paper describes what is known regarding alcohol measurement via various methods, and details which technologies deserve further study. Two approaches are identified that have considerable promise in measuring driver BAC non-invasively within the time and accuracy constraints: 1) Tissue Spectrometry, a touch-based approach allowing estimation of alcohol in tissue through detection of light absorption at a particular wavelength from a beam of near-infrared light reflected from within the subject's tissue, 2) Distant Spectrometry using part of the infrared light spectrum where the light is transmitted toward the subject from a source that receives and analyses the reflected and absorbed spectrum, thereby allowing assessment of alcohol concentration in the subject's exhaled breath.

INTRODUCTION

In response to concerns about limited progress in reducing alcohol-impaired driving in the United States during the last decade, attention is focused on technological approaches to address this problem. There are existing technologies, predominantly used by drivers convicted of Driving While Impaired or Intoxicated (DWI) that require drivers to provide breath samples before starting their vehicles. If a positive breath alcohol concentration (BrAC) is registered, the vehicle cannot be started. Studies indicate that when these devices are used on the vehicles of convicted DWI offenders, they can reduce recidivism by about two-thirds (Willis et al., 2004). Efforts are underway in the United States to increase the use of breath-alcohol ignition interlocks for convicted DWI offenders, both through passage of stronger state laws that will require them for first-time offenders, and through efforts to work within the criminal justice system to maximize their adoption (http://www.madd.org/Drunk-Driving/Drunk-Driving/Campaign-to-Eliminate-Drunk-Driving.aspx).

Even if efforts to get ignition interlocks into the vehicles of every convicted DWI offender are successful, they would only partially solve the problem of deaths and injuries due to alcohol-impaired driving. This is because a large proportion of the alcohol-impaired fatal crashes that occur every year involve drivers with no prior DWI convictions. In 2006, only 7 percent of drivers in fatal crashes with BACs 0.08 g/dL or higher (the threshold above which it is illegal to drive in every state in the United States) had previous alcohol-impaired driving convictions on their records for the prior three years (Insurance Institute for Highway Safety, IIHS, 2008). Moreover, it has been estimated that the actual risk of being arrested for alcohol-impaired driving is as low as 1 in 50 alcohol-impaired driving episodes (Hedlund and McCartt, 2002).

Wider deployment of current breath-based alcohol measurement technology as a preventative measure among the general public is not advisable because of the obtrusive nature of the technology. There are shortcomings associated with the current technology breath alcohol measurement systems that preclude their use on a more widespread basis. Drivers are required to provide a deep-lung breath sample, for which they have to blow long and hard, before starting the vehicle. Furthermore, nearly all units currently in use employ fuel-cell alcohol sensors. These sensors must be at the temperature of breath in order to meet the accuracy requirements, and measurements typically take about 30 seconds but can require as long as three minutes in very cold conditions (Pollard et al., 2007). In addition, current fuel-cell sensor measurements are known to drift over time and require relatively frequent calibration.

The performance standards for the adoption of in-vehicle alcohol detection devices among the general public, many of whom do not drink, let alone drink and drive, must be much more rigorous if they are to cause minimal inconvenience. The Automotive Coalition for Traffic Safety (ACTS, a group funded by vehicle manufacturers) and the National Highway Traffic Safety Administration (NHTSA) have commenced a five-year cooperative agreement to explore the feasibility of, and the public policy challenges associated with, widespread use of in-vehicle alcohol detection technology to prevent alcohol-impaired driving. This program, known as the Driver Alcohol Detection System for Safety (DADSS), seeks to develop technologies that are less-intrusive than the current in-vehicle breath alcohol measurement devices. Detection technology must be seamless with the driving task and quickly and accurately measure a driver's BAC in a non-invasive manner. These technologies will be a component of a system that may prevent the vehicle from being driven when the device registers that the driver's BAC exceeds the legal limit. Such devices ultimately must be compatible with mass-production at a moderate price, be durable, meet high levels of reliability, and require little or no maintenance.

The cooperative agreement seeks to assess the current state of detection technologies that are capable of measuring BAC, and to support the development and testing of prototypes and subsequent hardware that could be installed in vehicles. The goal, at the end of the 5-year program, is the practical demonstration of an alcohol detection subsystem, suitable for subsequent installation in a vehicle. The technical challenges are substantial; however the possible benefits to society are compelling, with the potential to prevent almost 9,000 motor vehicle deaths every year if all drivers with BACs at or above the legal limit of 0.08 g/dL were unable to drive (Lund et al., 2007).

The purpose of this paper is to outline the technological approaches taken in developing alcohol detection hardware. These approaches are founded on a clear understanding of the processes by which alcohol is absorbed into the blood stream, distributed within the human body, and eliminated from it. Not only must technologies under consideration quickly and accurately measure BAC, but the medium through which it is measured (e.g., breath, tissue, sweat, etc.) must provide a valid and reliable estimation of actual BAC levels. Alcohol absorption, distribution, and elimination measurement is a topic about which much has been written yet some large gaps in our understanding still remain. This paper will provide an overview of what is known regarding alcohol measurement via various methods and their implications for the decisions about which technologies deserve further study. The paper also will provide an overview of the current performance specifications developed to assess the in-vehicle advanced alcohol detection technologies and the rationale for them.

THE BASIS FOR DEFINING ALCOHOL-IMPAIRED DRIVING

It has long been known that ingested alcoholic beverages can result in the impairment of skills necessary to drive safely, so laws against driving while impaired or intoxicated by alcohol (DWI) have been in place in the United States for many years. At the time of the earliest state DWI laws there was no objective measure of BAC or impairment, thus, the practical effect of these laws was that only obviously drunk drivers - the so-called falling-down drunks - were likely to be arrested. Decades of research has established that driving impairment due to alcohol may be present even without obvious outward signs of impairment. Impairment is not indicated just by the appearance of gross physical symptoms, but involves the deterioration of judgment, alertness, and attention, loss of fine motor coordination, as well as reduced reaction times and diminishing sensory perceptions (Moscowitz et al., 1988, 2000). Furthermore, studies of drivers in crashes in which BAC measurements are taken have clearly established a dose-response relationship between BAC and risks of being in a crash. The data show unequivocally that as driver BAC goes up, the risk of being in a crash goes up. The probability of a fatal crash increases significantly when a driver's BAC is at 0.05 g/dL and climbs more rapidly after 0.08 g/dL (Peck et al., 2008; Zador et al., 2000).

Because of research establishing a dose-response relationship between blood alcohol concentration and crash risk, it was possible to define DWI offenses in terms of a BAC threshold above which drivers are considered impaired (so called per se laws). All states and the District of Columbia have per se laws defining the DWI offense as driving with a BAC above 0.08 g/dL. At a BAC of 0.08 g/dL drivers are about 5 times more likely to be in a fatal crash than sober drivers and these elevated risks grow even higher as BACs increase (Zador, et al., 2000). Per se laws have made it much easier to enforce alcohol-impaired driving laws in the United States and around the world, and have enabled progress to be made both in apprehending alcohol-impaired drivers and removing them from the road, as well as reducing alcohol-impaired crashes (Hedlund et al., 2002). Given such compelling research indicating increasing driver impairment with rising BACs, and laws throughout the United States and other countries specifying that it is illegal to drive with BACs at or above a given threshold, the most reasonable option for the DADSS subsystem is a device or devices that can accurately measure driver BAC.

PERFORMANCE SPECIFICATIONS

Based on input from the Blue Ribbon Panel (BRP), a group of experts formed by ACTS to help advise the DADSS program, ACTS has developed performance specifications to assess the in-vehicle advanced alcohol detection technologies that are being developed. The specifications are designed to focus the current and future development of relevant emerging and existing advanced alcohol detection technologies. In addition to requirements for a high level of accuracy and very fast time to measurement, the influences of environment, issues related to user acceptance, long-term reliability, and system maintenance also will be assessed. The resulting list of specifications with definitions, measurement requirements, and acceptable performance levels are documented in the DADSS Subsystem Performance Specification Document (http://dev.dadss.org/sites/default/files/dadss001-draft_100908.pdf). The accuracy and speed of measurement requirements adopted by the DADSS Program are much more stringent than currently available commercial alcohol measurement technologies are capable of achieving. As noted above, the devices would need to be seamless with the driving task and not inconvenience drivers. Translating that to appropriate performance specifications was approached by estimating the potential for inconvenience if reliability, accuracy, and time to measurement were set at various levels. Presented below are the processes used to derive them.

RELIABILITY

Developing an alcohol detection device as original equipment for the vehicle environment brings with it special challenges. Reliability is defined as the ability of a system or component to perform its required functions under stated conditions for a specified period of time. Levels of reliability that are too low would result in an unacceptable number of failures to operate the vehicle. It has been estimated that at the 3σ reliability (sigma - Greek letter σ - is used to represent the standard deviation of a statistical population) there could be the potential for 66,800 defects per million opportunities, where an opportunity is defined as a chance for nonconformance. The accepted level of reliability within the industry is 6σ. The term "six sigma process" comes from the notion that with six standard deviations between the process mean and the nearest specification limit, there will be practically no items that fail to meet specifications. In practice, 6σ is equivalent to 99.9997% efficiency. Processes that operate with "six sigma quality" over the short term are assumed to produce long-term defect levels below 3.4 defects per million opportunities.

ACCURACY AND PRECISION

Accuracy is defined as the degree of closeness of a measured or calculated quantity to its actual (true) value (also referred to as the Systematic Error - SE). Precision is the degree of mutual agreement among a series of individual measurements or values (also referred to as the Standard Deviation - SD). To limit the number of misclassification errors, accuracy and precision must be very high, otherwise drivers may be incorrectly classified as being over the threshold (false positives), or below the legal limit (false negatives). It has been determined that in order to assure that drivers with BACs at or above the legal limit will not be able to drive, while at the same time allowing those below the limit to drive unhindered, SE and SD requirements should be set at 0.0003.

SPEED OF MEASUREMENT

Another important performance requirement is that time to measurement be very short. Sober drivers should not be inconvenienced each and every time they drive their vehicle by having to wait for the system to function. Current breath-based alcohol measurement devices can take 30 seconds or more to provide an estimate of BAC. However, it was determined that the DADSS device should take no longer to provide a measurement than the current industry standard time taken to activate the motive power of the vehicle. Thus, the subsystem should be capable of providing a reading of the current BAC and communicating the result within 325 msec. It should be capable of providing a second reading, if necessary, within 400 msec.

ALCOHOL ABSORPTION, DISTRIBUTION, AND ELIMINATION IN HUMANS: WAYS IN WHICH ALCOHOL CAN BE MEASURED

The science of pharmacokinetics is concerned with the ways in which drugs and their metabolites are absorbed, distributed, and eliminated from the body (Jones, 2008). This is separate from pharmacodynamics which is the study of the physiological effects of drugs and their actions on the body (Jones, 2008). Ethyl alcohol or ethanol, more commonly referred to as alcohol, is only one of a family of organic compounds known as alcohols. Ethanol, referred to hereafter as alcohol, is highly soluble in the body's water, which makes up 50-60 percent of body weight. Even though alcohol is a central nervous system depressant, people perceive it as a stimulant and in the early stages it can produce feelings of euphoria (Jones, 2008). With the consumption of larger amounts of alcohol, performance and behavior can be impaired resulting in reduced coordination, loss of motor control, lack of good judgment, and at very high concentrations (greater than 0.4 g/dL) loss of consciousness and death.

Figure 1 portrays schematically the pathways by which alcohol is absorbed into the blood stream, is distributed throughout the body, and eliminated from it.

After ingestion, alcohol enters the stomach where it is partially absorbed through the stomach wall (about 20 percent), and then to the small intestines where most of the absorption takes place (about 80 percent). Alcohol is then transported to the liver and on to the heart before it is distributed by the arteries throughout all body fluids and tissues. Alcohol easily passes the blood-brain barrier where it affects central nervous system functioning. The time required for reaching equilibrium depends on the blood flow to the various organs and tissues, but over time alcohol mixes completely with all the water in the body and reaches into all fluid compartments within the body.

The characteristics of alcohol's distribution and elimination can point to potential ways in which BAC can be measured. There are two mechanisms by which alcohol is eliminated from the body, metabolism and excretion. The liver is the primary organ responsible for the elimination of alcohol and it is where about 95% of ingested alcohol is metabolized. The remainder of the alcohol, about 2-5 percent, is excreted unchanged wherever water is removed from the body; though the skin in sweat, from the lungs in breath, from the eyes in tears and from the kidneys in urine. As noted above, alcohol distributes completely into all the body's compartments so alcohol can be measured *in vivo* in bodily tissue.

(See Figure 1 after last section of paper)

METHODS USED TO MEASURE BLOOD ALCOHOL CONCENTRATION

For many years the only means to determine BAC was through blood and urine testing. As early as 1874 it was recognized that ingested alcohol can be measured in breath (Jones, 2008), and the smell of alcohol on breath is a well-known indication that someone has been drinking. Accurate measurement of alcohol in expired air has a physiological basis. Under normal lung function there is an efficient gas exchange between blood and gases, thus resulting in a close correlation between blood and gas concentrations of alcohol (Hök, 2006). Furthermore, a recent study (Lindberg et al., 2007) has established that the concentration of alcohol in breath is in very close agreement to that of alcohol in arterial blood (Figure 2), even though the gold standard for equating breath to blood alcohol is venous BAC.[1] Of note is that arterial BAC is a better indication of brain alcohol and hence impairment than venous blood, so BrAC is particularly well suited as a measure of driver impairment.

(See Figure 2 after last section of paper)

Dr. Robert Borkenstein is recognized as the inventor of the first system that measured alcohol on a person's breath. In 1954, he invented the first breathalyzer, which used chemical oxidation and photometry to measure alcohol concentration. Subsequently physiochemical methods were developed for the measurement of alcohol in breath such as gas chromatography, electrochemical oxidation, and infra-red analysis. Breath testing has flourished because it is non-invasive and, in contrast to urine and blood samples that have to be sent away for testing, provides on-the-spot results. As a result most countries have adopted breath testing both for roadside screening and evidential purposes to establish BAC.

In recent years a number of other approaches have been identified that could be used to measure alcohol in perspiration (either vapor phase or liquid phase) or from measurements of alcohol in a person's tissue. As noted below, these techniques have not yet been widely used to measure alcohol concentration.

DADSS TECHNOLOGICAL APPROACHES

One of the first tasks of the project team was to perform a comprehensive review of emerging and existing state-of-the-art technologies for alcohol detection. Technology scans were undertaken through patent and literature reviews. Based on these reviews four categories of technologies were identified with potential for measuring driver BAC within the vehicle environment:

1. Electrochemical/Transdermal Systems

Electrochemical Systems are chemical-reaction-based devices such as transdermal and breathalyzer-based systems. Alcohol in the presence of a reactant chemical will produce

---

[1] Venous BAC presently serves as a standard in forensic practice for the prosecution of drunk drivers and as a measure of their impairment and drunkenness.

colorimetric changes measured by spectral analysis or a semi-conductor sensor. In fuel cell systems (typically used in current technology breath-alcohol ignition interlocks), exhaled air containing alcohol passes over platinum electrodes which oxidize the alcohol and produce an electrical current; the more alcohol in the air sample, the greater the electrical current. The electrical current level permits accurate calculation of breath alcohol concentration (BrAC) which can be converted to blood alcohol concentration (BAC) using a standardized conversion factor.

2. Tissue Spectrometry Systems

Tissue Spectrometry Systems allow estimation of BAC by measuring the alcohol concentration in tissue. This is achieved through detection of light absorption at a particular wavelength from a beam of Near-Infrared (NIR) reflected from within the subject's tissue. As classified herein, they are touch-based systems and require skin contact. Variations of tissue spectrometry systems include Michelson, Raman, Fabry-Perot, Laser Diode and Light Emitting Diode (LED) based devices.

3. Distant/Offset Spectrometry Systems

Distant Spectrometry Systems use an approach that is similar to Tissue Spectrometry, except that no skin contact is required. Infrared (IR) is transmitted toward the subject from a source that also has a sensor to receive and analyze the reflected and absorbed spectrum to assess alcohol concentration in the subject's exhaled breath.

4. Behavioral Systems

Behavioral Systems detect impaired driving through objective behavioral measures. These include ocular, gaze, eye movement, and driving performance measures, as well as other performance measures believed to be related to driving performance.

In addition to the technology scans, a Request for Information (RFI) was published as a means by which the DADSS program was first communicated to potential technology developers. The goal of the RFI was to establish the level of interest among technology developers in taking part in the research, the kinds of technologies available, and their states of development relevant to in-vehicle applications. Based on information gleaned during the RFI process, a subset of technology companies was selected to receive a Request for Proposal. Detailed evaluation of the proposals that were received resulted in awards to three technology companies based on two of the technological approaches outlined above; tissue spectroscopy and distant spectroscopy. The reasons for proceeding with these preferred technological approaches, and not others, are outlined below.

Tissue Spectrometry Systems

Also known as near-infrared (NIR) spectrometry, this is a noninvasive approach that utilizes the near infrared region of the electromagnetic spectrum (from about 0.7 μm to 2.5 μm) to measure substances of interest in bodily tissue. NIR spectroscopy is the science that characterizes the transfer of electromagnetic energy to vibrational energy in molecular bonds, referred to as absorption, which occurs when NIR light interacts with matter. Most molecules absorb infrared electromagnetic energy in this manner. The specific structure of a molecule dictates the energy levels, and therefore the wavelengths, at which electromagnetic energy will be transferred. As a result, the absorbance spectrum of each molecular species is unique. Better-known applications include use in medical diagnosis of blood oxygen and blood sugar, but more recently devices have been developed that can measure alcohol in tissue (Ver Steeg et al., 2005).

TruTouch Technologies has developed a desk-top device that measures alcohol in tissue non-invasively. The measurement begins by illuminating the user's skin with NIR light which propagates into the tissue (the skin has to be in contact with the device). The beam of light can penetrate tissue at depths of up to 5 mm to reach the dermal layer where alcohol that is dissolved in water resides. A portion of the light is diffusely reflected back to the skin's surface and collected by an optical touch pad. The light contains information on the unique chemical information and tissue structure of the user. This light is analyzed to determine the alcohol concentration and, when applicable, verify the identity of the user. Because of the complex nature of tissue composition, the challenge is to measure the concentration of alcohol (sensitivity) while ignoring all the other interfering analytes or signals (selectivity).

Although the entire NIR spectrum spans the wavelengths from 0.7 - 2.5 μm, TruTouch has determined that the 1.25-2.5 μm portion provides the highest sensitivity and selectivity for alcohol measurement. The 0.7-1.25 μm portion of the NIR is limited by the presence of skin pigments such as melanin that can create large differences among people, particularly of different ethnicities. In contrast, the longer wavelength portion of the NIR, from 1.25-2.5 μm, is virtually unaffected by skin pigment (Anderson et al., 1981). One other advantage of using this part of the spectrum is that the alcohol signal in the 1.25-2.5 μm region is hundreds of times stronger than the signal in the 0.7-1.25 μm part of the NIR.

TruTouch has undertaken extensive *in vitro* testing - comprised of testing samples of alcohol, glucose, creatinine, urea, water, and polystyrene microspheres - as well as human subjects testing to determine the feasibility of this approach.

Extensive *in vitro* testing has demonstrated that the alcohol measurement can be done independently of other interfering analytes present in the sample set. In addition, hundreds of human subjects have participated in alcohol dosing studies where the noninvasive tissue alcohol measurements taken from the underside of the forearm have been directly compared to alcohol in samples of breath, capillary blood, and venous blood (*in vivo* testing). The findings of several published studies have demonstrated that tissue alcohol, as measured using the TruTouch approach, can be selectively identified and that its concentration corresponds reasonably to venous blood, capillary blood and breath during the elimination phase (Ridder et al., 2005a; Ridder et al., 2005b; Ridder et al., 2009).

The clinical data collected during the controlled drinking studies presented in these papers uses a protocol whereby human subjects drink a quantity of alcoholic beverage at one sitting (in a period of about 20 minutes) that is calculated to bring them to a BAC of 0.12 g/dL. This drinking is done after overnight fasting so results in a rapid excursion to a high blood alcohol concentration (absorption phase), and a larger difference in the blood and tissue compartments than would be expected during typical social drinking. Further, because of the rapidly changing BACs, individual compartment pharmacokinetic differences are more evident during the absorption phase when compared to the elimination phase. Research is underway to examine tissue alcohol concentration using a location on the finger rather than the forearm. At this location blood perfusion is greater and is expected to result in a better correspondence among the measures of the various compartments (blood and breath) as well as during absorption and elimination.

One question of interest is the length of time it takes for alcohol to reach the interstitial fluid in subdural tissue and the length of time to reach peak measurements relative to peak measurements of alcohol in blood. It should be noted that lag times for peak alcohol concentration in the blood and other compartments can vary based on a number of factors including the gastric emptying, dose of alcohol and time over which it is consumed, and factors that influence the distribution volume such as gender, age, and body mass index. As noted above, in the case of tissue measurements there can be additional differences based on where in the body the measurements are made. There currently are few published studies indicating differences in the timing of peak BACs in tissue versus blood or breath but based on discussions with TruTouch personnel, they estimate that the average time delay is on the order of 10-15 minutes (Personal Communication, Ben Ver Steeg, July 9, 2009).

DISTANT SPECTROMETRY SYSTEMS

Distant spectrometry systems use an approach similar to tissue spectrometry, in that the approach utilizes the infrared region of the electromagnetic spectrum, however no skin contact is required. Infrared light is transmitted toward the subject from a source that receives and analyses the reflected and absorbed spectrum to assess alcohol concentration in the subject's exhaled breath. There are a number of approaches under development that aim to remotely analyze alcohol in breath either within the vehicle cabin or around the drivers face without the driver having to provide a deep-lung breath sample.

As mentioned above, under normal lung function there is an efficient gas exchange between blood and gases, resulting in a close correlation between blood and breath alcohol concentrations (Hök, 2006) reflecting the very rapid equilibrium kinetics between pulmonary capillary blood and alveolar air (Opdam et al., 1986). In fact, as seen in Figure 2 (see page, 7), BrAC measurements (converted to units of BAC) track arterial BACs throughout the blood alcohol time curve; only slightly below during the ascending curve, then virtually identical on the descending limb of the BAC time curve (Lindberg et al., 2007).

Current breath-based alcohol measurement techniques require direct access to undiluted deep-lung air, and therefore employ a mouthpiece. The challenge in measuring alcohol in breath from around the driver's face or within the vehicle cabin is that the breath is diluted with the cabin air. With funding from the Swedish Road Administration, Autoliv, Hök Instruments AB, and SenseAir AB have collaborated in the development of a contactless method to measure alcohol in breath (http://www.hokinstrument.se/nyheter/KAIA%20Broschyr.pdf.)[2] The measuring principle of the sensor is to use measurements of expired carbon dioxide (CO_2) as an indication of the degree of dilution of the alcohol concentration in expired air. Normal concentration of CO_2 in ambient air is close to zero. Furthermore, CO_2 concentration in alveolar air is both known and predictable, and remarkably constant. Thus, by measuring CO_2 and alcohol at the same point, the degree of dilution can be compensated for using a mathematical algorithm. According to Hök (2006), the ratio between the measured concentrations of CO_2 and alcohol, together with the known value of CO_2 in alveolar air, can provide the alveolar air alcohol concentration.

The sensor technology under development by Autoliv and its partners uses infrared (IR) spectroscopy, which is superior to conventional fuel-cell devices in two ways. The IR-based sensors can be stable over the full product lifetime,

---

[2] Note that Delphi Research Labs also has been developing a passive method to detect alcohol in a driver's expired breath (Lambert et al. 2006) using similar principles.

eliminating the need for recurrent calibrations. Furthermore, the IR sensor is not as sensitive as the fuel-cell to major variations in ambient temperature. A patented optical device is used in which multiple reflections of the IR beam within a closed space enable the calculation of alcohol concentration with high resolution. The current device requires drivers to blow towards the sensor which is held at a distance of 10-15 centimeters. Future generations of such a device could employ multiple sensors placed strategically around the cabin of the vehicle close to the driver. The challenge will be to determine number and placement of sensors needed to measure alcohol quickly and accurately given the dynamics of the cabin air, and to ensure that there is no potential bias introduced as a result of passengers who may have been drinking.

Another method being pursued by Alcohol Countermeasure Systems (ACS), a Canadian company that produces alcohol interlocks and breath alcohol testers, employs a similar approach but relies on mid-IR (optical) detection methodology to detect alcohol in the cabin air. Like the Autoliv approach, the approach being used by ACS utilizes sensors that will not require frequent recalibration.

ELECTROCHEMICAL SYSTEMS

Electrochemical systems identified as potential technologies of interest include existing breath-based measurement systems as well as systems that measure alcohol concentration present in a person's sweat. Current breath-based measurement systems, such as breath testers used at the roadside by police officers as an enforcement tool, and after-market breath alcohol ignition interlocks, use electrochemical fuel-cells. Devices that measure alcohol in sweat (e.g., the Secure Continuous Remote Alcohol Monitoring (SCRAM) System, manufactured by Alcohol Monitoring Systems) also use fuel cell technology.

Fuel-cell based approaches to measuring breath alcohol are based on a mature technology that has been used successfully for many years, but there are some barriers to their use in a vehicle environment as original equipment. The fuel cells must be warmed up to breath temperature to meet the accuracy requirements. Thus, in a vehicle environment where widely fluctuating temperatures with changing weather conditions are the norm, alcohol measurement can take a long time. The time to determine BrAC accurately can be as low as 30 seconds at warmer temperatures but can take as long as 3 minutes in colder temperatures (Pollard et al., 2007).

Another concern is that current technology fuel cells are known to exhibit some drift over time, and require recalibration within a year or less. These shortcomings are acceptable inconveniences to drivers who have been convicted of DWI, and for whom the breath alcohol ignition interlock is installed as a condition of reacquiring a license to drive. Once their vehicles are equipped with ignition interlocks the devices are frequently monitored, data downloaded, and recalibrations done every 30, 60 or 90 days (TIRF, 2006). However, due both to the lengthy time of measurement and the recalibration needs, current technology fuel cells are not deemed suitable as original vehicle equipment.

About 1 percent of alcohol is eliminated from the body through sweat (Jones, 2008). It has been known for decades that some alcohol crosses the outer layer of the skin in the form of sweat in liquid phase (sensible perspiration) or vapor phases (insensible perspiration) (Hawthorne et al., 2006). This phenomenon has come to be known as transdermal excretion of alcohol (TIRF, 2006), and the approaches to measure it as transdermal monitoring. Early research specifically looked at alcohol concentration in liquid sweat, and was followed by research to examine alcohol concentration in vapors formed above the skin (Hawthorne et al., 2006). Research using both these approaches has indicated that the concentration of alcohol in sweat increases with increasing numbers of drinks and that the mean transdermal alcohol concentration (TAC) can be relatively close to that of the mean BAC but only after a long delay (Anderson et al., 2006; Davidson et al., 1997; Hawthorne et al. 2006; Swift, 2003; Webster et al., 2007). According to Hawthorne et al. (2006) and others, the TAC and BAC curves are very similar in amplitude and shape, but the peak is delayed by up to 120 minutes or more. For this reason, it is commonly recommended that it be used as a screening device to sense the presence of alcohol, but not as a device that is intended to accurately measure alcohol. In its screening mode, transdermal alcohol monitoring has been recognized as a useful means to determine whether episodes of drinking have occurred among substance abuse offenders who have been ordered to remain abstinent (Marques et al., 2009; Sakai et al. 2006; Vanlaar et al., 2006).

A number of studies have examined the validity of transdermal alcohol testing devices that were developed to be worn on the body for an extended period of time. Two devices have been developed, but only one is available for commercial use. The first wearable prototype transdermal alcohol sensor to be developed by Giner, Inc is known as the WrisTAS. First tested in the 1990s, it has been shown to produce BAC measurements that follow the patterns of the BAC curve but with a significant time delay (Vanlaar et al., 2007). However, this device is not currently available for commercial use.

The SCRAM device, developed by Alcohol Monitoring Systems, became available in 2003, and since that time has been the subject of numerous studies to examine its performance both in the laboratory and in the field. The SCRAM device is worn on the ankle continuously for up to several months. The general consensus is that such devices can consistently detect alcohol at the skin surface, but that they are not able to measure BAC accurately. Moreover, the delays between maximum TAC versus maximum BAC were extensive. Using laboratory dosing studies, Marques et al. (2009) reported a mean peak delay between BAC and TAC of 4.5 hours + 2.9 hours, and the WrisTAS device a delay of 2.3 + 1.5 hours; Sakai et al. (2006) reported mean peak delays of 2 to 3 hours.

It should be noted that both the SCRAM and WrisTAS devices were developed for the purpose of continuously monitoring the presence of alcohol and not to provide a quantitative measurement of alcohol concentration at a given point in time. That being said, if versions were to be developed for the purpose of providing a discrete measurement, the challenge would be to address the TAC/BAC time lag so as to provide a real-time estimation of BAC.

BEHAVIORAL SYSTEMS

There has been considerable research undertaken to establish objective behavioral measures of impairment, including efforts to develop in-vehicle behavioral measures of alcohol impairment. During the 1970s there was a NHTSA-sponsored effort to develop "Cars that drunks can't drive" (Stein et al., 1986). As part of that effort General Motors developed the "physiological tester" which involved punching a five digit code number in four seconds prior to starting the car. It was felt that drunks would be unable to accomplish the task in the time allotted. Another device under development was the Critical Tracking Task (CTT) - a test of psychomotor functioning that measures eye-hand coordination and delays in visual motor response. The CTT was first developed about 35 years ago to measure the ability of Air Force pilots to deal with aircraft malfunctions. In 1976, DOT contracted with the developers of the CTT to design and construct eleven prototype units which subsequently were tested in the laboratory and on vehicles in the field that were driven by second-time DUI offenders (Stein et al., 1986). The CTT devices were implemented in the vehicle as a steering competency test, in which the driver was required to balance a needle in a steering column mounted instrument by turning the steering wheel (Stein et al., 1986). Data indicate that failures on the test were correlated with increasing BACs. However, there is no evidence that performance on such tasks would be able to accurately predict BAC.

There also is evidence that alcohol can affect eye movements (see Pollard, 2007). Simulation studies have indicated that alcohol results in longer dwell times and reduces the frequency of eye movements. One well-known effect of alcohol on eye movements, known as Horizontal Gaze Nystagmus, often referred to as HGN, results in a type of jerk occurring in the movement of the eye whereby the eye gazing upon or following an object begins to lag and has to correct itself with a saccadic movement toward the direction in which the eye is moving or gazing. This phenomenon is used widely by police officers to determine that alcohol impairment may be present, and to indicate the need for further investigation (NHTSA, 2009). However, many ocular and eye movement measures may not be able to distinguish alcohol impairment from impairment from other sources, such as other drugs (Pollard, 2007). Furthermore, there currently are no proposed, non-intrusive and simple eye measurement technologies that appear potentially appropriate for vehicle use.

More recently attention has focused on vehicle-based measures for monitoring impairment such as lane position variability/lateral position, changes in driving speed and speed variability, pedal and steering control, vehicle headway, and delay in motor actions and responses such as braking reaction times. The most comprehensive research efforts have been undertaken in Europe, and often have employed multiple measures of driving behavior. For example, in the SAVE Project (System for effective Assessment of the driver state and Vehicle control in Emergency situations) researchers utilized a simulator and data from road tests to measure a large variety of driving behaviors, the goal of which was to identify driver impairment and classify it in one of a number of categories: fatigue or sleep deprivation, alcohol or drug abuse, sudden illness of the driver, and prolonged periods of inattention. Although some measures of driving behavior were consistent for individual drivers, some factors were more important than others and those factors varied among drivers. This finding indicated that the system would need to be trained for individual drivers (Peters et al., 1998). In the final report of the SAVE Project, Bekiaris (1999) reported only a 78 percent accuracy for detecting alcohol impairment at a BAC of 0.05 g/dL. However, no data are available regarding the ability of the system to discriminate alcohol impairment from that due to fatigue.

There are a few limitations that behaviorally-based devices for measuring alcohol impairment have in common that render them unsuitable for use as original equipment in new vehicles. First, in order to measure impairment there has to be some measure of "normal" abilities on the specific task that can act as a baseline measurement for comparison. Such a requirement would be hard to accomplish in an unobtrusive device. Another shortcoming is that changes in behavior on these tasks can result from any impairment, whether from fatigue, illness, alcohol, medications, illegal drugs, or other sources. There is no question that impairment from sources other than alcohol can be problematic for safe driving. However, the relationship between alcohol and crash risk has been well established, and the number of deaths and injuries resulting from impaired drivers has been tracked for many years. In contrast, the impairments from other drugs and fatigue for example, are often less well understood and the precise relationship between increasing "dose" and crash risk are undetermined. Another limitation specifically related to vehicle-based performance measures is that they can only be measured once the vehicle is moving. A preferable approach would be to prevent drivers at or above the legal limit from driving their vehicle.

SUMMARY/CONCLUSIONS

For the DADSS system to be installed as original equipment in new vehicles there are a number of critical requirements that would have to be met. Alcohol detection technology must be seamless with the driving task and must be able quickly and accurately to measure a driver's BAC in a non-invasive manner. Such devices must be compatible for mass-production at a moderate price, be durable, meet acceptable reliability levels, and require little or no maintenance. These requirements have been taken into account as decisions have been made as to which technology approaches to pursue.

The science of *in vivo* alcohol pharmacokinetics provides insight into the possible ways to measure blood alcohol concentration in the human body, as well as potential limitations of certain methods. Based on patent and literature searches, four technology categories have been identified for measuring driver BAC within the vehicle environment. These approaches are tissue spectrometry, distant spectrometry, electrochemical, and behavioral. These approaches encompass direct measurements of BAC through breath, tissue, and sweat, and indirect measurements through behavior.

Current breath-based measurement systems as well as transdermal systems that measure alcohol in vapor or liquid phase perspiration, utilize electrochemically-based fuel-cell technology which has several limitations. Fuel cell warm up time in colder temperatures, and measurement drift requiring calibration, render fuel cell technologies unsuitable for every-day use by the general public. An additional concern specific to transdermal fuel-cell based devices is the long lag time to reach peak alcohol concentration in sweat versus blood. It is not clear how future approaches to measuring TAC at a single point in time, rather than continuously, can address this fundamental physiological difference. Because of these limitations for the DADSS application, an electrochemical/transdermal approach is not being pursued at this time.

As noted above, there are a large number of measurable behaviors that are affected by alcohol. However, close correlations between BAC and changes in these behaviors have not yet been established. Another concern is that behavioral task performance may change as a result of a variety of impairments, whether from fatigue, illness, alcohol, medications, illegal drugs, or other sources. Furthermore, in order to measure impairment there has to be some measure of "normal" abilities on the specific task that can act as a baseline measurement for comparison. It should be noted that other sources of impairment can result in unsafe driving, and research continues to identify those risks and determine potential countermeasures. However, the limitations outlined above would be hard to deal with in an unobtrusive device to measure alcohol. Because of these limitations for the DADSS application, a behavioral approach is not being pursued at this time.

Two technological approaches have been identified that have considerable promise in measuring driver BAC unobtrusively; tissue spectrometry and distant spectrometry, and, three contracts have been awarded to develop prototype devices. TruTouch Technologies is developing a touch-based prototype sensor that uses tissue spectrometry to measure alcohol concentration in tissue. Autoliv and ACS have been awarded contracts to develop prototype sensors that will measure breath alcohol in the vehicle cabin air.

Although none of the devices under development currently can meet the speed of measurement and accuracy specifications outlined in the DADSS performance specifications, there are indications that with additional research and development, they could meet them.

The new methods under development do not require breath samples to be provided by blowing through a tube, but sense the air around the driver passively. In addition, the approaches being taken do not rely on fuel cells, but use other technologies that will not require recalibration over the life of the vehicle. Further development will be required to meet the speed and accuracy but there are good indications that this could be achieved.

Tissue spectrometry also has the potential to provide a non-invasive approach to alcohol sensing. Such sensors could readily be incorporated into driver functions already required before driving the vehicle. The viability of this approach to measuring BAC has been demonstrated there is a commercial device available which is being used in judicial systems and workplace applications. Further research and development will address improvements in the optical sensor to improve speed and accuracy to meet the DADSS requirements, as well as engineering needs to suit a vehicle environment.

In summary, promising technologies have been identified that with additional development could meet the needs of an in-vehicle alcohol measurement system. It is possible that additional technologies will emerge as potential candidates for the DADSS system as development proceeds. The DADSS team is committed to continue to review the science and evolving technologies to assure that any potential technologies are given due consideration.

REFERENCES

Anderson, R.R., Parrish, J.A. 1981. The Optics of Human Skin. The Journal of Investigative Dermatology, 77, 13-19.

Anderson, J.C., and Hlastala, M.P. 2005. The kinetics of transdermal exchange. Journal of Applied Physiology, 100, 649-655.

Bekiaris, E. 1999. System for effective Assessment of the driver state and Vehicle control in Emergency Situations. The Telematics Applications Program. Final Report TR 1047 http://www.regione.piemonte.it/trasporti/prss/biblioteca/dwd/progetti/uomo/save8.pdf Accessed 7/20/09.

Hawthorne, J.S. and Wojcik, M.H. 2006. Transdermal alcohol measurement: A review of the literature. Alcohol Monitoring Systems Inc.

Hedlund, J.T. and McCartt, A.T. 2002. Drunk Driving: Seeking Additional Solutions. Washington, DC: AAA Foundation for Traffic Safety. Available: http://www.aaafoundation.org/pdf/DrunkDriving-SeekingAdditionalSolutions.pdf.

Hök, B., Pettersson, H., Andersson, G. 2006. Contactless measurement of breath alcohol. Proceedings of the Micro Structure Workshop. May, 2006, Västerås, Sweden.

Insurance Institute for Highway Safety. 2008. Q&As. Alcohol:General. Arlington, VA. http://www.iihs.org/research/qanda/alcohol_general.html.

Lambert, D.K., Myers, M.E., Oberdier, L., Sultan, M.F. et al., "Passive Sensing of Driver Intoxication," SAE Technical Paper 2006-01-1321, 2006.

Lindberg, L., Brauer, S, Wollmer, P., Goldberg, L., Jones, A.W., Olsson, S.G. 2007. Breath alcohol concentration determined with a new analyzer using free exhalation predicts almost precisely the arterial blood concentration. Forensic Science International. 168: 200-207.

Lund, A.K., McCartt, A.T., and Farmer, C.M. 2007. Contribution of Alcohol-Impaired Driving to Motor Vehicle Crash Deaths in 2005. Proceedings of the T-2007 Meeting of the International Council on Alcohol, Drugs, and Traffic Safety. Seattle, Washington.

Marques, P.R., McKnight, S.A. 2009. Field and laboratory alcohol detection with 2 types of transdermal devices. Alcoholism: Clinical and Experimental Research. 33, 1-9.

Moskowitz, H., and Robinson, C.D. 1988. Effect of Low Doses of Alcohol on Diving Related Skills: A Review of the Evidence, report # DOT HS-807-280, US Department of Transportation, NHTSA, Washington DC, July 1988.

Moskowitz, H. and Fiorentino, D., 2000. A Review of the Literature on the Effects of Low Doses of Alcohol on Driving-Related Skills, Report # DOT HS 809-028, U.S. Department of Transportation, NHTSA, Washington, DC, April 2000.

National Highway Traffic Safety Administration. 2009. Horizontal Gaze Nystagmus: The science & the law: A Resource Guide for Judges, Prosecutors and Law Enforcement. (http://www.nhtsa.dot.gov/people/injury/enforce/nystagmus/hgntxt.html#one, Accessed 07/20/09)

National Highway Traffic Safety Administration. 1997. Federal Register, Vol. 58, No. 179, pp 48705-48710. Washington, DC: U.S. Department of Transportation. Friday, September 17, 1993.

Opdam, J.J., Smolders, J.F. 1986 Alveolar sampling and fast kinetics of tetrachloroethene in man. I Alveolar sampling. British Journal of Industrial Medicine, 43, 814-824.

Peck, R.C. Gebers, M.A. Voas, R.B. and Romano, E. 2008. The relationship between blood alcohol concentration (BAC), age, and crash risk. Journal of Safety Research 39:311-19.

Peters, B, Van Winsum, W. 1998. SAVE: System for effective assessment of the driver state and vehicle control in emergency situations. R&D Program Telematics: EU

Pollard, J. K., Nadler, E. D., and Stearns, M. D. 2007. Review of Technology to Prevent Alcohol-Impaired Crashes (TOPIC). DOT HS 810 833, Washington, DC: National Highway Traffic Safety Administration, U.S. Department of Transportation.

Ridder, T.D., Hendee, S.P., Brown, C.D. 2005. Noninvasive alcohol testing using diffuse reflectance near-infrared spectroscopy. Applied Spectroscopy, 59, 181-189.

Ridder, T. D., Brown, C.D., Ver Steeg, B. J. 2005. Framework for multivariate selectivity analysis, Part II: Experimental applications. Applied Spectroscopy, 59, 804-815.

Ridder, T.D., Ver Steeg, B. J., Laaksonen, B.D. 2009. Comparison of spectroscipally measured tissue alcohol concentration to blood and breath measurements. In press.

Sakai, J.T., Mikulich-Gilbertson, S.K., Long, R.J., and Crowley, T.J. 2006. Validity of transdermal alcohol monitoring: Fixed and self-regulated dosing. Alcoholism: Clinical and Experimental Research. 30, 26-33.

Stein, A.C. and Allen, R.W., "The Use of In-Vehicle Detectors to Reduce Impaired Driving Trips," SAE Technical Paper 860360, 1986.

Traffic Injury Research Foundation. 2006. Continuous transdermal alcohol monitoring: A Primer for criminal justice professionals. Ottawa, Canada.

Vanlaar, W. and Simpson, H. 2007. Monitoring alcohol use through transdermal alcohol testing. The Journal of Offender Monitoring.

Webster, G.D., and Gabler, H.C. 2007. Feasibility of transdermal ethanol sensing for the detection of intoxicated drivers. Proceedings of the 51st Annual Meeting of the

Association for the Advancement of Automotive Medicine. Melbourne, Australia.

Willis, C, Lybrand, S, Bellamy, N. 2004. Alcohol ignition interlock programmes for reducing drink driving recidivism. Cochrane Database of Systematic Reviews 18(4): CD004168.

Zador, P.L., Krawchuk, S.A., and Voas, R.B. 2000. Alcohol-related relative risk of driver fatalities and driver involvement in fatal crashes in relation to driver age and gender: an update using 1996 data. Journal of Studies on Alcohol 61:387-95.

CONTACT INFORMATION

Susan Ferguson, Ph.D.
Ferguson International LLC
508 Bay Villas Lane
Naples
FL 34108
USA
fergsusan@gmail.com

ACKNOWLEDGEMENTS

The authors thank members of the Blue Ribbon Panel for their careful review of the draft paper and insightful feedback.

DEFINITIONS/ABBREVIATIONS

ACS
    Alcohol Countermeasure Systems

ACTS
    Automotive Coalition for Traffic Safety

BAC
    Blood alcohol concentration

BrAC
    Breath alcohol concentration

CO_2
    Carbon dioxide

CTT
    Critical Tracking Task

DADSS
    Driver Alcohol Detection System for Safety

DWI
    Driving While Impaired or Intoxicated

HGN
    Horizontal gaze nystagmus

IIHS
    Insurance Institute for Highway Safety

IR
    Infrared

LED
    Light emitting diode

NHTSA
    National Highway Traffic Safety Administration

NIR
    Near-Infrared

RFI
    Request for Information

SAVE Project
    System for effective Assessment of the driver state and Vehicle control in Emergency situations

SCRAM
    Secure Continuous Remote Alcohol Monitoring

SD
    Standard deviation

SE
    Systematic error

TAC
    Transdermal alcohol concentration

TIRF
    Traffic Injury Research Foundation

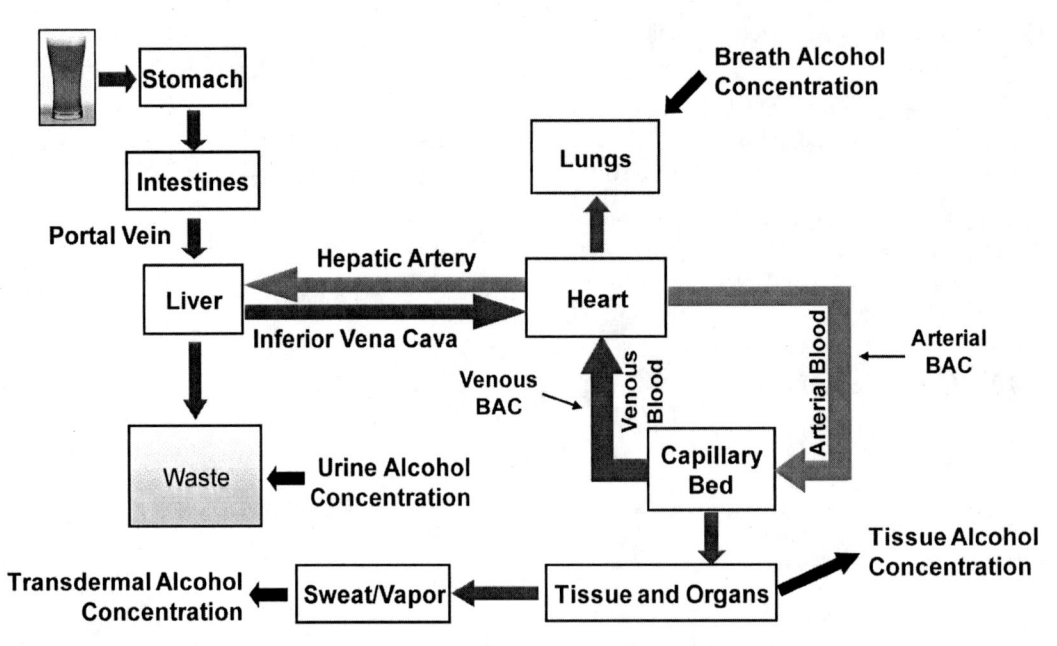

Figure 1. Alcohol absorption, distribution and elimination though the body

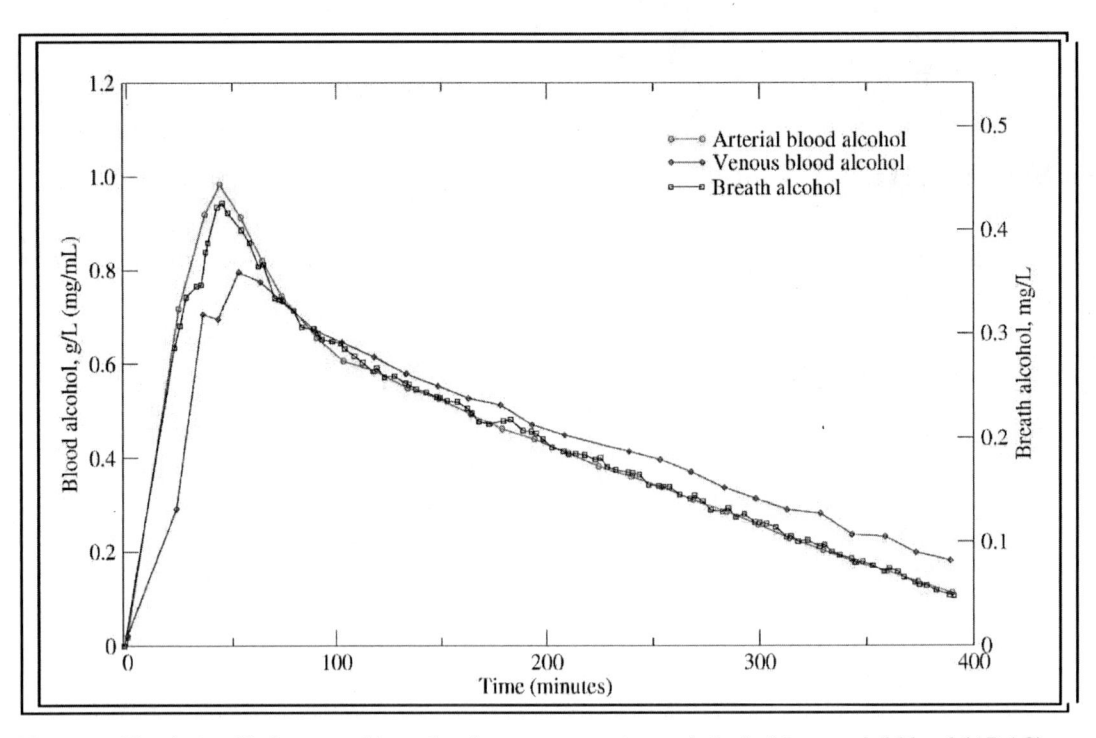

Figure 2. Pharmacokinetic profile in one subject showing concentrations of alcohol in arterial blood (ABAC), venous blood (VBAC) and breath after oral ingestion of 0.6 g of alcohol per kg body weigh

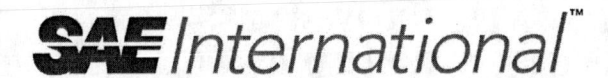

Drowsiness Detection Using Facial Expression Features

2010-01-0466
Published
04/12/2010

Satori Hachisuka, Teiyuu Kimura, Kenji Ishida, Hiroto Nakatani and Noriyuki Ozaki
Denso Corporation

Copyright © 2010 SAE International

ABSTRACT

This paper presents the method of detecting driver's drowsiness level from the facial expression. The motivation for this research is to realize the novel safety system which can detect the driver's slight drowsiness and keep the driver awake while driving.

The brain wave is commonly used as the drowsiness index. However, it is not suitable for the in-vehicle system since it is measured with sensors worn over the head. We precisely investigated the relationship between the change of brain wave and other drowsiness indices that can be measured without any contact; PERCLOS, heart rate, lane deviation, and facial expression. We found that the facial expression index had the highest linear correlation with the brain wave. Therefore, we selected the facial expression as the drowsiness-detection index and automated the drowsiness detection from the facial expression.

Three problems need to be solved for automation; (1) how to define the features of drowsy expression, (2) how to capture the features from the driver's video-recorded facial image, and (3) how to estimate the driver's drowsiness index from the features. First, we found that frontalis muscle, zygomaticus major muscle, and masseter muscle activated with increase of drowsiness in more than 75 percents of participants. According to the result, we determined the coordinates data of points on eyebrows, eyelids, and mouth as the features of drowsiness expression. Second, we calculated the 3D coordinates data of the features by image processing with Active Appearance Model (AAM). Third, we applied k-Nearest-Neighbor method to classify the driver's drowsiness level. Eleven participants' data of the features and the drowsiness level estimated by trained observers were used as the training data. We achieved the classification of the drivers' drowsiness in a driving simulator into 6 levels. The average Root Mean Square Errors (RMSE) among 12 participants was less than 1.0 level.

INTRODUCTION

Although active safety systems in vehicles have contributed to the decrease in the number of deaths occurring in traffic accidents, the number of traffic accidents is still increasing. Driver drowsiness is one reason for such accidents and is becoming an issue. The National Highway Traffic Safety Administration (NHTSA) estimates that approximately 100,000 crashes each year are caused primarily by driver drowsiness or fatigue in the United States [1]. In Japan, attention lapse, including that due to driving while drowsy, was the primary reason for traffic accidents in 2008. The Ministry of Economy, Trade and Industry in Japan reports that the number of such accidents has increased 1.5 times in the 12-year period from 1997 to 2008 [2].

One solution to this serious problem is the development of an intelligent vehicle that can predict driver drowsiness and prevent drowsy driving. The percentage of eyelid closure over the pupil over time (PERCLOS) is one of the major methods for the detection of the driver's drowsiness [3]. We developed a method for the detection of driver drowsiness using the whole facial expression, including information related to the eyes. This method is based on the results of observational analyses. The results of such analyses revealed that features of drowsiness appear on the eyebrows, cheeks, and mouth, in addition to the eyes [4]. The aim of using the facial expression is to detect drowsiness in the early stages, on the basis of the many minute changes in the facial parts. Our goal is to develop an intelligent safety vehicle that can relieve drivers from struggling against drowsiness by detecting their drowsiness and keeping them awake naturally. Two systems are necessary for the establishment of such an intelligent vehicle; one is a system for detecting the information predictive of driver drowsiness in the early stages, and the other is a system for providing feedback to keep the driver awake. In this paper, we discuss a method for detecting driver drowsiness in the early stages. Our method detects drowsiness with accuracy equivalent to that of brain waves, which is the general index of drowsiness. The method does not require any attachment of sensors. We developed the drowsiness detection method with a system comprising a camera set on the dashboard, an image processing algorithm, and a drowsiness detection algorithm.

This method categorizes drowsiness into 6 levels using features of facial expression based on the mechanism of facial muscle activities. This paper presents a novel drowsiness detection method and assesses its effectiveness.

EARLY-STAGE DROWSINESS DETECTION

The changes in brain waves, especially alpha waves, are one of the indices used to detect changes in the level of drowsiness [5]. Although change in brain waves is an effective index for detecting drowsiness, it is not feasible to apply this index in a vehicle because of the electrodes that are used as contact-type sensors. However, it is recognized in the field of cerebral neuroscience that the facial nerve nucleus is contained in the brain stem, which is defined as an organ of drowsiness [6]. Therefore, we adopted facial expression as the index of drowsiness as an alternative to brain waves. In addition, it is apparent from our experience that we can recognize drowsiness in others from their facial expressions.

In Japan, Kitajima's trained observer rating is a commonly used method for the detection of driver drowsiness on the basis of appearance [7]. The method divides drowsiness into 5 levels with criteria such as "slow blink", "touching the face with the hand", "frequent yawning", and so on. Since these criteria are qualitative, the method is not appropriate for automatic detection of drowsiness. In general, percentage of eye closure time, heart rate, and lane deviation are considered to be suitable drowsiness measurement indices that do not require the attachment of sensors [3, 8,9,10,11,12,13,14]. To determine the best index as an alternative to brain waves, we examined the correlation between brain waves and other indices such as PERCLOS, heart rate, lane deviation, and facial expression [15]. Figure 1 indicates that facial expression has the highest correlation with brain waves (correlation coefficient = 0.90) and it detects drowsiness at an earlier stage than other indices. This indicates that facial expression is the most appropriate index to use for the detection of driver drowsiness in the early stages. Therefore, to be able to predict and prevent drowsy driving, the development of a method that detects driver drowsiness from facial expression is necessary.

<figure 1 here>

AUTOMATIC DROWSINESS-DETECTION SYSTEM USING FACIAL EXPRESSION

It was necessary to solve 3 problems for the development of an automatic drowsiness-detection system:

(1). How to define the features of drowsy expression.

(2). How to capture the features from the driver's video-recorded facial image.

(3). How to estimate the driver's drowsiness index from the features.

Our approaches to solving these problems are explained in this chapter.

FEATURES OF DROWSY EXPRESSION

We clarified the particular features of drowsy expression by comparing the facial muscle activities of the waking expression with those of the drowsy expression [16]. We measured 9 facial muscles of each of 17 volunteer participants during the task of monotonous driving in the driving simulator for 1 hour. Figure 2 shows the 9 facial muscles; inner frontalis, upper orbicularis oculi, lower orbicularis oculi, zygomaticus major, masseter, risorius, upper orbicularis oris, lower orbicularis oris, and mentalis. We divided the reference states of drowsiness into 6 levels, i.e., "not sleepy", "slightly sleepy", "sleepy", "rather sleepy", "very sleepy", and "sleeping" (Table 1) by adding the "sleeping" level to Kitajima's trained observer rating scale.

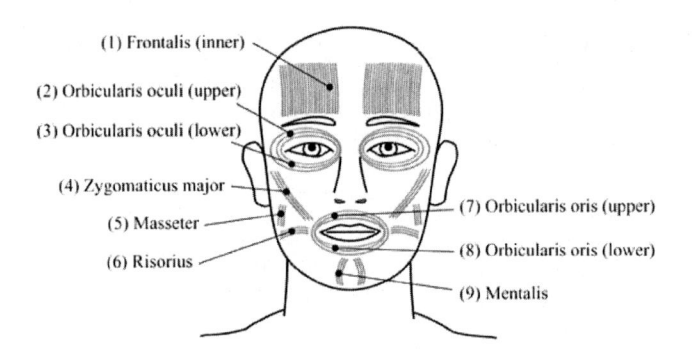

Fig. 2. Nine facial muscles. Facial muscles were measured by facial electromyograph.

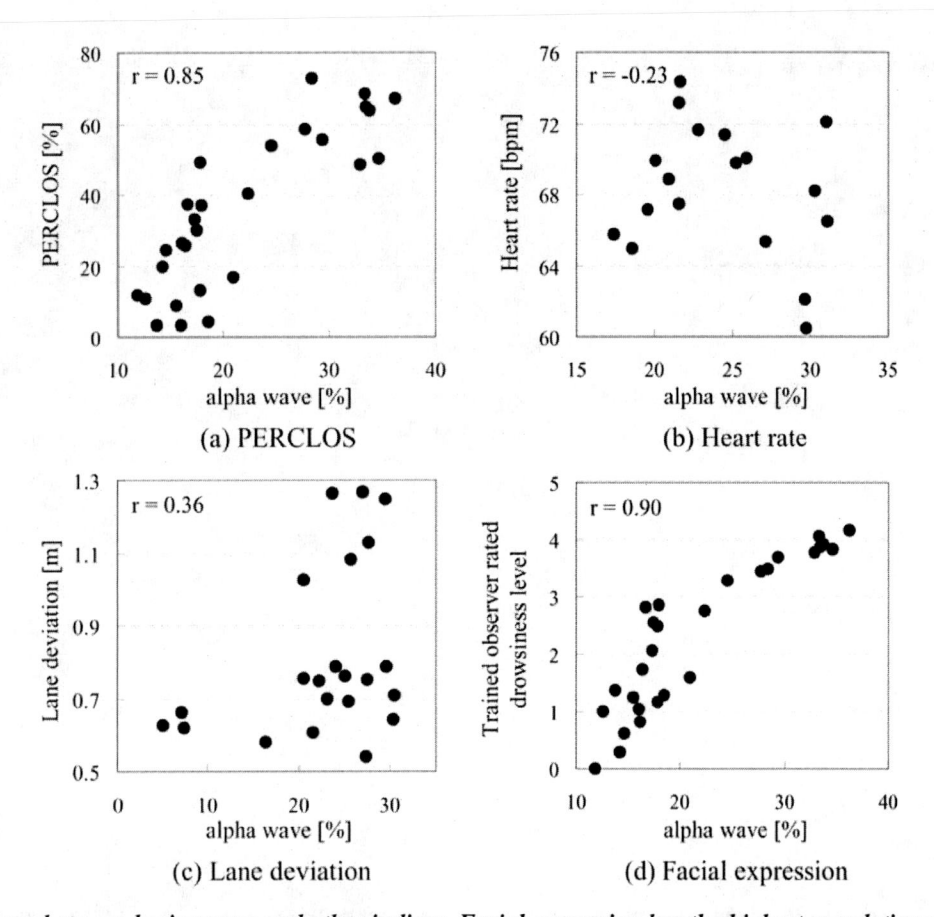

Fig. 1. Correlations between brain waves and other indices. Facial expression has the highest correlation with brain waves.

Table 1. Drowsiness levels by trained observer rating

Category (Drowsiness Level)	
0	Not Sleepy
1	Slightly Sleepy
2	Sleepy
3	Rather Sleepy
4	Very Sleepy
5	Sleeping

Figure 3 shows the comparison results of the drowsiness levels and the facial muscle activities. The contractions of the frontalis and the relaxation of the zygomaticus major were detected in more than 75 percent of participants, and the relaxation or contraction of the masseter muscle was detected in 82 percent of participants. In addition, contraction of the frontalis, which was detected in 94 percent of participants, was the characteristic expression of resisting drowsiness. This characteristic expression does not appear during the natural drowsy state without any struggle against drowsiness. According to the result, we chose the eyebrows, edges of the mouth, and the lower lip as the facial features related to the frontalis, zygomaticus major, and masseter, respectively, in addition to the eyelids, which are the general features of the drowsiness expression (Fig. 4).

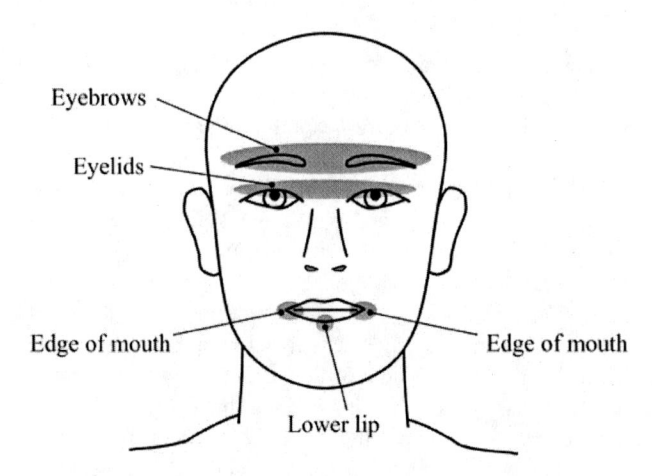

<figure 3 here>

Fig. 4. Features for detecting drowsiness

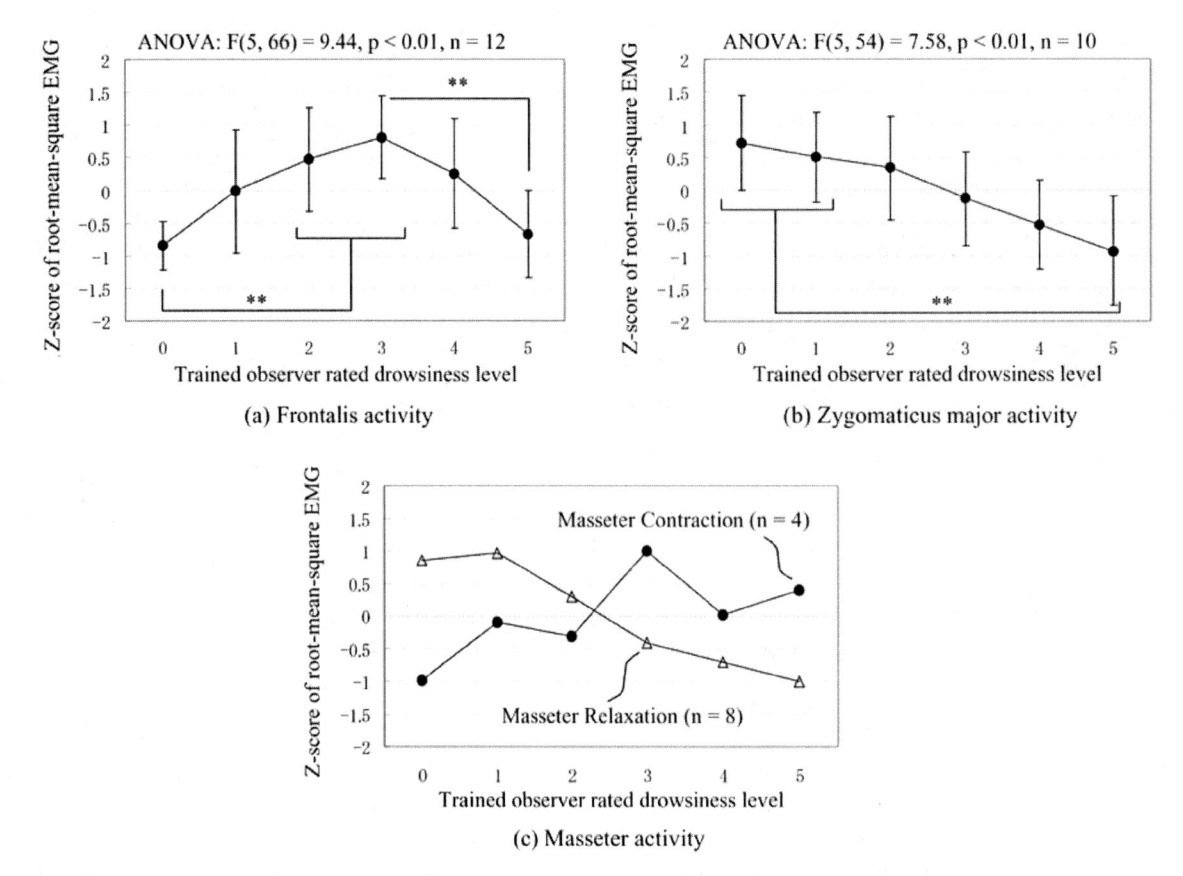

Fig. 3. Typical comparison results of the drowsiness levels and the facial muscle activities. The contractions of frontalis and the relaxation of zygomaticus major were detected in more than 75 percent of participants, and the relaxation or contraction of the masseter muscle was detected in 82 percent of participants.

IMAGE PROCESSING FOR MEASURING FEATURES OF DROWSY EXPRESSION

We developed a method of image processing for measuring the features of drowsy expression, without any sensor contact, from a driver's video-captured facial image [17]. The method, which is based on the Active Appearance Model (AAM) [18], detects three-dimensional coordinates of 68 points on the driver's face per frame (Fig. 5). Our AAM consists of the specific 2-dimensional model (Fig. 6 (a)) and the generic 3-dimensional model (Fig. 6 (b)). The specific 2-dimensional model has information relating to the shape and texture of the individual driver's facial image. The generic 3-dimensional model has the 3-dimensional vectors of each of the 68 points. We developed a method that extracts change in facial expression without individual differences in the shape of each driver's face by using the generic 3-dimensional model. This method is an effective way of detecting the coordinates of the points on the face in the vehicle, which is expected to be driven by an unspecified number of drivers. The process of this method is shown in Fig. 7. First, the specific 2-dimensional model is generated by a captured static facial image of the driver. This process is performed once for each driver. Next, the specific 2-dimensional model is fitted on each frame of the driver's facial image and the 2-dimensional coordinates of the 68 points are output. Finally, the generic 3-dimensional model is deformed based on the 2-dimensional coordinates and the 3-dimensional coordinates of each of the 68 points are output per frame. We employed the method of steepest descent to the fitting of the specific 2-dimensional model and the deforming of the generic 3-dimensional model.

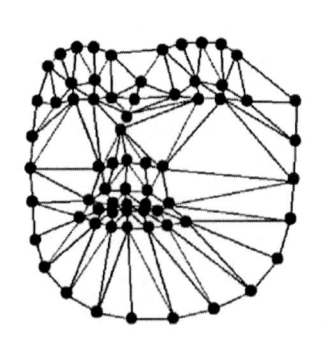

Fig. 5. Sixty-eight points on face

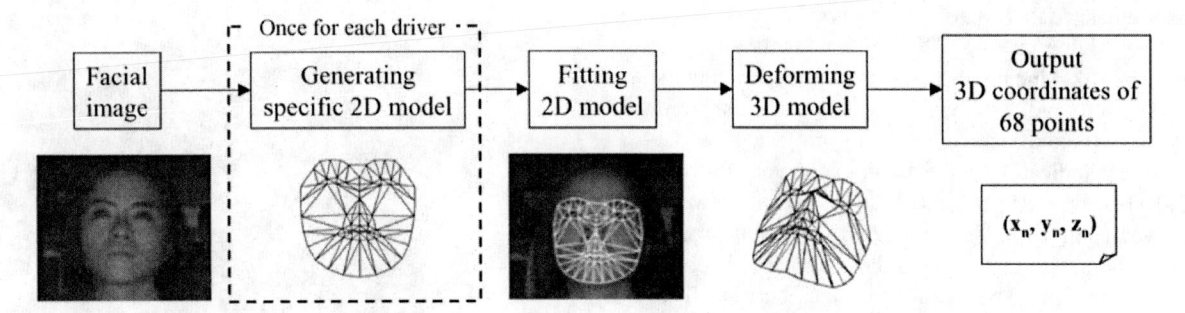

Fig. 7. Flow of the image processing with AAM

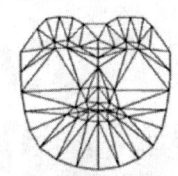

(a). 2-dimensional model

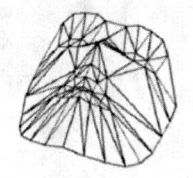

(b). 3-dimensional model

Fig. 6. Specific 2-dimensional model and generic 3-dimensional model for AAM

<figure 7 here>

METHOD OF DETECTING DROWSINESS LEVEL

We adopted 17 points as the measurement objects to detect the drowsy expression. Figure 8 shows the 17 points: 10 points on the right and left eyebrows (5 points on each side), 4 points on the right and left eyelids (2 points on each side), 2 points on the right and left edges of the mouth (1 point on each side), and 1 point on the lower lip. As the features of drowsy expression we used scalar quantities of the change in the 17-point positions, which were measured from the positions on the waking-state expression. The individual differences of the waking-state expressions are reduced by defining the positions on the waking-state expressions as reference positions. According to this normalization, it is possible to detect the changes in the driver's facial expression based on drowsiness. We employed the k-Nearest-Neighbor method, which is one of the pattern classification methods, for detecting the drowsiness level. This decision was based on the result of a preliminary experiment in which we compared the results of the drowsiness levels detected by the trained observer with other estimation methods: multiple regression analysis method, subspace method, and k-Nearest-Neighbor method. The drowsiness level estimated by the k-Nearest-Neighbor method had the highest correlation with the drowsiness level as estimated by the trained observer. In accordance with this result, we regarded the facial change of drowsy expression as multidimensional data that were nonlinear, nonparametric, and unpredictably distributed. Our method uses the prebuilt database that consists of the 6-level drowsy expression features of several individuals. The driver's features are compared with the whole database, and the similarities of each comparison are applied to detect the drowsiness level. The similarity-based method is able to detect drowsiness with higher time resolution than the method using trends in the change in the facial expression at a specific time interval, such as 30 seconds [19].

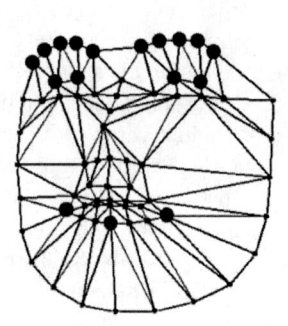

Fig. 8. Seventeen measurement points for detecting drowsy expressions

The every 5-second (150-frame) average features are used as the feature data in this method. The 5-second block time is applied as the bare minimum sampling time for the trained observer rating for facial expressions [7]. According to the averaging, it is possible to detect the difference between "eye closure based on blinking" and "eye closure based on drowsiness", which is difficult to distinguish from a still frame. Therefore, it is possible to reduce erroneous detection of the drowsiness level.

We investigated the accuracy of our drowsiness detection off-line. We used the driving simulator in a sound-proof room (Fig. 9). Motion system was excluded from the driving simulator to induce drowsiness in the participants efficiently and to measure basic data of participants' drowsy expressions accurately. The driving task was also designed monotonously for the purpose of inducing drowsiness in the participants efficiently. The longitudinal flows of two sine curves, from the top to the bottom of the screen, were projected on the screen. The circle indicating the position of the vehicle from an overhead view was also projected on the screen between the sine curves (Fig. 10). We instructed the participants to operate the driving simulator with the steering wheel to maintain their position between the sine curves. The participants' facial images were recorded by the digital video camera (480 × 640 pixels, 30 fps, progressive scan) on the dashboard. The participants were instructed to remain awake during the driving and maintain the same position they would adopt while driving a real vehicle, even if they became drowsy. As a reference for the 6 drowsiness levels, we used the results of ratings for the drivers' recorded facial images from 2 trained observers. The 12 volunteer participants had drivers' licenses and were aged in their 20s to 40s. They were informed of simulator sickness before the experiments and required to sign an informed consent document. During the experiments, at least one examiner observed the participant's appearance from outside the sound-proof room. After the experiment, the participant rested with the examiners for approximately 10-15 minutes. Drowsiness detection was performed off-line using the leave-one-out cross validation procedure with the features, which were calculated by referring to the 12 participants' recorded facial images. All of the 12 participants fell asleep during the experiment. In the leave-one-out cross validation, data for 11 arbitrarily chosen participants were used as training data and used to detect the drowsiness of the remaining participant, who was excluded from the training data; this was performed repeatedly. Figure 11 shows the training data image. The training data consist of the 5-second average features and the reference of the drowsiness levels, which are labeled on the features. The flow of drowsiness detection is shown in Fig. 12. The entered driver's features are compared with all of the training data. The top 80 training data points, which have a strong similarity to the driver's features, are picked up. The driver's drowsiness level is estimated based on a majority decision of what the referential sleeping levels are, which are labeled on the 80 training data points. We employed the Euclidean distance as the index of similarity between the driver's features and the features of the training data. A small distance indicates strong similarity.

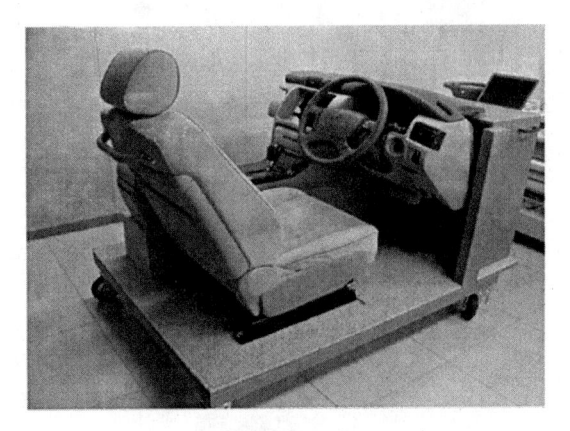

Fig. 9. Image of the driving simulator

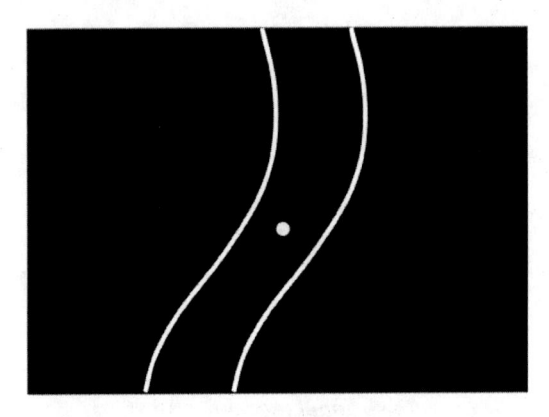

Fig. 10. Projected sine curves and circle on the screen

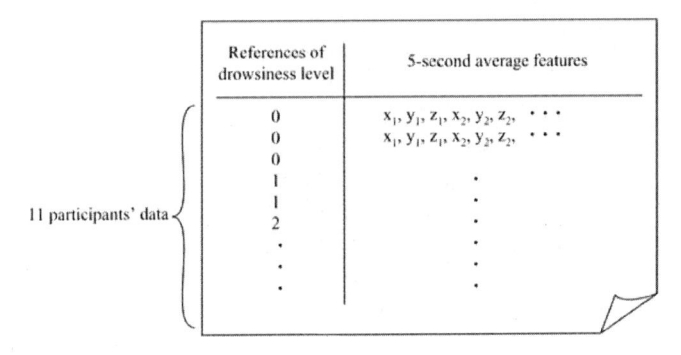

Fig. 11. Image of training data. The training data consist of the 5-second average features and the reference of the drowsiness levels labeled on the features.

<figure 12 here>

3D coordinates data of feature points → Comparing with training data → Pickup 80 training data similar to feature points → Majority decision of 80 labeled drowsiness levels → Output drowsiness level

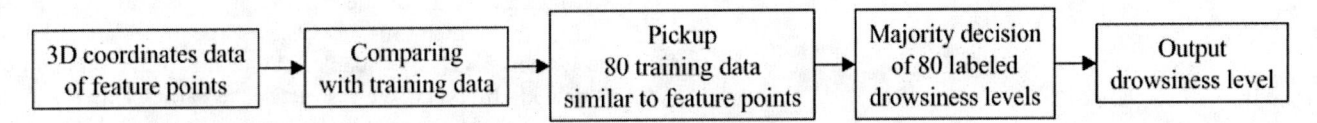

Fig. 12. Flow of the drowsiness detection with k-NN

We investigated the effectiveness of the method by comparing the detected drowsiness level with the referential drowsiness level. To detect the facial change based on the drowsiness accurately, we clipped the parts of the video of the participants' facial images before the comparison, and detected the drowsiness levels from those partial videos. The 3 criteria for clipping the partial videos were as follows.

(1). The facial image of the driver in a front-facing position.

(2). The facial image without any occlusion such as a steering wheel and/or a hand.

(3). The facial image without any actions that cause facial change, such as yawning, smiling, or laughing.

Figure 13 shows the one participant's result of the detected drowsiness levels which were detected by our method, and the referential drowsiness levels which were estimated by trained observers. The Root Mean Square Errors (RMSE) of 12 participants are shown in Table 2. These results demonstrate that our method detects the drowsiness with a RMSE of less than 1.0 for 8 participants. The average RMSE among 12 participants is 0.91. On the other hand, RMSE was increased when we used fewer or more feature points than 17 such as 10 points on eyebrows, 4 points on eyelids, or 68 points on whole face to detect the drowsiness. Therefore, 17 points, which described in chapter "METHOD OF DETECTING DROWSINESS LEVEL", were the best features for our drowsiness-detection method. In addition, it is shown that our method detects the drowsiness level of the participant who does not fall asleep during the examination with the same level of accuracy (Fig. 14). The detected drowsiness levels were indicated as being below level 3 ("rather sleepy") for the participant who does not fall asleep during the examination and the referential drowsiness levels of the participant were below level 2 ("Sleepy"). These results indicate that our drowsiness detection method is able to detect the drowsiness level as accurately as that of the trained observer in 5-second time resolution.

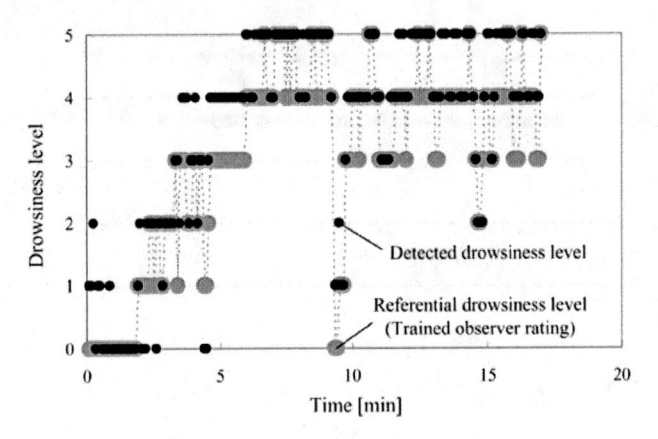

Fig. 13. Comparison result of the detected drowsiness levels and the referential drowsiness levels of participant No. 12.

Table 2. Root Mean Square Errors of drowsiness detection

Participant #	Root Mean Square Error (RMSE)
1	1.11
2	0.90
3	1.16
4	0.90
5	0.78
6	1.08
7	0.69
8	1.02
9	0.82
10	0.81
11	0.94
12	0.77
Average	0.91
SD	0.15

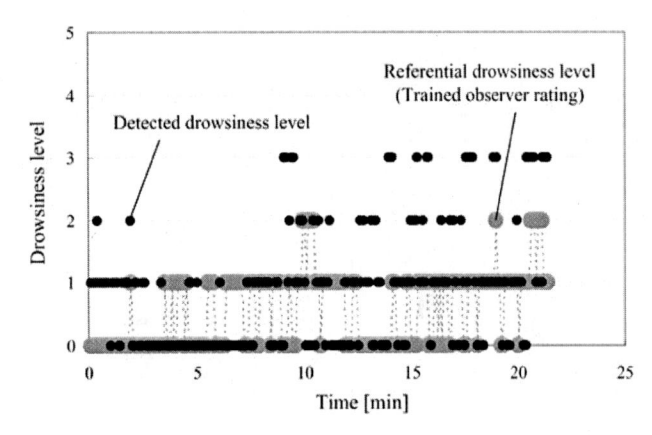

Fig. 14. Comparison result of the participant who does not fall asleep during the examination.

REAL-TIME DROWSINESS DETECTION

As described in chapter "AUTOMATIC DROWSINESS-DETECTION SYSTEM USING FACIAL EXPRESSION", we developed an off-line detection method that determines the driver's drowsiness level from the driver's facial expression. To apply this method in practice, it is imperative to perform image processing and drowsiness detection in real time. Therefore, we developed a real-time drowsiness detection method and tested the accuracy of this method for detecting drowsiness.

The experimental set-up is shown in Fig. 15. The Complementary Metal-Oxide Semiconductor (CMOS) camera (IEEE-1394a, 640 × 480 pixels, 60 fps, progressive scan) was placed on the dashboard of the driving simulator in the sound-proof room. Motion system was excluded from the driving simulator to induce drowsiness in the driver efficiently and to measure basic data of participants' drowsy expressions accurately. The band pass filter with a center wave-length of 845 nm (half width: 52 nm) was placed in front of the lens on the CMOS camera. The CMOS camera takes the driver's facial image through the filter. Two infrared Light-Emitting Diodes (LEDs) with a secondary lens (peak wavelength: 850 nm, half-value angle: 15 deg.) were placed at the right and left side of the camera, respectively. The camera was able to record the driver's facial image without any interference from ambient light by using the filter and the infrared LEDs. The artificial daylight (standard illuminant D65) was set up in the sound-proof room to simulate the outdoor lighting environment. The programs of the image processing and the drowsiness detection were executed on the Linux-based computer (Pentium D, 2.8 GHz CPU, 3GB memory). As the training data for the k-NN method, the 12 participants' data (3004 data sets) that had been recorded off-line (mentioned in chapter "AUTOMATIC DROWSINESS-DETECTION SYSTEM USING FACIAL EXPRESSION")

were used. The 2 volunteer participants had drivers' licenses and were in their 30s to 40s. As the reference of the 6 drowsiness levels, we adopted the results of the ratings from 2 trained observers. We instructed the participants to operate the driving simulator with the steering wheel held in such a manner as to maintain the position of the projected circle between the sine curves and, to remain awake during the experiment while adopting the same stance employed when driving a real vehicle, even if they became drowsy. The participants were informed of the possibility of simulator sickness before the experiments and were required to sign an informed consent document. During the experiments, at least one examiner observed the participant's appearance from outside the sound-proof room. After the experiment, the participant rested with the examiners for approximately 10-15 minutes.

<figure 15 here>

To detect facial change based on the drowsiness accurately, we clipped the video and compared the detected drowsiness levels from those partial videos with the referential drowsiness levels. The 4 criteria for clipping the partial videos were as follows:

(1). The facial image of the driver in a front-facing position.

(2). The facial image without any occlusion such as a steering wheel and/or a hand.

(3). The facial image without any actions that cause facial change, such as yawning, smiling, or laughing.

(4). The facial image with appropriate fitting position of 2-demensional model.

The Root Mean Square Errors (RMSE) of 2 participants are shown in Table 3. The results of our study demonstrate that our method detects the drowsiness level with an average RMSE approximately 1.0 in real time.

Table 3. Root Mean Square Errors of drowsiness detection

Participant #	Root Mean Square Error (RMSE)
1	1.24
2	0.83
Average	1.04

CONCLUSION

In this paper, we presented the driver's drowsiness detection method using facial expression, and we established the effectiveness of this method experimentally. Our method is executed according to the following flow: taking the driver's facial image, tracing the facial features by image processing,

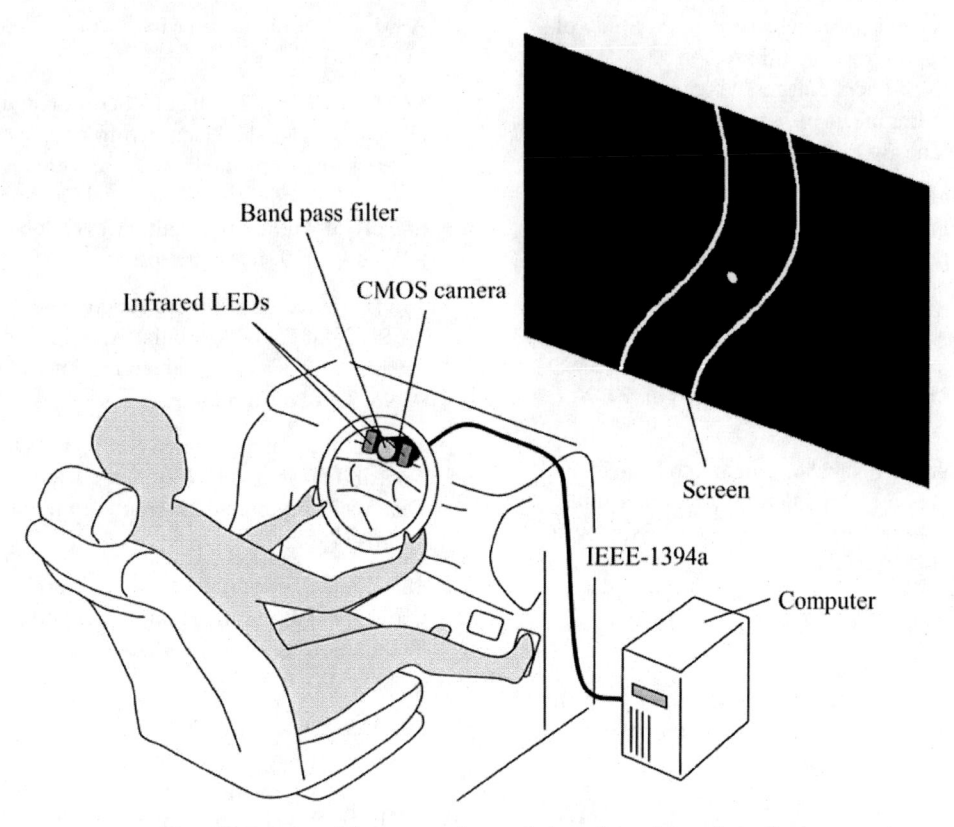

Fig. 15. Image of the set-up for real-time drowsiness detection

and rating the driver's drowsiness according to a 6-level scale from the features by pattern classification, in real time. The results of the drowsiness detection correspond to the drowsiness reference as estimated by a trained observer with an average RMSE of less than 1.0 level. The distinguishing feature of our method is that it uses 17 facial features based on the activities of facial muscles.

The limitations of this paper were the reality of driving environment and the number of the participants. In future work, we will verify practical effectiveness of our drowsiness detection method using motion-based driving simulator and/or real car. Additionally, we will increase the number of participants in our experiments and develop the effective training data for detecting drowsiness of a large number of drivers.

On the other hand, we tested the partial videos to detect the drowsiness expression accurately. The criteria from the clipped partial videos were to use the driver's facial image in a front-facing position, without any occlusion such as a steering wheel or a hand, and without any actions such as yawning, smiling, or laughing. However, in the real vehicle, these events will occur frequently. Therefore, we have started to develop an artifact detection and cancellation method. It will also be possible to develop methods for the detection of other expressions in addition to drowsiness, and apply these to a novel Human-Machine Interface (HMI) such as emotion estimation and agent communication in the vehicle. In addition, the integration of personal verification into our method will lead to the development of a highly precise drowsiness-detection method.

It is also necessary to develop a feedback system to achieve an intelligent safety vehicle that can relieve drivers struggling against drowsiness. We have now started to develop a feedback system that keeps the driver awake effectively and naturally.

REFERENCES

1. U.S. Department of Transportation, "Saving lives through advanced safety technology. Intelligent Vehicle Initiative 2002 annual report," http://www.itsdocs.fhwa.dot.gov//JPODOCS/REPTS_TE//13821.pdf, May 2003.

2. The Ministry of Economy, Trade and Industry, "Technological Strategy Map 2009," http://www.meti.go.jp/policy/economy/gijutsu_kakushin/kenkyu_kaihatu/str2009/7_1.pdf, April 2009 (in Japanese).

3. Wierwille W. W., Ellsworth L. A., Wreggit S. S., Fairbanks R. J., and Kirn C. L., "Research on Vehicle-Based Driver Status/Performance Monitoring; Development, Validation, and Refinement of Algorithms For Detection of Driver Drowsiness," National Highway Traffic Safety Administration Final Report: DOT HS 808 247, 1994.

4. Ishida Kenji, Ito Akiko, and Kimura Teiyuu, "A Study of Feature Factors on Sleepy Expressions based on Observational Analysis of Facial Images," Transactions of Society of Automotive Engineers of Japan, Vol. 39, No. 3: 251-256, 2008 (in Japanese).

5. Eoh Hong J., Chung Min K., and Kim Seong-Han, "Electroencephalographic study of drowsiness in simulated driving with sleep deprivation," International Journal of Industrial Ergonomics, Vol. 35: 307-320, 2005.

6. Saper Clifford B., Chou Thomas C., and Scammell Thomas E., "The sleep switch: hypothalamic control of sleep and wakefulness," TRENDS in Neurosciences, Vol. 24, No. 12: 726-731, 2001.

7. Kitajima Hiroki, Numata Nakaho, Yamamoto Keiichi, and Goi Yoshihiro, "Prediction of Automobile Driver Sleepiness (1st Report, Rating of Sleepiness Based on Facial Expression and Examination of Effective Predictor Indexes of Sleepiness)," The Japan Society of Mechanical Engineers Journal (Series C), Vol.63, No. 613: 93-100, 1997 (in Japanese).

8. Sato Shuji, Taoda Kazushi, Kawamura Masanori, Wakaba Kinzou, Fukuchi Yasuma, and Nishiyama Katsuo, "Heart rate variability during long truck driving work," Journal of Human Ergology, Vol.30: 235-240, 2001.

9. Kozak Ksenia, Pohl Jochen, Birk Wolfgang, Greenberg Jeff, Artz Bruce, Blommer Mike, Cathey Larry, and Curry Reates, "Evaluation of lane departure warnings for drowsy drivers," Proceedings of the Human Factors and Ergonomics Society 50th annual meeting: 2400-2404, 2006.

10. Liu Charles C., Hosking Simon G., and Lenne Michael G., "Predicting driver drowsiness using vehicle measures: Recent insights and future challenges," Journal of Safety Research, Vol. 40: 239-245, 2009.

11. Ueno A., Manabe S., and Uchikawa Y., "Acoustic Feedback System with Digital Signal Processor to Alert the Subject and Quantitative Visualization of Arousal Reaction Induced by the Sound Using Dynamic Characteristics of Saccadic Eye Movement: A Preliminary Study," Proceedings of the 2005 IEEE Engineering in Medicine and Biology 27th Annual Conference: 6149-6152, 2005.

12. Bergasa Luis M., Nuevo Jesus, Sotelo Miguel A., Barea Rafael, and Lopez Maria Elena, "Real-Time System for Monitoring Driver Vigilance," IEEE Transactions on Intelligent Transportation Systems, Vol. 7, No. 1: 63-77, 2006.

13. Ji Qiang, Zhu Zhiwei, and Lan Peilin, "Real-Time Nonintrusive Monitoring and Prediction of Driver Fatigue," IEEE Transactions on Vehicular Technology, Vol. 53, No. 4: 1052-1068, 2004.

14. Arimitsu Satori, Sasaki Ken, Hosaka Hiroshi, Itoh Michimasa, Ishida Kenji, and Ito Akiko, "Seat Belt Vibration as a Stimulating Device for awakening Drivers," IEEE/ASME Transactions on Mechatronics, Vol. 12, No. 5: 511-518, 2007.

15. Ishida Kenji, Hachisuka Satori, Kimura Teiyuu, and Kamijo Masayoshi, "Comparing Trends of Sleepiness Expressions Appearance with Performance and Physiological Change Caused by Arousal Level Declining," Transactions of Society of Automotive Engineers of Japan, Vol. 40, No. 3: 885-890, 2009 (in Japanese).

16. Ishida Kenji, Ichimura Asami, and Kamijo Masayoshi, "A Study of Facial Muscular Activities in Drowsy Expression," Kansei Engineering International Journal, Vol. 9, No. 2, 2010 (in press).

17. Kimura Teiyuu, Ishida Kenji, and Ozaki Noriyuki, "Feasibility Study of Sleepiness Detection Using Expression Features," Review of Automotive Engineering, Vol.29, No. 4: 567-574, 2008.

18. Cootes Timothy F., Edwards Gareth J., and Taylor Christopher J., "Active appearance models," IEEE Transactions on Pattern Analysis and Machine Intelligence, Vol. 23, No. 6: 681-685, 2001.

19. Vural Esra, Cetin Mujdat, Ercil Aytul, Littlewort Gwen, Bartlett Marian, and Movellan Javier, "Automated Drowsiness Detection For Improved Driving Safety," Proc. 4th International conference on Automotive Technologies, Turkey, Nov. 13-14, 2008.

CONTACT INFORMATION

Research Laboratories, DENSO CORPORATION
500-1, Minamiyama, Komenoki-cho
Nisshin-shi, Aichi-ken, 470-0111, Japan
arimitsu@rlab.denso.co.jp

Modeling of Individualized Human Driver Model for Automated Personalized Supervision

2010-01-0458
Published
04/12/2010

Xingguang Fu and Dirk Söffker
University of Duisburg-Essen

Copyright © 2010 SAE International

ABSTRACT

In contrast to driver assistance systems focused on the vehicle-navigation or stabilization problems, the study of the loop-oriented interaction between driver, vehicle, and environment has been focused in the last years. The core of the proposed approach is the Situation-Operator-Modeling (SOM), which assumes that changes in the parts of the real world to be considered are understood as a sequence of effects modeled by scenes and actions. Based on SOM approach, the logic of interaction between the driver, the vehicle, and the environment can be formalized for supervision of the drivers' behaviors in a real car. Based on a general model of driving in combination with driver-vehicle-interaction developed in previous works, the personalization and individualization of the human driver model is focused. Therefore the analysis of multidimensional probability distributions, which depend not only on the measurements from the vehicle dynamics and the driving environment, but also on the perception and mainly on observed drivers decision during interactions is used for detailing and refining elements of the global model. By implementing a human driver model, a closed-loop algorithm, consisting of a number of fuzzy elements, has been developed for the task of highway driving including passing maneuvers etc. The paper repeats the experimental previous works and focuses on the development of the closed-loop approach of model refinement realizing personalized drivers model. The results of the proposed approach allow the cognitive supervision and also autonomous driving.

INTRODUCTION

Driver safety and driving assistance systems have been focused in automotive research and development for the last few decades. Among the technologies of intelligent transportation system (ITS), the vehicle-navigation systems, vehicle stability control systems like anti-lock braking system (ABS) or Electronic Stability Control (ESC) have been used with great success in the automotive industry. Furthermore, adaptive cruise control (ACC), automated lane-changing (ALC) and lane-keeping (ALK) are also developed realizing partially components necessary for full automated driving system, which may assist drivers by automatically controlling vehicles by using the signals of vehicle dynamics and environmental positions, such as absolute or relative velocity, following distances and lane positions.

With the goal of protection against potential unsafe situations or accidents, a number of research activities have been put forward to predict the future status of the vehicle. In [1] driver's actions can be predicted and recognized using hidden Markov model (HMM)-based pattern recognition without the environmental information at the accuracy 0.632. Besides employing those driving signals, some other research contributions show that the behaviors of the driver should also be observed for the prediction of the future operations and status. From this point of view, the driver activities can be classified into multiple levels [2] and analyzed for the sake of safety [3, 4].

Although driving education is given under similar instructions, drivers drive differently with their own styles, e.g., how fast driving on the highway, how hard to press the gas pedal for accelerating, and how far to keep when they are following a vehicle [5]. Therefore, one approach to improve the driving assistance systems, is to personalize different drivers according to their driving styles. These driving styles representing the individual driver patterns can be described by a number of driver models, which are acquired by supervised off/on-line training process. The typical parameters of the individual driving styles from each driver are learned and classified into more detailed categories and saved in a individual knowledge base of driver behaviors during the training process, so that the driving safety of a specific driver can be supervised by comparing his/her behaviors with the normal patterns. Furthermore, the controlling of the vehicle can also be taken by the personalized assistance systems autonomously without changing the styles of the driver.

In this paper, the underlying structure and concept of the approach, the Situation-Operator-Modeling (SOM), is briefly introduced, then the principle rules of driving cars on highways are mapped into a driving algorithm. By specifying several main parameters of driving signals, the method of Multivariate Normal Distribution (MND), gives the core for driver individualization and assistance. As an example to illustrate this approach, the relative velocity v_r and the distances to the followed vehicle in front d_f are selected as two variables for MND in bivariate case.

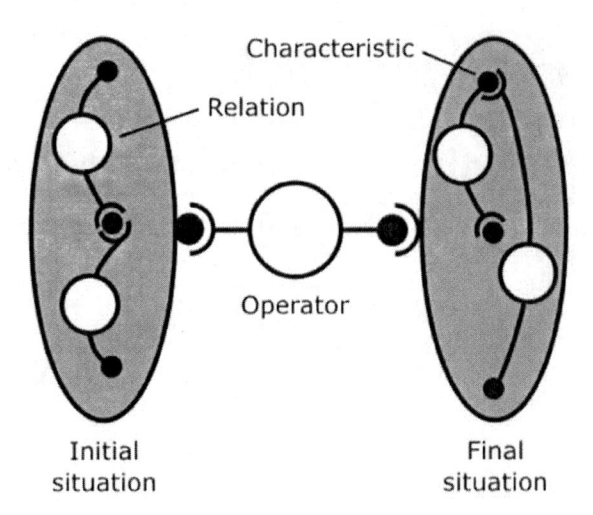

Fig. 1. Graphical notation of SOM [7]

SOM-BASED CONCEPT OF COGNITIVE SUPERVISION AND ASSISTANCE

THE UNDERLYING STRUCTURE OF ACTION LOGIC: SITUATION-OPERATOR-MODELING (SOM)

In [6, 7] a systemtheoretic modeling approach was introduced, dealing with a special situation-operator modeling kernel (calculus) called SOM, which will be repeated here briefly to illustrate the contributions extensions and related use. The approach combines ideas of the situation and event[8] and leads to a uniform and homogenous modeling approach allowing to describe human learning, planning, acting, and also the formal description of human errors. The introduced SOM approach gives the modeling framework, which means the structure of changeable scenes, and therefore maps the structured 'reality' of the real outside world of a system into a formalizable representation.

The core of the approach is the assumption that changes in the parts of the real world to be considered are understood as a sequence of effects. The items scenes and actions of the real world are used to model these changes. The item scene denotes a problemfixed moment in time but independent from time, and the item operator denotes the action changing the scene in Fig. 1. Both are connected by the underlying logic of action, the operator follows the scene, the following scene follows the operator, etc. The definition of the items scene and action are coordinated in a double win. They are related to each other and therefore can also be used to relate the assumed structure of the real world to the structure of the database - called the mental model - of an intelligent system. Humans (as human operators) and intelligent systems are included in the real world.

THE UNDERLYING STRUCTURE OF THE ACTION LOGIC: SOM APPROACH

Using the Situation-Operator-Modeling, a related concept for automated supervision is developed and proposed in [9]. For the representational level as part of the cognitive approach, the SOM is used to model and structure the complex scene of the driver-vehicle-environment interaction. First, the scene is modeled as situation. Therefore, its characteristics have to be defined. Then, the set of actions the driver can perform, is specified and modeled as operators. The passing maneuver is chosen as an example because it provides enough complexity to demonstrate the approach, but on the other hand it is not too complex to get lost in details.

In Fig. 2, the whole concept of automated supervision of the passing maneuver as an example is illustrated. On the sensory level, the relevant facts of the real world are perceived to set up a characteristic vector as situation description on the processing level. The operator representing the current action of the driver has to be identified from the existing basis operator library. With the situation description and the actual operator, a meta-operator representing the goal-oriented action of the driver can be chosen from the metaoperator library. On the analysis level, the assumptions of the operator are checked, whether the operator is - related to the actual situation - is applicable or not. This can be realized by checking the consistency between operators and situations as well as the interaction logic. The operator and error libraries can be replaced or extended for automated supervision of other systems, while the overall structure remains unaltered. With this formalization possibilities the action sequence of the driver can be checked for consistency (checking assumptions), typical human errors, and goal conflicts so that the driver can be informed, if an error or conflict is detected.

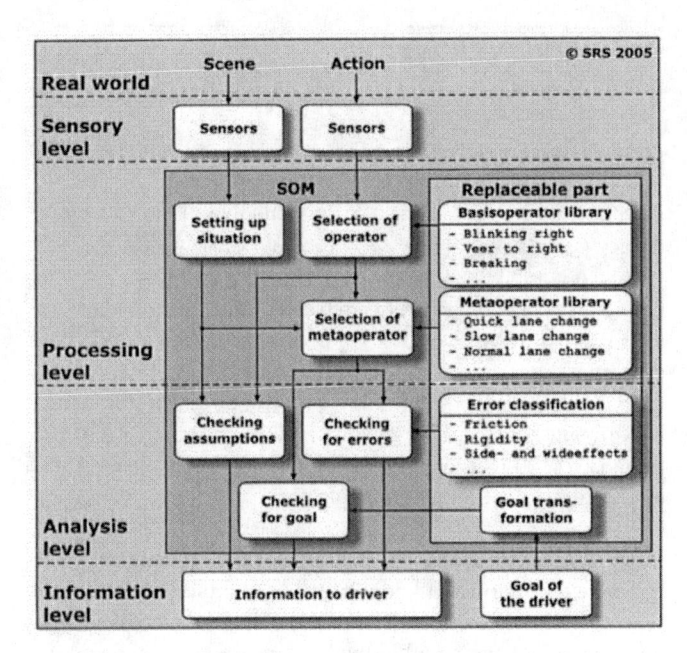

Fig. 2. Concept of automated supervision [9]

DATA COLLECTION FROM EXPERIMENTAL ENVIRONMENTS

According to the SOM approach, the collection of the driving signals can be categorized as

1). status of vehicle dynamics and of environment situation, which are used for setting up the situations vectors, e.g., velocity, acceleration, lane position, distances, and relative velocities to other vehicles, and

2). signals describing driving behaviors, which denote the operation of the driver, e.g., turning signal status, steering angles and pressures of gas/brake pedals.

In the personalized model focused in this contribution based on former work [10], the data from both groups were selected and combined for analysis. The driving signals were collected both from a driving simulator (Dynamic Driving Simulator®) and a real car (ViewCar®) of German Aerospace Center (DLR), Braunschweig, Germany. Both of the driving signals and driver actions were recorded and transmitted via CAN-bus.

The ViewCar® equipped with data collecting devices was also used to gather data about traffic information, state of driver, and vehicle guidance. For the experiment, 15 groups of signals were sampled at 100 Hz including vehicle velocity, engine speed, pressure on the gas/brake pedals, accelerations, steering angles, lateral deviation, lane positions, and turning indicator signals. Furthermore, the driver behaviors and environmental information were observed by 5 digital cameras mounted at different positions in the ViewCar®. The on-line data collecting devices and the setting of the cameras are shown in Fig. 3.

<figure 3 here>

Examples of driving signals collected from the ViewCar® are shown in Fig. 4 a) velocity of the vehicle, b) position of the gas pedal, c) yaw rate, and d) lateral deviation from the vehicle center to the midline of the lane.

<figure 4 here>

MODELING AND PERSONALIZATION OF HUMAN DRIVER MODEL

RULE-BASED MODELING: GENERAL DRIVING ALGORITHM FOR HIGHWAY-DRIVING

The modeling of the driving process on highway is illustrated by the flowchart of a driving algorithm in Fig. 5. The algorithm, which is based on the general driving rules (here: in Germany), separates the whole driving processes by a cluster of nodes representing the main actions or observations of the driver. The elements with white background indicate the non-individual description of the vehicle status, such as the confirmation of lane position with a vector (n, i) standing for driving currently on the i-th lane of a highway which consists of n lanes. Within the whole loops of driving algorithm, the lane vector (n, i) is checked for the decisions of the driver's next actions. Some of those depend on the individual behaviors of the driver, which are represented by the colored blocks and some others depend on a mix of information and desires.

In the following algorithm, the individual elements (colored) can be classified as:

1). driver's estimations for the occupation of the current or nearby lanes, and

2). lane changing actions depending on the driving styles.

<figure 5 here>

RULE-BASED MODELING INCLUDING EXPERIMAL-BASED VARIABLES DESCRIBING REGULAR AND UNUSUAL HUMAN DRIVE BEHAVIOR

The driving process includes a number of maneuvers. According to the SOM approach they can be considered as a series of scenes and actions modeled by interconnected situations and operators. The individual driving behaviors appear in typical maneuvers such as driving into/off the highway, following patterns, overtaking scenarios, and lane changing maneuvers. The modeling of personalization process is built by catching the related measurements when the corresponding maneuver or events occurred and recording into the individual knowledge base about driver's behaviors. The cyclic structure for developing the individual knowledge base in this case is illustrated in Fig. 6.

The driver interacts with the vehicle by perceiving information from the environment and acting operations to the vehicle. All the measurements are observed by transmitting the data via UDP-connections. The occurrence of a lane changing maneuver is detected by the lateral deviation ld. In the experiments, the width of each lane is 3m and a lane changing maneuver is recognized in this case if $|ld| > 1$, which means the vehicle is touching the lane marks and ready for passing.

<figure 6 here>

MODELING WITH MULTIVARIATE NORMAL DISTRIBUTION

The MND method is widely used in pattern recognition and classification [12, 13]. This method is able to generate random values and vectors from the distribution of choice given its sufficient statistics or chosen parameters. Before changing lanes, the driver needs to estimate the relative velocity v_r to the nearest vehicle on the desired lane and the distances to the leading vehicle in front d_f. In this chapter, v_r and d_f are considered as two independent standard normal random variables within a certain range. The probability density function (PDF) of the lane changing event depending on v_r and d_f can be expressed with a bivariate normal distribution by

$$\varphi(v_r, d_f) = \frac{1}{2\pi\sigma_1\sigma_2\sqrt{1-\rho^2}} \exp\left\{-\frac{1}{2(1-\rho^2)}\left[\left(\frac{v_r-\mu_1}{\sigma_1}\right)^2 - 2\rho\left(\frac{v_r-\mu_1}{\sigma_1}\right)\left(\frac{d_f-\mu_2}{\sigma_2}\right) + \left(\frac{d_f-\mu_2}{\sigma_2}\right)^2\right]\right\}$$

(1)

where μ_1, μ_2 are respective means of v_r and d_f, and σ_1^2, σ_2^2 are their respective variances. The parameters of biavariate normal distribution were determined using maximum likelihood parameter estimation (MLE) from the collected data for its robustness and good statistical properties [14]. According to MLE, means and variances of collected vectors (d_f, v_r) are estimated by

$$\hat{\mu}_1 = \frac{1}{N_1}\sum_{i=1}^{N_1} v_{r_i}, \quad \hat{\mu}_2 = \frac{1}{N_2}\sum_{i=1}^{N_2} d_{f_i}$$

(2)

and

$$\hat{\sigma}_1 = \sqrt{\frac{1}{N_1}\sum_{i=1}^{N_1}(v_{r_i}-\hat{\mu}_1)^2}, \quad \hat{\sigma}_2 = \sqrt{\frac{1}{N_2}\sum_{i=1}^{N_2}(d_{f_i}-\hat{\mu}_2)^2}$$

(3)

Examples of MND-based on MLE estimated of lane changing maneuvers are shown in Fig. 7 for 2 drivers. In Fig. 7 a) and b) the recorded vectors v_r, d_f were marked when the lane changing maneuvers were detected. The parameters are estimated with MLE by Eq. (2) and (3) and used to build the corresponding biavariate normal distribution, which are plotted in 3-dimentional form in c) and d). In the initial experiment 2 borders are marked to illustrate the 5% boundaries shown clearly in e) and f) by projecting the distributions on 2-dimentional planes.

After 2 hours test driving, data of each driver are taken to generate the individual knowledge base by the distribution of lane changing behaviors. The probabilities of successful lane changing maneuvers within the borders are generated and the results are compared among different borders. Within the 5% borders, 87.2% of lane changings and 51.6% of fitting vectors without lane changing are detected. However, within the 10% borders the two probabilities are 82.6% and 47.1%.

<figure 7 here>

SUMMARY

In this paper, a driving algorithm is developed representing a rule-based mapping to realize a complete human driver model being constructed by a series of individual elements specified by personalized parameters. Lane changing maneuvers are analyzed and illustrated by the relationship between relative velocity to nearby vehicle and the distance to the leading vehicle in front. As approaches to map the individual driver behaviors into related cognitive models MND and MLE approaches are used to generate parameters from collected driving data and generate the probability of lane changing maneuvers. The results show that better detection of events can be achieved by extending the boundary of the occurrence distribution, but the probability of non-occurrence of the event arises simultaneously.

For the future work, the personalization of driver's behaviors needs to be focused on, e.g. the following patterns, overtaking maneuvers and acceleration styles. Other driver modeling approaches besides MND, such as Gaussian mixture model (GMM) and hidden Markov model (HMM) are supposed to be applied for comparison. Furthermore, it is also intended to improve to improve the model for recognition of differences in the driving styles of one driver from a long period of time.

REFERENCES

1. Itoh, T., Yamada, S., Yamamoto, K., Araki, K., "Prediction of Driving Actions from Driving Signals," In-Vehicle Corpus and Signal Processing for Driver Behavior, Springer US, New York, USA, 197-210, 2009.

2. Park, S., Trivedi, M., "Driver activity analysis for intelligent vehicles: issues and development framework," In IEEE Intelligent Vehicles Symposium, 644-649, 2005.

3. Riener, A., Ferscha, A., "Driver Activity Recognition from Sitting Postures," Mensch und Computer 2007, Workshop Automotive User Interfaces, Weimar, Germany, Verlag der Bauhaus-Universität Weimar, 55-63, 2007.

4. Veeraraghavan, H., Atev, S., Bird, N., Schrater, P., Papanikolopoulos, N., "Driver activity monitoring through supervised and unsupervised learning," In IEEE Intelligent Transportation Systems, 580-585, 2005.

5. Ohta, H., "Individual differences in driving distance headway," Proceedings of Vision in Vehicles IV Conference, Amsterdam, Elsevier, pp. 91-100, 1993.

6. Söffker, D. "Systemtheoretic modeling of knowledge-guided human-machine-interaction," Habilitation thesis, Faculty of Safety Engineering, University of Wuppertal, Germany, 2001 (in german), also by Logos Verlag, Berlin, 2003.

7. Söffker, D., "Interaction of Intelligent and Autonomous Systems - Part I: Qualitative Structuring of Interactions," *MCMDS-Mathematical and Computer Modelling of Dynamical Systems*, 14(4): 303-318, 2008.

8. McCarthy, J., "Situations, actions, and causal laws", Memo 2, Stanford University Artificial Intelligence Project, 1963.

9. Ahle, E. and Söffker, D, "Interaction of Intelligent and Autonomous Systems - Part II: Realization of Cognitive Technical Systems," *MCMDS-Mathematical and Computer Modelling of Dynamical Systems*, 14(4): 319-339, 2008.

10. Fu, X. and Söffker, D., "Cognitive Awareness of Intelligent Vehicles," Proceedings of SAE 2010 World Congress and Exhibition, 2010, submitted.

11. Fu, X., Gamrad, D., Mosebach, H.; Lemmer, K. and Söffker, D., "Modeling and implementation of cognitive-based supervision and assistance," Proc. 6th Vienna Conference on Mathematical Modeling on Dynamical Systems MATHMOD 2009, Vienna, Austria, 2009.

12. Pratt, J. W., Raiffa, H., Schlaifer, R., "Introduction to Statistical Decision Theory," MIT Press, MA, USA, ISBN 0-262-16144-3, 1995.

13. Kachigan, S. K., "Multivariate Statistical Analysis: A Conceptual Introduction," Radius Press, New York, USA, ISBN 0-942-15491-6, 1991.

14. Aldrich, J., "R. A. Fisher and the Making of Maximum Likelihood 1912-1922," *Statistical Science*, 12(3):162-176, 1997.

CONTACT INFORMATION

Fu, Xingguang
Ph. D. Student
Chair of Dynamics and Control (SRS)
University of Duisburg-Essen
Lotharstrasse 1-21
47057 Duisburg, Germany
xingguang.fu@uni-due.de

Söffker, Dirk
Professor, Head of the Chair of Dynamics and Control (SRS)
University of Duisburg-Essen
Lotharstrasse 1-21
47057 Duisburg, Germany
soeffker@uni-due.de

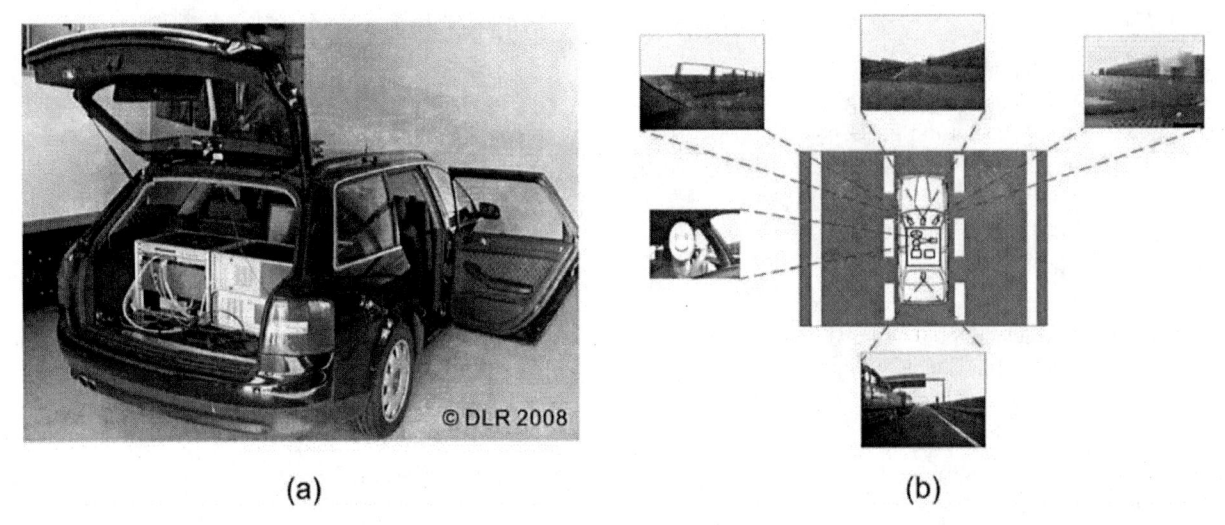

Fig. 3. a) The ViewCar® and b) digital cameras mounted inside the ViewCar® showing different views [11]

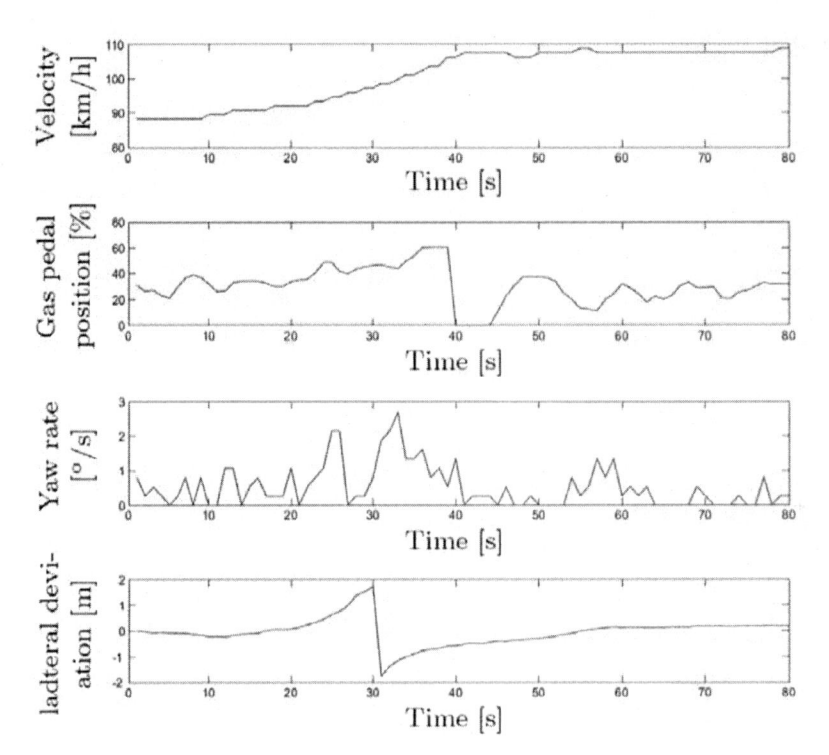

Fig. 4. Driving signals collected from the ViewCar®

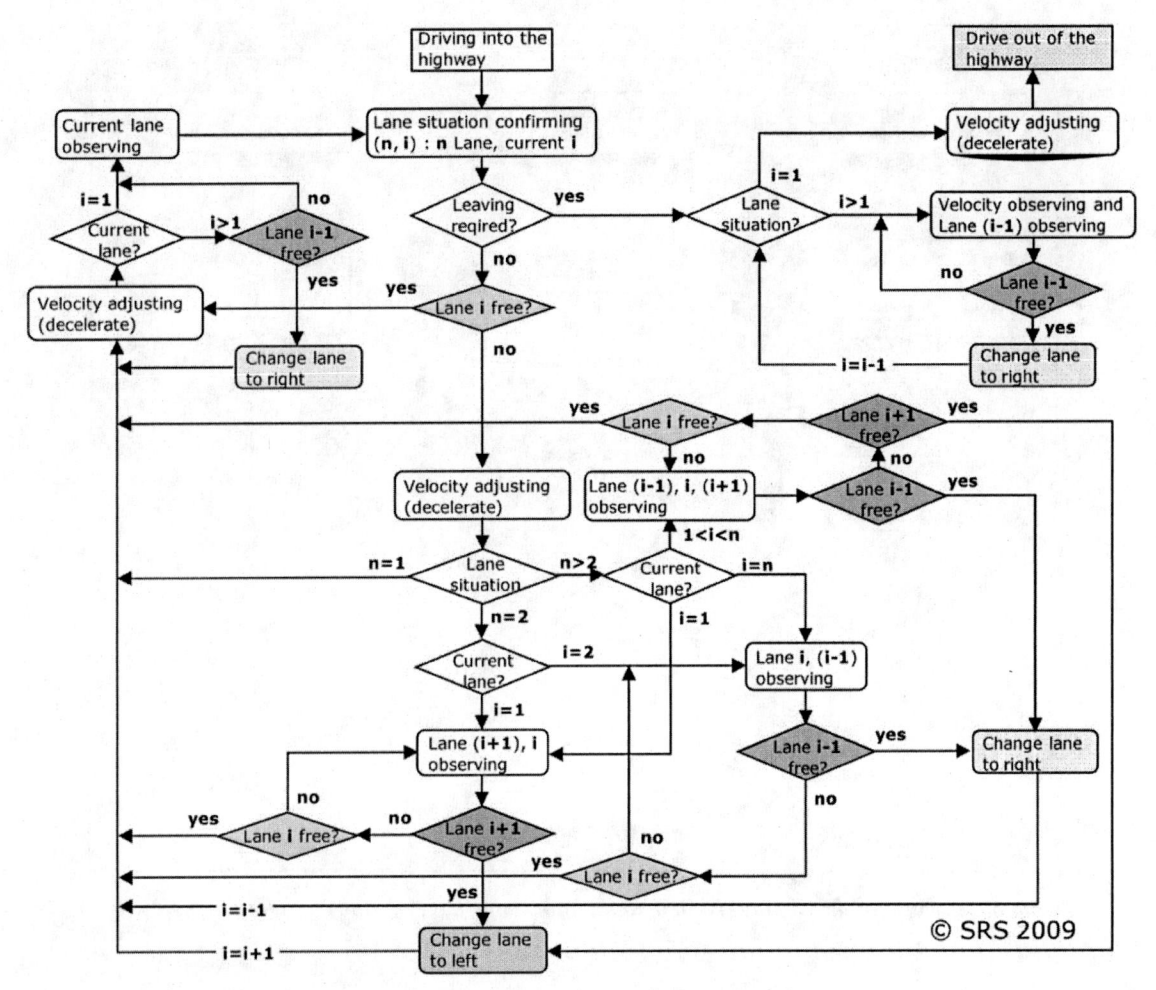

Fig. 5. Flowchart of the driving algorithm on highway

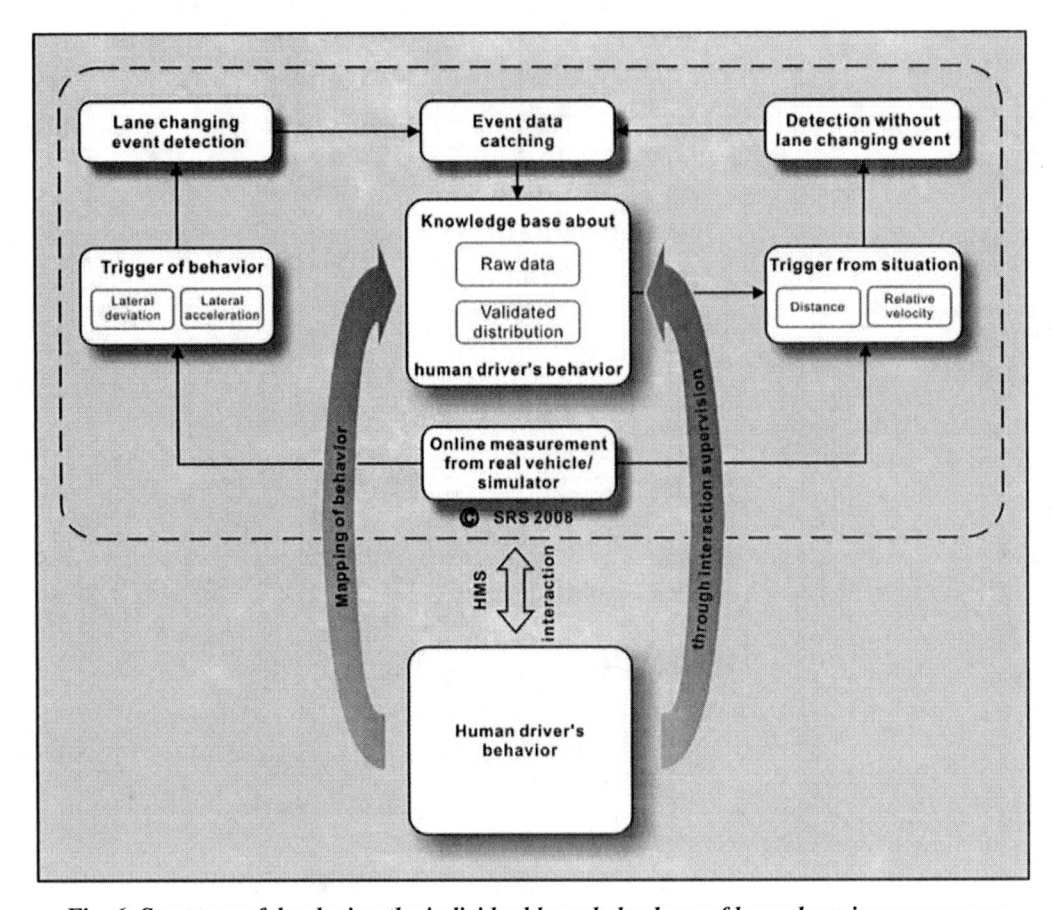

Fig. 6. Structure of developing the individual knowledge base of lane changing maneuvers

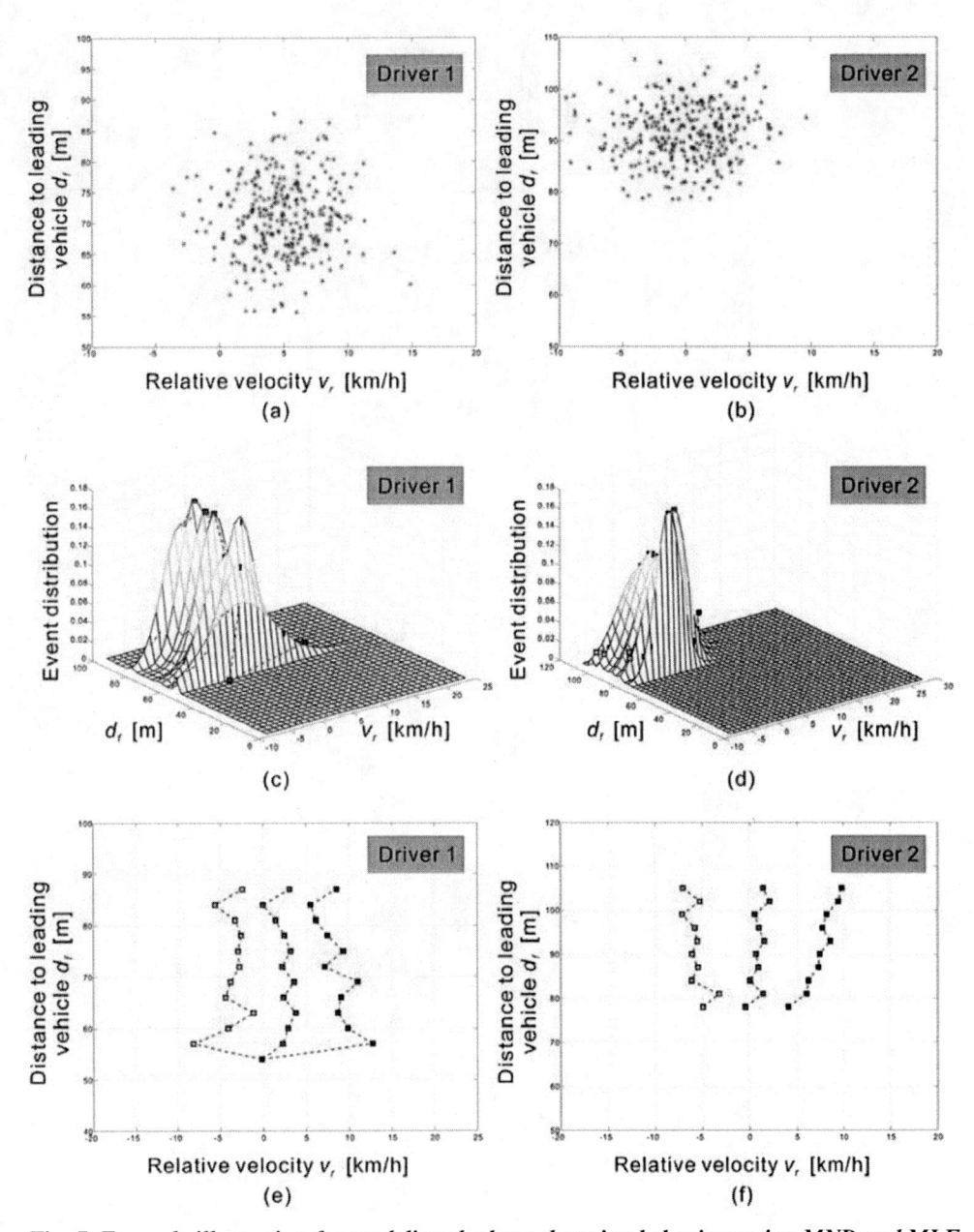

Fig. 7. Example illustration for modeling the lane changing behaviors using MND and MLE

Designing Reusable and Scalable Software Architectures for Automotive Embedded Systems in Driver Assistance

2010-01-0942
Published
04/12/2010

Dirk Ahrens, Andreas Frey and Andreas Pfeiffer
BMW Group, Munich, Driving Dynamics

Torsten Bertram
Technische Universität Dortmund

Copyright © 2010 SAE International

ABSTRACT

In this paper a model based design approach is described helping to improve the process of automotive embedded software development. The methodology contains a prototypical development process which shows great improvement towards the more and more upcoming non-functional demands faced by automotive software development like reusability, scalability and resource efficiency. This is achieved by use of increased formalism, continuousness and open standards. The whole hereby proposed and surrounding development process is introduced including all essential artifacts, process steps and formal or automated transitions between them.

Significantly important is the Abstract Automotive Software Architecture (ABSOFA), a meta model based software architecture modeling concept for structuring automotive software fully abstracting from realization, implementation and platform details. The meta model as well as the modeling language itself are based on the Unified Modeling Language (UML). This concept was particularly developed for the specific needs of complex software systems found e.g. in the domain of driver assistance.

The ABSOFA is also the backbone for several stages of expansion of the process. One of them is the concept of evaluating software architectures automatically and objectively. For doing so criteria for 'good' software architectures have been identified, formalized and worked out into quantitative algorithms that can be applied to the architecture calculating one or many quality values. For obtaining concrete variant and platform specific software architectures model transformations can also be applied automatically to any ABSOFA configuring and later on transferring them to common software implementation tools. For both concepts, architecture evaluation and model transformations, a common and formal XML (Extensible Markup Language) Schema based data model was required and developed which is also described.

The whole process works for top down as well as well as bottom up development. The article concludes with a practical example where the whole longitudinal dynamics software architecture of BMW's driver assistance systems is reengineered, restructured and optimized.

INTRODUCTION

Automotive mechatronic systems nowadays are facing a big variety of challenges. The complexity of these systems is determined by rising functionality of the single system as well as the steady increase of interactions and interconnections between different systems. On the other hand systems have to be developed faster by using fewer (financial) resources for a big variety of product lines, models and variants. Turning to safety critical systems these challenges become even greater.

This paper deals in particular with developing modern driver assistance systems. There is a multitude of different systems in the market today with either unique or overlapping characteristics. Many systems exist in different alternative variants depending on which product line or vehicle type has to be handled with and which system characteristic the customer has ordered. Depending on the platform one or many of the following specifics may vary: *sensors* (e.g. radar, lidar, ultrasound), *actuators* (e.g. engine, brake system, steering system), *manual control elements* and/or *displays* (HMI), *electronic control units* (ECU). As driver assistance systems interact on a very intense level and share certain functionalities (e.g. usage of identical interfaces to actuators and therefore needing coordination) an overall and common systems and software architecture is needed which allows modeling and implementing of all offered driver assistance systems for any possible configuration as automated as possible. If you just focus on the currently offered longitudinal dynamics systems of BMW the following systems or functions have to be considered in this architecture:

• Cruise Control (CC)

• Dynamic Cruise Control (DCC)

• Active Cruise Control (ACC)

• Active Cruise Control with Stop&Go-Function (ACC Stop&Go)

• Speed Limitation Device (SLD)

• Adaptive Brake Assist (iBrake)

• Hill Descent Control (HDC)

To be able to deal with the complexity of the highly and complexly interacting systems in all possible configurations and environments a new and efficient process for developing, modeling and implementing driver assistance systems was needed. The key concepts is a flexible and scalable software architecture offering innovative means to deal with the rising complexity and empower the user to do formal and automatic transitions between process steps and artifacts to minimize manual work. This paper therefore presents a methodology containing a prototypical development process helping to face and solve the major challenges for developing driver assistance systems (Figure 1). One of the main concepts is the newly created and from now on referred to as *Abstract Automotive Software Architecture (ABSOFA)* [1,2] which allows modeling of complete software systems disregarding and abstracting from concrete realization, implementation and deployment details. This software architecture is modeled with the Unified Modeling Language (UML) [3] and bases on a formal meta model. By this abstraction and a separation of general system properties from realization dependent properties as long as possible, **reusability** as well as **portability** are strongly increased. Configuration and vehicle type variants can also be modeled and managed. Additionally *Software Architecture Metrics* are deployed that can be applied automatically to any given architecture. They help evolving and improving it by measuring an objective 'quality value' at each evolution step. After that the optimized software architecture can be transformed - via *Automated*

Model Transformations - into common software development tools for embedded systems, i.e. ASCET [4] and Simulink [5]. Therefore code generation almost seamlessly follows the architecture design. This shows the benefit of the development process regarding **continuousness** and **formalism**.

<figure 1 here>

The field of model based software development for automotive applications is currently under keen investigation. Many promising efforts inside of concluded and still ongoing work can be found. For instance there are different modeling languages or Architecture Description Languages (ADL) especially designed for automotive efforts like AML [6,7], EAST-EEA [8,9], SAE-AADL [10]. All these languages typically do not go 'far enough' for being immediately used in real automotive serial development. The languages themselves are powerful, sometimes even too mighty and therefore complex, and highly adapted to the branch's needs. But there is no intersection point to current and approved development processes. Additionally these single concepts usually are not compatible among each other, they are commercialized and use proprietary formats. On the other hand there are other efforts like CAR-DL [11] and COLA [12] introducing whole and completely new development processes and thereby going 'too far'. As the automotive branch is quite conservative regarding the introduction of new processes to replace approved ones, even if they are known to be imperfect, replacing old processes with unapproved innovations is often referred to as too risky.

All concepts pointed out in this paper are designed to be easily integrated into currently used development processes. They support and ease a goal-oriented, formal and efficient development of automotive software and mechatronic systems. All relevant ideas of the AUTOSAR-Initiative [13] are supported as well, AUTOSAR even is strongly extended by the structuring, implementation and variation handling concepts of this work. Furthermore only open and freely exchangeable and compatible standards are used which basically enhance world-wide utilization and make a continuous development possible.

The paper contains a brief survey of all key facts and steps mentioned above. The adapted development process, the ABSOFA, architecture metrics and model transformations are introduced to the reader in their own sections and form the backbone of this article. Additionally we give a little insight in current work where the whole methodology is applied in practice by restructuring and refactoring our current driver assistance software architecture. Conclusions and future prospects for ongoing work end this article.

THE DEVELOPMENT PROCESS

To obtain the here presented final development process (see Figure 1) it was necessary to analyze the currently existing one. This section deals with this very topic. At first we point out this analysis of the process concerning the given non-functional requirements and demands towards the development. The goal was to examine:
• Which weaknesses can be identified inside the process?
• Where is potential for and necessity of improvement?
• Which process steps have to be dealt with to fulfill and satisfy the requirements?

ANALYSIS OF THE DEVELOPMENT PROCESS

Basis for our work was the currently practiced process of development according to an internal BMW Group standard for developing embedded software. This standard is similar to commonly known sources like the VDI standard 2206 [14]. The process is primarily described by a so called 'phase model' that in fact is a tailored V-Model [15]. Its left branch is depicted in Figure 2. Without being able to describe every step of this process or every detail of the analysis inside this document the following aspects where identified as major weaknesses and potential points of improvement:
• Transition from technology and platform independent details to concrete realization way too early.
• Unwanted mixing of hardware independent (functional) logic and implementation.
• Hindered reuse because most steps have to be traversed for each realization variant over and over again.
• Deployment done on basis of functional modules, no consideration of software demands.
• Concrete wiring system is prerequisite for functional development.
• No real software architecture design as functional modules become software modules one-to-one.
• No formal or automatable criteria for the transitions between the process steps and their artifacts.

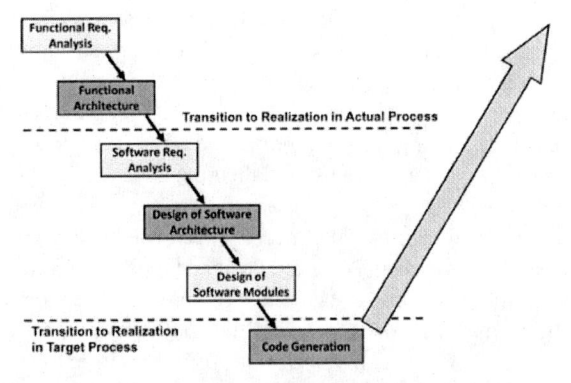

Figure 2. Actual Development Process before Modification

ADAPTATIONS AND CHANGES TO THE PROCESS

The previous subsection has identified the key points and steps where improvement is needed and also pointed out which way to act. The main target is to move the line marking the transition to realization details (see Fig 2) as downwards as possible. As Source Code is always particularly generated and suited for a certain platform this point is definitely the only possible and likewise optimal target position for this line. Obviously just moving a line inside a diagram does not improve anything. Therefore in order to really change and advance the process certain adaptations to various process steps and adding of new steps were necessary. All changes can be integrated into the process any time as optional upgrade. That is why they can be applied instantly to any new or even running project.

First of all we introduced an outer frame offering three views on the technical system 'vehicle' (Fig 1 left). BMW traditionally is very good on the first and third layer of this pyramid. This means that on the one hand it is clear which function shall be developed on a very abstract level meeting all customer requirements. On the other hand the implementation of this function for specific platform is also done very well. Therefore customer satisfaction is very good but the path from specifying the function to implementation requires a lot of resources as the same steps have to be traversed for each implementation variant repeatedly. Studies show that functional details change way less frequently than platforms and implementation technologies. That is why **the medium layer of the pyramid is the key to resource efficient development.** Only by modeling the mechatronic system in a reusable i.e. platform and realization independent way supporting automatic transitions the amount of non-functional requirements can be fulfilled. This means that the process steps two to five of the current development process (Fig. 2) have to be newly defined and interpreted to fit in the medium layer. Functional and software architecture from now on are platform independent and therefore reusable. Deployment and implementation is done as final step traversing to the technical architecture. Moving resources from multiple implementations to an intensified platform independent modeling of technical systems is often referred to as **frontloading**. Concerning the large amount of different platforms and variants that have to be dealt with the little additional effort pays off very quickly when noteworthy resources are saved for every single realization.

Reckoning software architectures already at this platform independent and therefore abstract level is a new perspective in the automotive branch. Software architectures need to be considered separately from functional architectures as their scope is completely different. Functional architectures (or functional nets) structure a technical system according to logical, physical or technical demands and constraints while software architectures shall focus on quality as well as non-functional issues, an outline can be found in [16]. The software system needs to be structured primarily according to these demands which differ from technical constraints very much. The ABSOFA contains information for any possible variant (software) configuration and thus maybe be referred to as '150% software architecture'. The introduction of this abstract software architecture is conclusive and required to fulfill the big variety of non-functional demands. More details can be found in the next section.

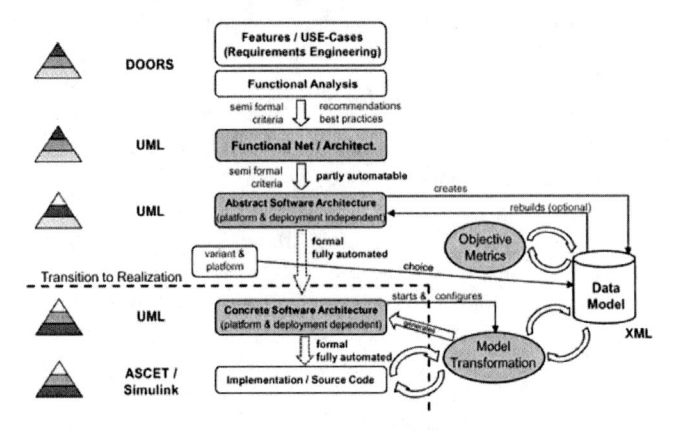

Figure 3. Transitions inside the new Development Process

Figure 1 (right hand side) already showed all process steps of the final version of the acquired development process. You can see the classification of every process step according to the views pyramid. Figure 3 shows these process steps with additional information. You can see how the adaptations of the old process steps and introduction of the new abstract software architecture have made formal and automated transitions between them possible. This is done by formalized criteria, a consistent and continuously used data model combined with compatible, universal and freely available data formats. The continuousness derived from this method guarantees the efficiency of the newly developed process. Crucial for the success of this process apart from the abstract software architecture are the two concepts of objective software architecture evaluation and automated model transformations which are explained in the following main sections.

ADVANTAGES OF THE NEW DEVELOPMENT PROCESS

This subsection summarizes the advantages of the adapted development process compared to the original one. Many of these advantages are at hand, others can be found at second sight.

• Easy integratability into current development process.

• Continuous and consistent model based development starting from the functional architecture.

• Explicit modeling and intentional separation of functional and software architecture.

• Meta model based abstract software architecture containing variant information as '150%-model'.

• Objective evaluation and iterative, goal-oriented evolution of software architectures.

• Partly or fully automatable transitions between process steps because of formal criteria.

• Usage of open source and open standards instead of proprietary and incompatible third party tools.

ABSTRACT AUTOMOTIVE SOFTWARE ARCHITECTURE

Software architectures are used for documentation and structuring of software systems. Many different definitions of the term can be found in literature e.g. in [17]. The common sense of most publications concerning software architecture definition is this:

'A software architecture describes the structures of a software system which comprises their structural elements (e.g. components, modules), their externally visible characteristics (e.g. attributes, methods) as well as their interactions and interrelations.'

The goal of designing software architectures typically is to improve the system towards one or many (unfortunately mainly competing) non-functional characteristics. This means that the user, i.e. customer, does not benefit from these efforts directly. Maybe that is why struggles dealing with these topics have been neglected recently; more probable however is the fact that software systems were very uncomplex and therefore quite manageable in the beginning with no real need for these structural issues. This fact has changed extremely during the past years making new approaches and concepts for dealing with the complexity at hand necessary. Depending on the specific domain these characteristics vary strongly. For driver assistance systems the following requirements usually occur: *reusability, scalability, extensibility, maintainability, resource efficiency, clarity, portability*. For further interest you may find a complete list and their dependencies in [18].

Software architectures have recently been used mainly on the implementation level. ECUs are designed with an architecture - typically designed as layered structure - to separate the applications (also referred to as 'high level software') from hardware dependent parts of the software like operating system, drivers, bus communication (also referred to as low level software) (Figure 4). The Figure shows these typical layers and classifies them relating to the familiar OSI

Reference Model [19]. AUTOSAR proposes a similar architecture.

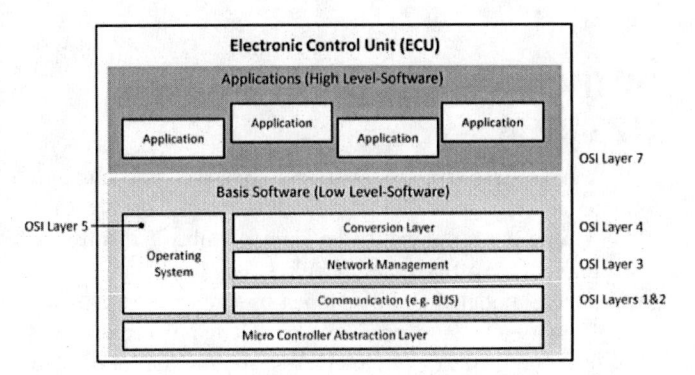

Figure 4. A Typical ECU Software Architecture

The high level software (OSI Layer 7) until today however has not been structured according to the addressed needs above. The internal software structure typically was derived from the functional structure one-to-one. That is why the structure performed very poorly when it comes to non-functional demands. As these demands become more and more important - the reasons have been mentioned in the beginning of this article - a different and independent software structuring process is needed after the functional net creation. The already mentioned ABSOFA is designed to accomplish these goals. It abstracts from realization, implementation and platform specific details thus guaranteeing full reusability. Additionally it also forms the basis of all further process steps like architecture evaluation and evolution, variant configuration and generation of platform specific software architecture as well as transition to implementation/code generation tools. The generated high level software at the end of this process equates to the one shown in Fig. 4 and therefore fits into the common concept.

The key facts of the ABSOFA may be summarized as follows:

• Depiction based on formal UML meta model (UMP Profile) highly adapted to automotive software needs.

• Design and modeling of whole embedded software systems abstracting from concrete realizations (i.e. ECU platforms, wiring system, variant configuration, sensors, actuators).

• Abstraction from concrete implementation tools, usage of open source and free standards only.

• Arbitrary and flexible composition and decomposition of the system.

• 150%-models possible to show full functionality across all platforms.

• Optional integration of functional and technical variants for automatic variant configuration.

THE META AND DATA MODEL

The ABSOFA's meta model combines and abstracts from typical implementation tools like ASCET and Simulink. All relevant concepts for structural elements and interfaces are merged and converted to an abstract form. Also by interviewing developers and software architects towards their own personal needs some additional modeling elements have been introduced. Therefore the UML depiction is very powerful and offers all required means of modeling automotive embedded software architectures. Special focus was set on a simple and easy to learn modeling language with the ability of arbitrary hierarchical (de)composition. On different architecture (detail) levels the same modeling elements can be used. Variant information can optionally be added at any level to express the dependencies and interrelations of the model elements in different contexts (Figure 7). Other approaches totally lack this type of information which indeed is decisive for dealing with complex automotive software systems. Some of the main concepts of the meta model shall be explained shortly. Further information can also be obtained from [2].

<figure 5 here>

The whole modeling concept and language only needs five different structural elements (Figure 5 left) to fully model the structure of a software system. For detailed information some of these elements can be specialized into different versions of the basis element (e.g. Figure 5 right). The meta model is established as an independent UML-Profile using several stereotypes to formally redefine and extend the standard UML elements. To assure the characteristics and attributes of the stereotypes typically tagged values are used (see. Figure 6).

<figure 6 here>

For modeling interfaces two different depictions are possible. Depending on the hierarchical level and the intention of the user, interfaces can be modeled in detail as single and unique interface or be combined to abstracted signal groups. The latter allow very clearly arranged depictions for getting a quick overview, the first allows full detail modeling. The user can decide freely where to use which depiction, but the meta model demands that every interface/signal is modeled in detail somewhere in the architecture. Therefore unambiguous and formal modeling is combined with the possibility of abstraction and increase of clarity. Signal group interfaces always communicate via UML ports enhancing their abstract nature. Figure 6 also shows which metaclass the interface elements are derived from and which subtypes are needed to model all relevant communication mechanisms inside the ABSOFA.

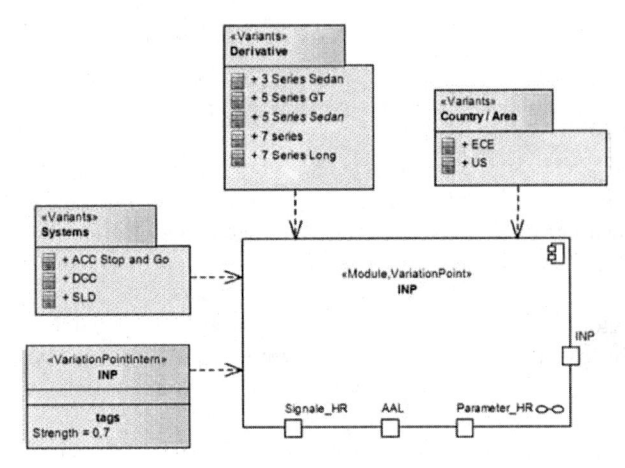

Figure 7. Modeling of a Module with Variant Information

AUTOSAR focuses on standardization and abstraction of interfaces. The primary target is to increase interchangeability between OEMs and suppliers. AUTOSAR does not offer any means of internally structuring or implementing software components. Also concepts for handling or configuring different variants of software components, like Figure 7, are not supported. Therefore the approach of this paper closes this gap but still is compatible to the essential AUTOSAR concepts like AUTOSAR Software Components and their Interfaces.

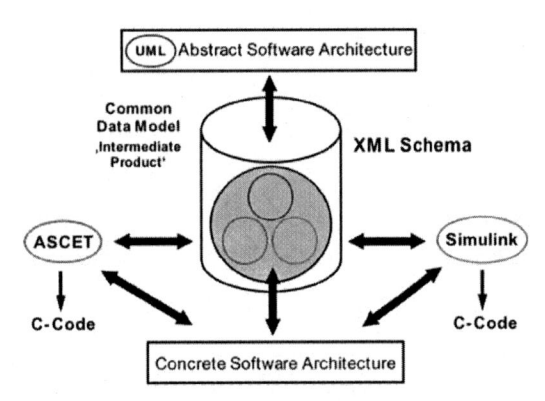

Figure 8. The Common Data Model 'Intermediate Product' and the Connectable Tools

Equivalently to the UML meta model a formal data model has been developed to save any information of the architecture in an independent database. The database cannot only save the information of the UML model but of all tools that take part in the development process (Figure 8). Therefore consistent data handling is guaranteed. As the data model mediates between these tools it is also referred to as *'Intermediate Product'*. The data model does not only contain modeling elements or pieces of information belonging to all three tools in common but can save any tool specific information as well. Therefore no information is lost at any time. Furthermore this data model is the backbone for architecture evaluation and the various model transformations (see following sections). It is defined by a formal XML Schema [21].

OBJECTIVE ARCHITECTURE METRICS

Until today there are no available means for evaluating software architectures. This is not only a specific problem for BMW Group but for anybody developing embedded software of all branches. Software architectures, if regarded at all, are documented, modeled and developed by single experts based on their own personal experience, knowledge or preferences. Therefore the software architectures are on the one hand dependent on subjective design decisions, on the other hand knowledge concerning building 'good' software architectures can hardly be passed from these single individuals to other employees, e.g. for job training. Software architectures of new software systems thus are designed according to commonly known or company specific best practices, feelings and intuitions about what is good or by the principle 'we always did it like this'.

That is why we focused on ways for *objective evaluation of automotive software architectures*, not exclusively for, but especially adapted to the well-known concept of the ABSOFA. Evaluation of whole software architectures cannot be found in today's literature although there are different approaches and concepts of evaluating software in general (i.e. source code or implementations), often based on criteria found in [16]. Our approach also bases on this norm choosing and adapting capable criteria for objective software architecture evaluation. The key question is: How is 'good' software architecture characterized and how can it be evaluated by objective, i.e. quantitative and reproducible, manners? This was solved by the following steps:

• Finding, adapting and defining criteria (Figure 9).

• Defining objective algorithms to 'measure' a quantitative value of each criterion (metrics).

• Implementing all metrics to be automatically applicable to the ABSOFA and the affiliated data model.

• Designing a prototypical tool with a Graphical User Interface (GUI).

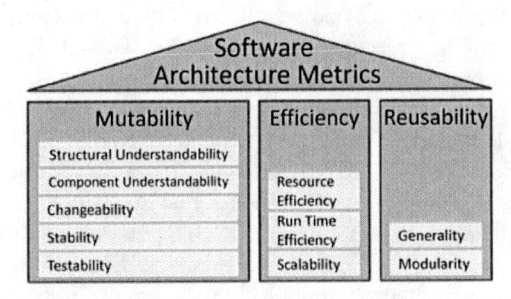

Figure 9. Objective Software Architecture Metrics - Groups and Criteria

For example the metric for the criterion 'modularity' is introduced in short.

Idea: Good modularity of software architectures is characterized by strong cohesion and few couplings of participating software components or modules. This means that well defined and encapsulated software units exist for certain and distinguished functionalities. Coupling and interrelations between those units should be limited as much as possible because they create hardly controllable dependencies. The metric works as a recursive algorithm and calculates modularity for every hierarchical level. The relation from internal to external communication is measured.

$$M_{mod} = \frac{\frac{\sum Mess_{ext,i}}{\sum Mess_{ext,i} + \sum Mess_{int,i}} + \frac{\sum_{j=1}^{n} M_{mod,j}}{n}}{2}$$

$1 \leq j \leq n$
i = current component
n = quantity of all sub-components of i
$Mess_{mod,j}$ = metric value of sub-component j
$Mess_{ext,i}$ = external communication between component i and another component
$Mess_{int,i}$ = internal communication between sub-components of component i among each other

All other metrics are defined in a similar way and can objectively be calculated to single quality values. The user, i.e. software architect or developer, can use the GUI of the implemented tool to configure the evaluation. He can choose which criteria shall be calculated and how they are weighted. The tool itself is realized as executable java file (.jar) and can therefore be run on any pc. The algorithms have been implemented using JAXB technology [20]. Typically the user compares two architecture versions to each other, but also a single evaluation is possible. The tool needs the software architecture (given by the ABSOFA) as input artifact in its defined XML format, outputs are the metrics values. Appendix A shows the GUI of the tool and a calculation example. Detailed information on this peculiar and all other metrics can be obtained from [22].

MODEL TRANSFORMATIONS

To make an integration of all the concepts into the development process as easy as possible, model transformations are needed. Because of these transformations the newly introduced artifacts and views can work hand in hand with the current and established process steps and, even more important, the actual tool chain. Model transformations are useful and already implemented for the following cases:

• Configuration/Generation of a concrete software architecture for a certain vehicle type, platform or variant.

• Transformations between UML and the data model (bidirectional).

• Transformations between implementation tools (ASCET, Simulink) and the data model (bidirectional).

Heart of all of these transformations is the common data model of the software architecture. This shows the consistency and continuousness of the new process. Similar to the architecture metrics a prototypical tool was implemented which enables the user to easily configure and execute the desired transformation. JAXB is also the technology for all transformations. The developed transformation concepts are fully compatible to the Model Driven Architecture (MDA) approach [23]. MDA is a very abstract idea of formal modeling of software architectures allowing several transformations between models as well as model-to-code; it does not explicitly provide specific tools, concrete (meta) models or transformation rules, therefore it can be regarded as a general concept of modeling and transforming software architectures.

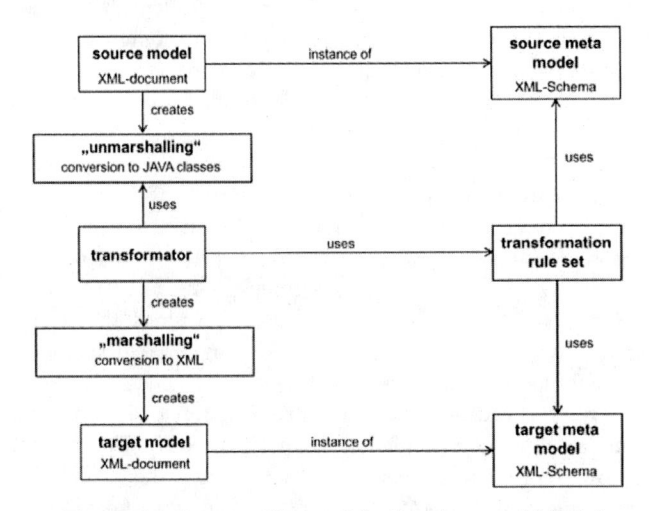

Figure 10. Process Chain of the Implemented Model Transformations

Our approach and software prototype is a concrete implementation of the MDA concept for a specific domain offering a specific problem solution and tool chain. The process of a typical model transformation is depicted in Figure 10. Meta (data) models for both source and target sides are provided by XML Schemata. JAXB reads the XML document (instance of the data model) and converts it into java classes ('unmarshalling'). The transformation rules are applied by common java programming constructs to manipulate the attributes of the classes thus transforming the structure inside java from source to target. In the end the new structure is written back to XML ('marshalling'). Platform specific details are included in the particular rule set that is chosen for the desired transformation direction by the user. The concept can be referred to as meta model based model-to-model-transformation regarding MDA-terminology. Summing up the main advantages of the realized transformations are the following:

- Full reuse of only once to be created ABSOFA 150% model. Automatic transformation into specific variant configurations with nearly no effort.

- Singular creation of transformation rules for infinite times of transforming ABSOFA software architecture to implementation tools.

UML is an object-oriented modeling language and therefore is often being used for designing object oriented software systems. This (ABSOFA) UML-profile however is compatible to the non object-oriented tools ASCET and Simulink which can generate C-Code. This is no antagonism as the meta model as well as the data model only represent the structure of the software system not the software itself, i.e. the implementation. On this abstract level corresponding structural elements and interfaces can be found in all tools and have to be united and abstracted inside the data model. The challenge is to fully cover all relevant elements and an unambiguous as well as tool-independent storage of their attributes. The transformation rules can then transform the tool-specific construct into the universal data model construct and vice versa.

Nonetheless UML is no convenient tool for modeling parts of the implementation of the software because the available constructs inside UML do not allow proper modeling of processes, sequences and algorithms. As only structural aspects are in focus of the presented methods this is no problem at all. To support these structural issues the transformations between the listed tools can be realized entirely. The data model can later on be extended towards implementation information easily if needed. This would allow full transformations including any behavior and implementation information between all possible directions.

AUTOSAR models are formalized via a meta-model, i.e. UML-profile, as well. As mentioned before the ABSOFA reuses the AUTOSAR structural elements and extends them e.g. towards variant handling and internal structure information. The following AUTOSAR constructs, among others, can be represented in ABSOFA as well as the embedded data model: AUTOSAR (Atomic) Software Component (SWC), Sensor/Actuator-Component, Calibration Parameter Component. This means that AUTOSAR models are compatible to ABSOFA models and can be transformed vice versa. This concrete transformation however has not been implemented yet but due to the extensibility of this approach a set of rules to do so can easily be integrated. All infrastructural prerequisites (interfaces, data formats, tools) exist and can be reused to full extent. The extensibility of the concept, resulting from the use of only free and open standards and tools is one of its key benefits for the user.

RESTRUCTURING AND REFACTORING OF LONGITUDINAL DYNAMICS DRIVER ASSISTANCE SOFTWARE ARCHITECTURES

One of the big advantages of the whole methodology, apart from its easy integratability into the development process, is the capability to be used for both, top down and bottom up, development. Top down development is suitable for totally new driver assistance systems or functions. As the dominating part of driver assistance development is characterized by improving and extending existing systems we chose the bottom up approach for our practical example to be much closer to real life's needs.

The task was to restructure the complete existing software architecture of the longitudinal dynamics driver assistance systems. Due to further development during the past years structural efforts have been neglected resulting in a software architecture which is full of structural dependencies and unneeded interrelations. This makes functional extensions as well as scaling of the overall architecture very difficult. Many software modules which are not really needed for sub-functionalities cannot be removed as there are dependencies because of the often quickly and structurally imperfect included extensions. A typical problem is the extraction of the DCC-Function out of the ACC Stop&Go-Function. Much overhead wasting precious ECU resources is generated. By restructuring the architecture according to the dominating non-functional criteria DCC implementation has to be made much more efficient. The whole process of restructuring with all constraints and steps is shown in Figure 11.

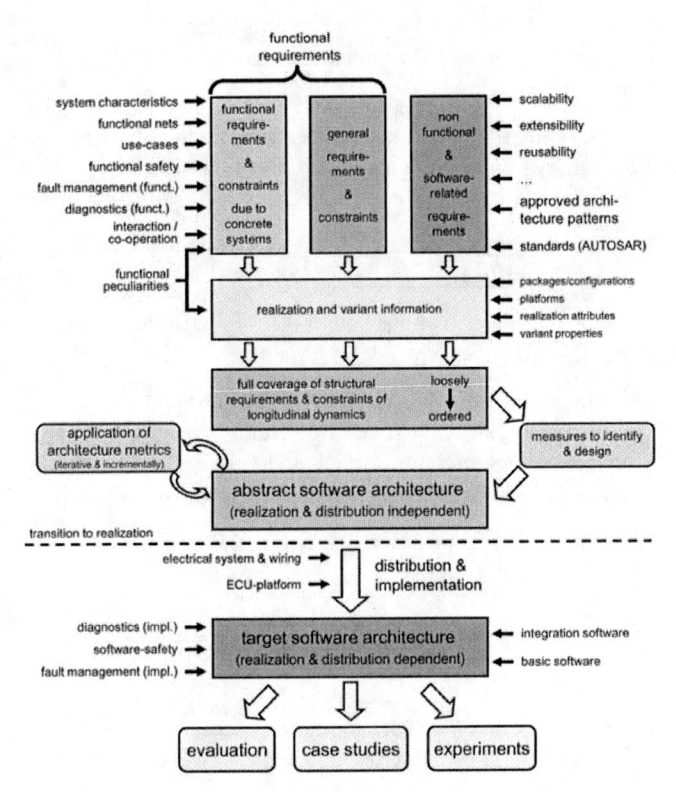

Figure 11. The Process of Restructuring Longitudinal Dynamics Driver Assistance Systems

Estimations resulting from this work show an approximate 50% resource saving. This is not to total extent to be owed to the process optimizations. The architecture offered a lot of potential which could have been made accessible by other approaches too. Nonetheless only the continuousness and formalism of the process steps aiding the user with mighty utilities at any time made the goal-oriented and efficient optimization and restructuring of the architecture possible in this short amount of time.

Additionally the architecture was therefore reengineered and depicted in UML for the first time giving new ideas and room for improvement of further versions of the UML meta model and the data model as well.

SUMMARY/CONCLUSIONS

This paper introduces to a new methodology containing a prototypical development process for an efficient and resource optimized development of driver assistance systems. The peculiar characteristic of this process is its easy integratability into existing embedded software development processes. One or more of the newly added or differently defined process steps can be included. The process focuses on formal and partly or fully automated process steps increasing continuousness.

Core feature is the so called Abstract Automotive Software Architecture (ABSOFA) allowing the user to model the structure of the high level software independent and therefore abstracting from realization or platform details. This dramatically increases reusability of the software system. Variant information can be depicted as well helping to model not only vehicle specific architectures but complete software systems valid for any product line (150% models). The approach supports the user in fulfilling the large variety of non-functional demands.

The ABSOFA is defined by a formal UML meta model and supported by a likewise formal XML Schema based data model which allows storing any architecture information inside a database. The data model is named "Intermediate Product". Attached to the data model are different automated process steps optimizing the development. Objective software architecture metrics have been acquired which can be applied to the architecture calculating an objective and quantitative 'quality value' of any given software architecture. This allows goal-oriented and iterative architecture evolution. For creating concrete variant architectures and transferring them to established implementation tools like ASCET and Simulink automated model transformations have been developed and realized.

Practical evaluation of the methodology is done by a bottom up approach. The whole longitudinal dynamics driver assistance systems software architecture was reengineered and restructured. The aim was to improve greatly towards non-functional criteria like scalability, extensibility and reusability. First results show potential for up to 50% resource saving depending on the peculiar task.

Our future work will be further evaluation and optimization of the process. The practical example will be concluded shortly offering additional results and awareness of possible weaknesses. Additionally in the near future new systems have to be implemented using the top down approach to evaluate this way of using the process as well. Due to the modeling of the software architecture in UML some details to be improved and extended in the ABSOFA were found which will be integrated soon. Mentionable potential for improvement is the complete depiction of variant information and their automated further processing.

The model transformations will be entirely available for all possible directions shortly, including transformations involving AUTOSAR models.

REFERENCES

1. Ahrens D., Pfeiffer A. and Bertram T., "Comparison of ASCET and UML - Preparations for an Abstract Software Architecture," Forum on specification and Design Languages (FDL) 2008, Proceedings. Stuttgart, Germany, 2008

2. Ahrens D., Frey A., Pfeiffer A. and Bertram T., "Entwicklung einer leistungsfähigen Darstellung für komplexe Funktions- und Softwarearchitekturen im Bereich Fahrerassistenz," VDI Mechatronik 2009, Wiesloch, Germany, 2009 (in German)

3. Object Management Group: UML - Unified Modeling Language, online informations, www.uml.org, 2009

4. ETAS GmbH: Ascet SD/SE, online informations, www.etas.com, 2009

5. The MathWorks: Matlab&Simulink, online informations, www.mathworks.com, 2009

6. Freund, U., von der Beeck, M., Braun, P., and Rappl, M., "Architecture Centric Modeling of Automotive Control Software," SAE Technical Paper 2003-01-0856, 2003.

7. Braun P., von der Beeck M., Rappl M. and Schröder C., "Automotive UML", UML for Real, Lavagno L., Martin G., and Selic B. (eds.), Kluwer Academic Publisher, ISBN-1402075014, 2003

8. Embedded Architecture and Software Tools, the EAST-EEA project, online informations, www.east-eea.net, www.atesst.org, 2009

9. Lönn H., Tripti S., Törngren M. and Nolin M., "FAR EAST: Modeling an Automotive Software Architecture Using the EAST ADL," ICSE 2004 workshop on Software Engineering for Automotive Systems (SEAS), p. 43-50, Edinburgh, 2004

10. SAE: SAE-AADL, online informations, www.aadl.info, 2005.

11. Wild, D., Fleischmann, A., Hartmann, J., Pfaller, C. et al., "An Architecture Centric Approach Towards the Construction of Dependable Automotive Software," SAE Technical Paper 2006-01-1222, 2006

12. Kugele S. et al., "COLA - The Component Language," Institut für Informatik, Technische Universität München, TUM-I0714, 2007

13. AUTOSAR: Automotive Open Systems Architecture, online informations, www.autosar.org, 2009

14. VDI 2206, "Design Methodology for Mechatronic Systems," Beuth Verlag, Berlin

15. Bundesministerium des Inneren, Koordinierungs- und Beratungsstelle der Bundesregierung für Informationstechnik in der Bundesverwaltung: V-Modell Entwicklungsstandard für IT-Systeme des Bundes, Bonn, 1997

16. ISO - International Standards Organization, "Software engineering - Product quality," ISO 9126, 2001

17. Clements P. et al., "Documenting Software Architectures: Views and Beyond", Addison-Wesley, Boston, USA, 2005

18. Ahrens D., Pfeiffer A. and Bertram T., "Entwicklung einer flexiblen und skalierbaren Funktions- und Softwarearchitektur im Bereich Fahrerassistenz," 2. Dortmunder Autotag, Dortmund, Germany, 2007 (in German)

19. ISO - International Standards Organization, "Information Technology - Open Systems Interconnection - Basic Reference Model: The Basic Model," ISO/IEC 7498-1, 1994

20. SUN Microsystems: JAXB - Java Architecture for XML Binding, https://jaxb.dev.java.net, 2009

21. Object Management Group: XML - Extensible Markup Language. http://www.w3.org/XML

22. Ahrens D., Frey A., Pfeiffer A. and Bertram T., "Entwicklung eines objektiven Bewertungsverfahrens für Softwarearchitekturen im Bereich Fahrerassistenz," Software Engineering 2010, Paderborn, Germany, 2010 (in German)

23. Object Management Group: MDA - Model Driven Architecture, http://www.omg.org/mda/, 2009

24. Broy M., von der Beeck M., Braun P. and Rappl M., "A fundamental critique of the UML for the specification of embedded systems"

25. Hatley D. and PirbhaiI., "Strategies for real time system specification," Dorset House Publishers, New York, 1988.

CONTACT INFORMATION

Dipl.-Ing. Dirk Ahrens
BMW Group
Driving Dynamics
Systems Development Driver Assistance and Active Safety
80788 Munich, Germany
dirk.ahrens@bmw.de
+49-89-382-60776

Prof. Dr.-Ing. Prof. h.c. Torsten Bertram
Chair for Control and Systems Engineering
Technische Universität Dortmund
44221 Dortmund, Germany
torsten.bertram@tu-dortmund.de
+49-231-755-2760

ACKNOWLEDGMENTS

During the past two years several committed students contributed to our work who we are obliged and pleased to thank for their successful efforts. To mention here are especially (in alphabetical order): *S. Grunow, J. Heinichen, E. Resvoll, C. Vodermair* and *A. Zeidan*.

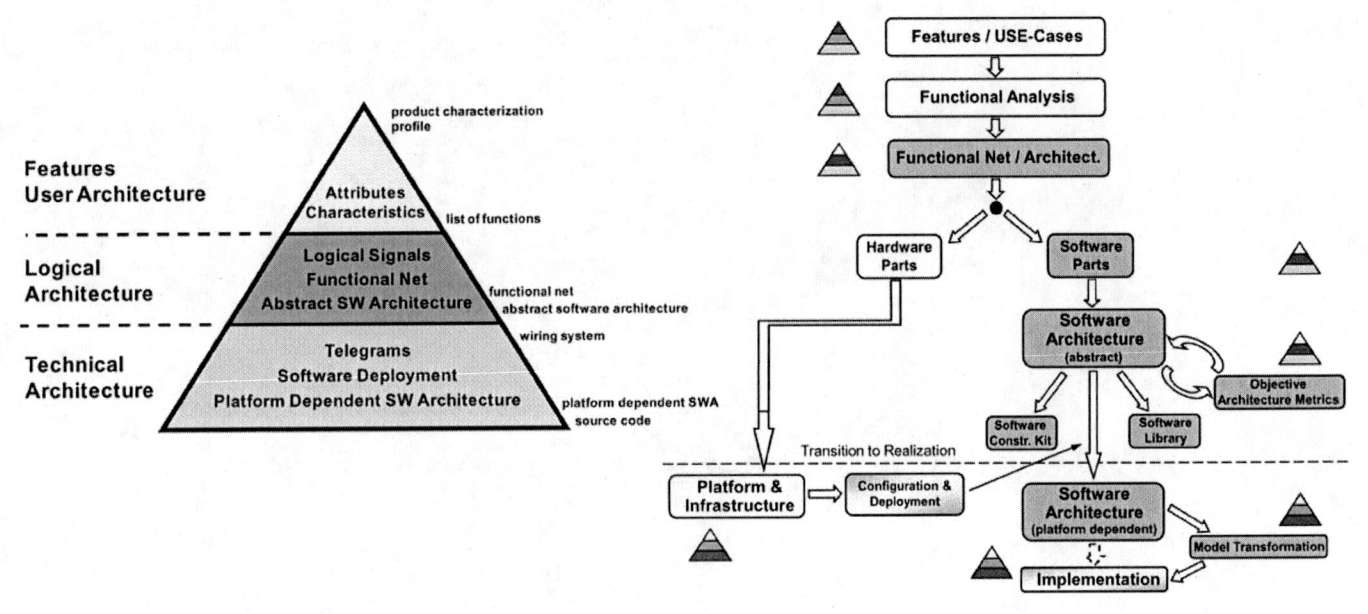

Figure 1. left: Views on the System 'Vehicle'; right: Adapted Development Process

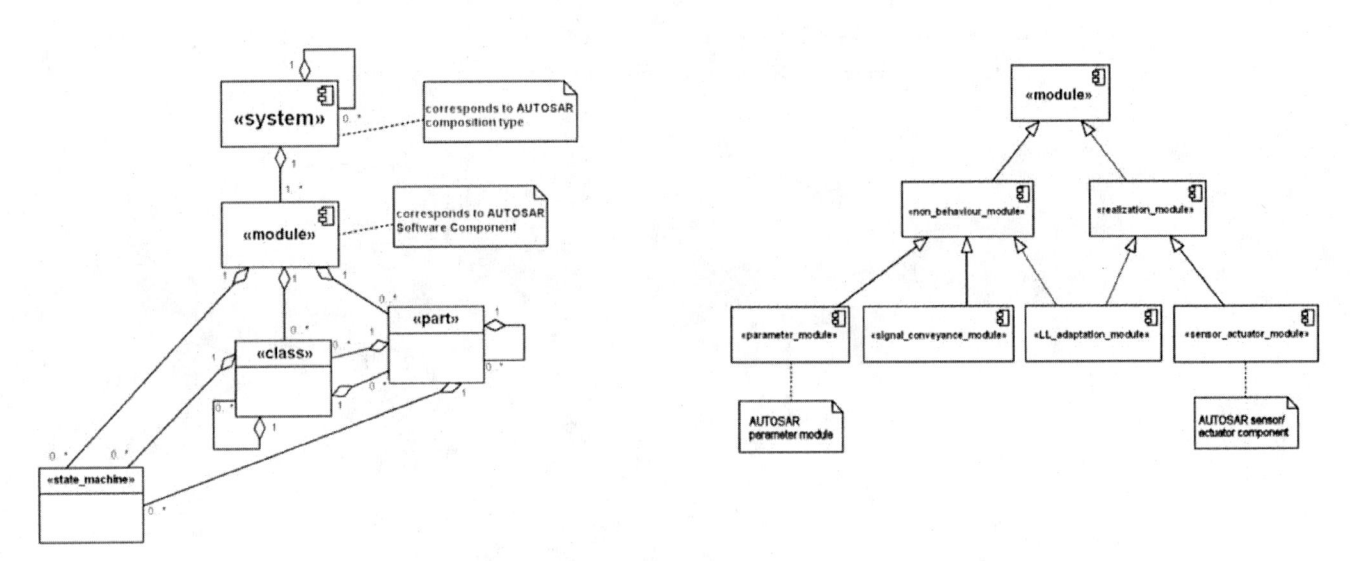

Figure 5. Structural Elements of the ABSOFA meta model

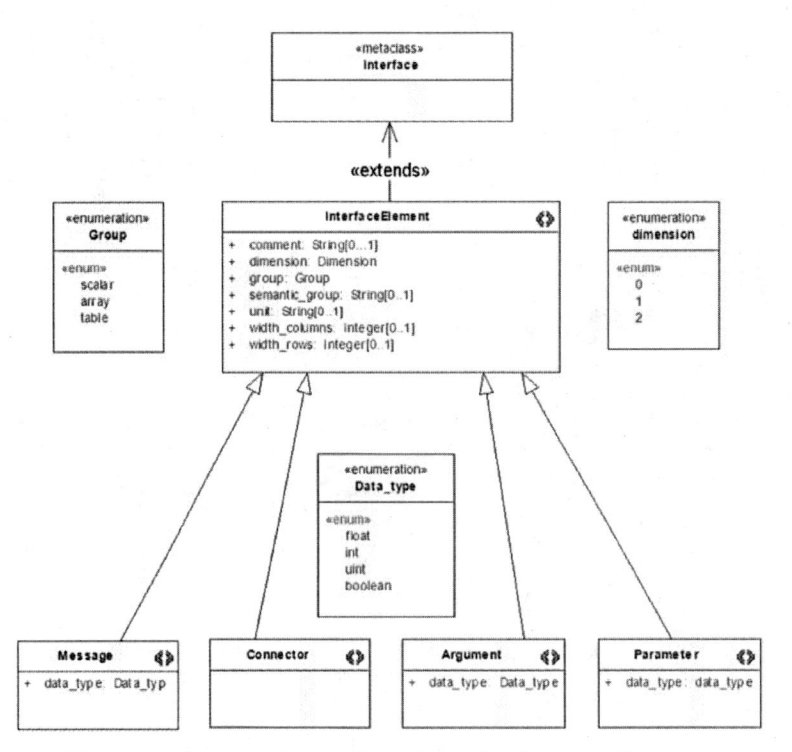

Figure 6. Extension Mechanism shown for Interface Elements

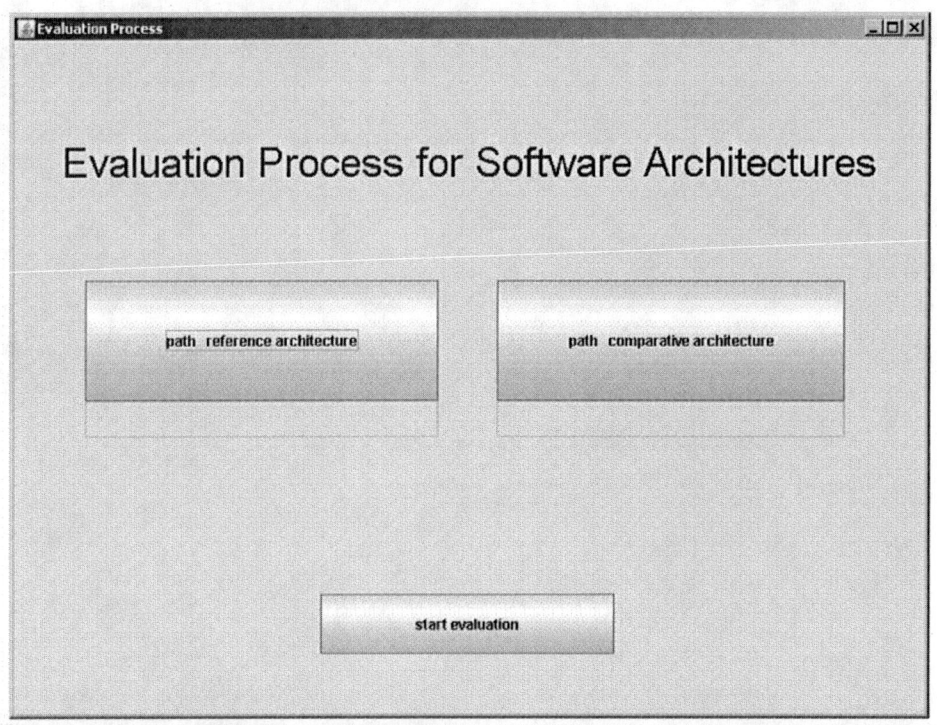

APPENDIX A
(SOFTWARE ARCHITECTURE EVALUATION)

Figure A1. GUI and Practical Example for Objective Software Architecture Evaluation

COMMUNICATIONS

2009-01-1479

Integration of Car-to-Car Communication into IAV

Tae-Kyung Moon, Jun-Nam Oh, Hyuck-Min Na and Pal-Joo Yoon
MANDO Corporation

Copyright © 2009 SAE International

ABSTRACT

In this paper, we present the integration of C2C communication into IAV. Traditional IAV detects target vehicles only they are in visible area; however this integration makes IAV to sense target vehicles even they are blocked by obstacles. In this system, C2C ECU keeps monitoring the target vehicles on the road and sends a warning to IAV controller when it detects any event or risky situation. Finally IAV avoids the collision with the target vehicle by reducing the moving speed or generating a new path.

INTRODUCTION

IAV (Intelligent Autonomous Vehicle) is an intelligent vehicle, which automatically drives in a given track according to driving rules. This system monitors around the vehicle by using various sensors. Normally radar, laser, and camera are used. When the obstacles are detected, it generates a new path and applies control to avoid the situation like a human driver does. However on-vehicle sensors can only detect limited objects in the visible area. Vehicles should cooperate with other vehicles to reduce accidents, since road safety cannot be achieved by a single vehicle. So, we need a communication based safety system in the future. [2,3,4,5]

In our previous study, we built a simple C2C (Car-to-Car) communication system for two vehicles to exchange their driving data wirelessly. [1] We also defined five issues to apply C2C communication for safety purpose: Localization, Communication, Tracking, Perception, and Decision. In this study, we will show how we integrated both systems, and how they worked. Traditional IAV detects target vehicles when they are in visible area [6], however this integration makes IAV to sense targets even they are blocked by obstacles. C2C function gives early warning to handle more complicated situations (e.g. moving vehicle from blind spot, traffic signal) by increasing an information domain.

SYSTEM CONCEPT

Our objective was to make IAV to be driven safely and dynamically within many vehicles. And this could be achieved by integrating C2C communication. Figure 1 shows a road traffic example and a block diagram of each vehicle. There are two kinds of vehicles on the road. Red car is the IAV and blue cars are normal vehicles. Every vehicle has C2C ECU to communicate with each other. Also there are roadside devices. These devices collect the environmental data (e.g. traffic signal, traffic jam, emergency signal) and relay wireless data packets from nearby vehicles to faraway.

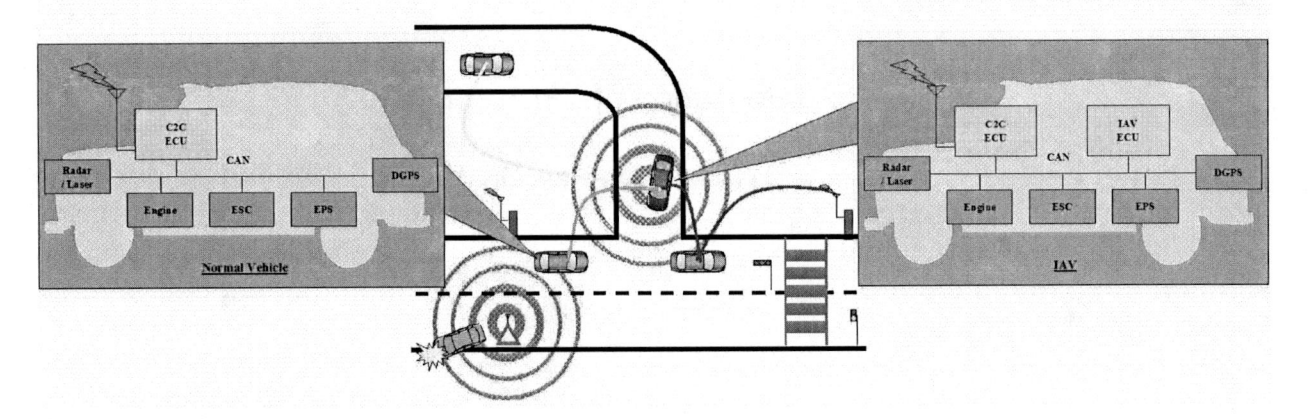

Figure 1. Road traffic example and block diagram of IAV, Normal vehicle

IAV consists of several electrical control units. IAV ECU is the main controller, which serves novel tasks for autonomous driving: map generating, path planning and motion controlling. Radar and laser sensors are used for detecting obstacles around the vehicle. However its direction and range is limited, redundant sensors are needed. Additional vision sensor can be applied to detect obstacles in visual. Engine ECU, ESC, and EPS serves low-level vehicle control according to the motion control command: acceleration, deceleration, and steering angle. These ECUs also collect the driving information: vehicle speed, applied steering angle, yaw rate, longitude/latitude G force, moving distance, and so on. DGPS measures the absolute location of the vehicle. C2C ECU is the communication interface to talk with other vehicles or to the road equipments. This ECU measures the driving data of its own vehicle and periodically broadcast its information to nearby vehicles. This ECU also keeps listening to other vehicles' voice and checks the relationship and the risk level of an accident. Then C2C ECU notifies the situation to IAV ECU and the driver to avoid the accident. Also this ECU periodically broadcast its own driving information to nearby vehicles. The normal vehicles have the same C2C ECU except the IAV ECU, which drives the vehicle automatically. Human drivers operate these vehicles manually. C2C ECU operates same in IAV, so that every vehicle can perceive other vehicles in this area.

In this environment, IAV can make the proper decision and the operation according to the road traffic. When the road situation is as Figure 1, IAV senses that the traffic signal is turned to red. And one vehicle is stopped in front of the signal before it enters into the highway. IAV can also detect another vehicle which is approaching to the intersection. (This IAV slows down the other vehicle's speed and notifies to the following vehicle to avoid the sudden stop.)

C2C ECU CONTROL ARCHITECTURE

Our C2C control logic consists of 5 elements; Localization, Communication, Tracking, Perception, Decision. Figure 2 shows the control architecture of our C2C ECU.

Localization serves measuring vehicle's position correctly to define interrelation between the vehicles on the road. Higher precision, which can detect every single lane, is needed. DGPS, dead reckoning, map matching to the e-Map can be applied for this purpose. Dead reckoning can improve the response rate and map matching can reduce the heading error.

Communication serves establishing a network link (called VANET; Vehicular Ad-Hoc Network). There are protocol handler and network manager for this module to share driving information wirelessly with other vehicles within certain area. Since every end-device in this network has a high mobility, link cost changes at anytime. New study for physical media, MAC, network topology, and routing method is needed.

Tracking serves monitoring and managing information of nearby vehicles continuously. Normally, communication based control system has a high latency, and the performance is affected by a communication period. Furthermore, packets can be lost by collision or signal loss if the wireless media is used. Therefore in tracking, we need to estimate the movement of every vehicle from the last receiving data to keep the system working even though we lost the network link for a while. Vehicle dynamics model and driver behavior model is needed for estimation.

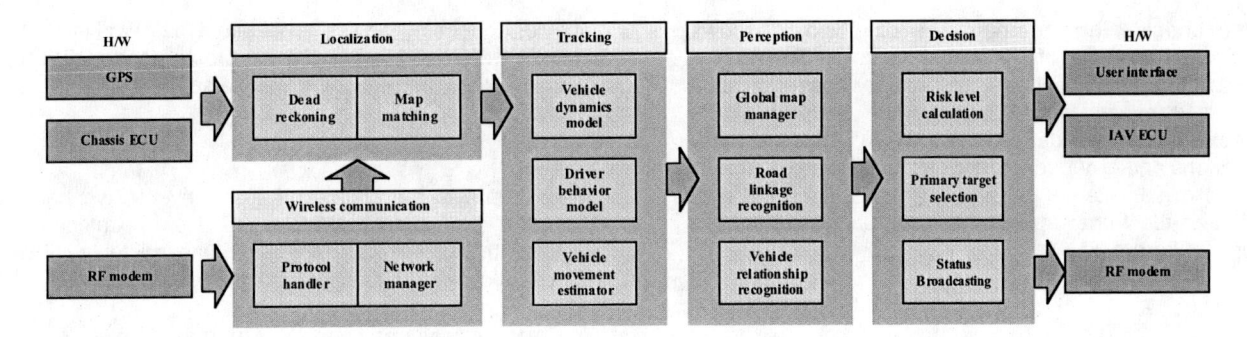

Figure 2. C2C ECU logic diagram

Perception serves defining the relationship between my own vehicle and other vehicles and generates the global map. This map contains not only the physical route information but also the dynamic vehicle movement and relationship. Road linkage recognition module finds the road topology using nodes and link information on e-Map,

and distinguishes which road is related to current driving road. Vehicle relationship recognition module picks up the valid vehicles, which is highly related to itself from many candidates on the nearby road. The types of the relationships are depending on the road linkage and vehicle' moving direction. The map manager coordinates whole information into the global map.

Decision serves calculating a risk level of an accident with nearby vehicles, and making a final decision for an action. Several factors (i.e. position, heading angle, speed, distance, TTC (Time To Collision) of two vehicles, TTC to node) are selected based on the relationship to define a risk level. Then C2C ECU chooses a primary target and notify to driver or IAV ECU to avoid a risky situation. Sometimes emergency reactions can be applied in IAV ECU, and C2C ECU broadcasts them immediately.

This system is communication based control system; therefore C2C ECU should not directly control the actuator. It is recommended to use a superior ECU with on-vehicle sensor to reactive control faster then several hundred milliseconds. And C2C ECU has slower then several hundred milliseconds and up to several seconds for deliberative control.

IMPLEMENTATION AND EXPERIMENT

SYSTEM IMPLEMENTATION

We prepared one IAV and two normal vehicles for target, and installed C2C ECU to every vehicle.

Our C2C ECU is consisted of CPU, GPS, 2.4 GHz wireless transceiver, and is interfaced to the chassis through CAN (Figure 3). This ECU collects and broadcasts its driving data (GPS position, vehicle speed, heading angle, and status flags), and listens driving data from other vehicles, and notifies the warning situation to IAV ECU.

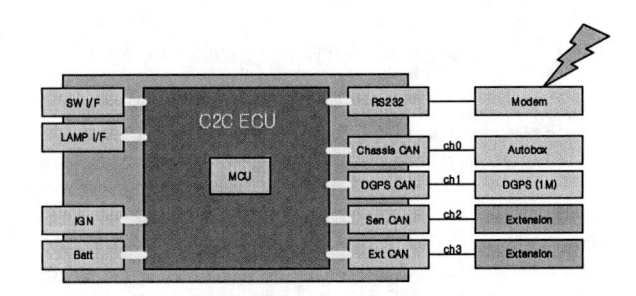

Figure 3. C2C ECU HW block diagram

We used DGPS and map matching for localization. One link, which exists in the nearest distance and the smallest heading error from vehicle location, was chosen to a current road from e-Map. Considering heading angle helps to identify the current road when we pass the crossroad. Between the every GPS measurement, dead reckoning is used to compensate the vehicle position and to increase the system response. While C2C communication is occurred in the limited range of area (200~1000m), we used IEEE 802.15.4 wireless modem [7] with additional PA and LNA. Data rate was set to 38,400 bps, and 60mW (18dBm) of the power covered up to 800m areas. Data packet, which consists of 14 bytes of payload, is transmitted every 200ms, and contained GPS data (Longitude, Latitude), Vehicle speed, Heading angle, and Status flag, Simple estimator, which calculates the linear vehicle trajectory via vehicle speed and heading angle from the last received packet, is used for tracking. Packet decay time is applied to check the target whether it is alive or not. We eliminate the target vehicles from the list when the signal is lost for a while.

Perception served checking the road linkage and identifying the relationship between itself and nearby vehicle Figure 4 shows the road connection and the vehicle arrangement. The dot and black line means the node and link on the map. The blue vehicle shows available locations of the target vehicles. During the experiment, we supposed an ordinary road condition with and every vehicle, which keep their lane and direction. Then we defined (1), (5), (8) cases are risky situations and grouped into 2 classes. ((1), (5) → Class I. IAV follows the target on the same road. (8) → Class II. IAV and the target approaches to a same intersection from a different way.)

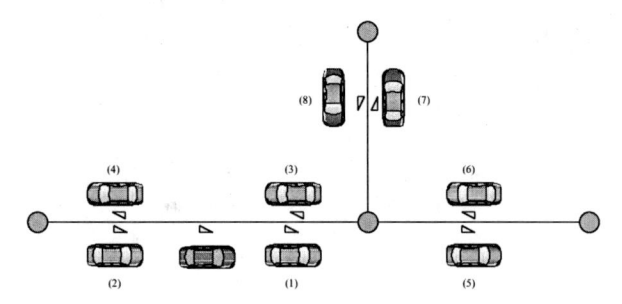

Figure 4. Road connection and vehicle arrangement

Decision logic calculates the risk level of every vehicle and figures out a primary target vehicle. Different parameters were used to define a risk level depending on the relation. When the relation is class I, TTC between two vehicles were used. In class II, a difference of two TTC from each vehicle to the intersection was used. We defined 16 levels of risk, and it is linearly increased according to the risk. After every risk level has calculated, we figures out the most risky target and notified to the driver and IAV ECU.

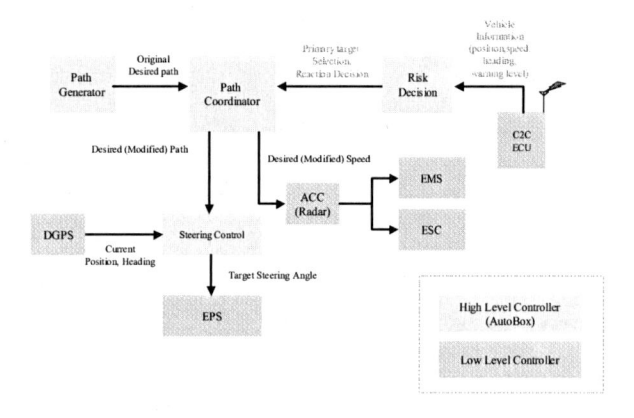

Figure 5. IAV ECU logic diagram

IAV that we used in the experiment was a simple path tracer with forward collision avoidance. We used Micro AutoBox for IAV ECU. Figure 5 shows the block diagram of IAV controller. With controlling the Engine, EPS, and ESC, this ECU drives vehicle autonomously to follow the adequate trajectory. Radar sensor is used to detect and avoid linear forward obstacles. When the risk level of 7 or higher notified from C2C ECU, IAV chose avoidance behavior depending on the target situation. Two kinds of behavior were used to prevent the crash; generating an alternative path, and reducing vehicle speed.

EXPERIMENT

Figure 6 shows the track and traffic events we used during the experiment. IAV followed a given trajectory, and the traffic events happened unexpectedly. We supposed following situations for traffic events.

1. Emergency stopped vehicle is in front of IAV.
2. Sudden stopped vehicle in front of curved way.
3. Vehicles are approaching to same intersection.

C2C ECU monitored whether emergency lamp is turned on, and this is used to identify the event 1) and event 2). When the IAV detected the event 1) and risk level was high, IAV generated an alternative path which avoids the obstacle, and followed the new path. When the event was 2), 3), IAV slow downed its moving speed. The deceleration rate was increased until the vehicle stops when the risk level went.

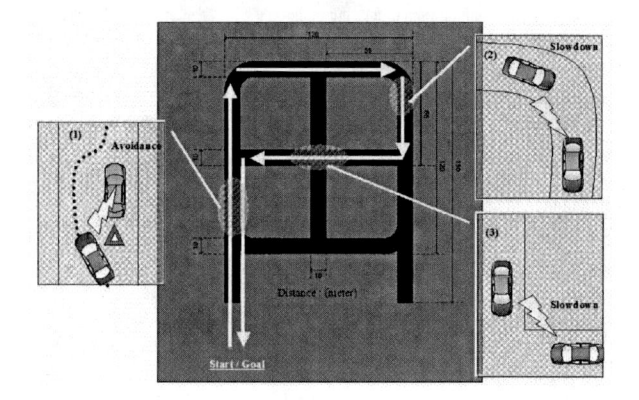

Figure 6. Test track and traffic events

We prepared one IAV and two target vehicles. Figure 7 shows ECU and monitoring SW that is installed into vehicles. One target vehicle played the event 1), 3) and the other one played the event 2). Figure 8 shows the driving log. Green quadrilateral indicates the IAV, and blue quadrilateral indicates the target vehicle. This quadrilateral is called 'virtual drive area', which deduced from vehicle speed and heading direction, and means the probabilistic area of a vehicle in a certain future. IAV avoided every risky situation and traveled all given trajectory successfully.

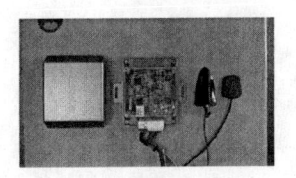

Figure 7. System installation

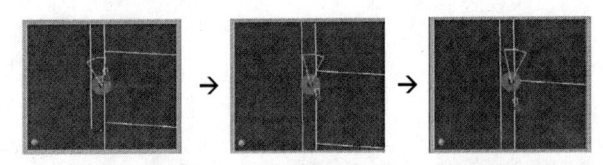

(< C2C ECU >,  < GPS & Modem antenna >, 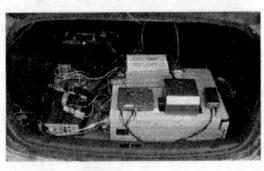 < IAV system >, 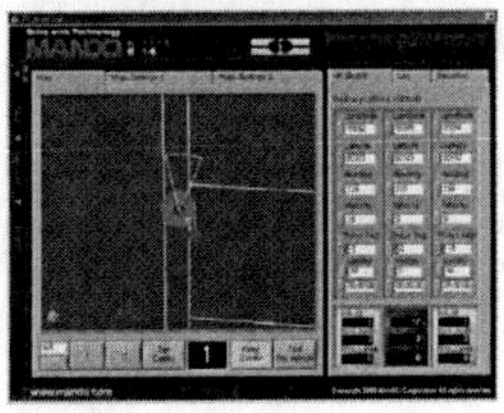 < Monitoring SW >)

Event 1. Avoiding emergency stopped vehicle

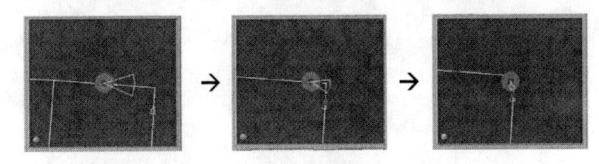

Event 2. Stopped vehicle in front of curve

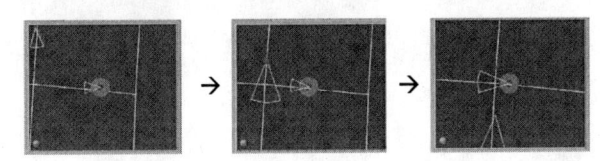

Event 3. Intersection collision avoidance

Figure 8. Driving log

CONCLUSION

In this study, we integrate C2C communication into IAV. Our objective was to make IAV to be driven safely within many vehicles, however traditional IAV detects vehicles only when they are in visible area. And this was achieved by installing the wireless vehicular network. C2C ECU was installed in every vehicle, and it served Localization, Wireless communication, Tracking, Perception, and Decision-making. This ECU figured out the most risky vehicle around itself, and notified to the driver and IAV controller. Our system could detect the target vehicles through the C2C communication; even if they are located where the on-vehicle sensors can't detect. Then the vehicle avoided severe accidental situation and traveled all given trajectory successfully.

REFERENCES

1. Tae-Kyung Moon, Seong-Hee Jeong, Jun-Nam Oh, "A Study of Vehicle Safety System - An Integration of Wireless Communication and Chassis Control", The 14th Asia Pacific Automotive Engineering Conference, August 2007
2. Honda ASV-3, http://world.honda.com/ASV/
3. Matthias Schulze, Gerhard Nöcker, Konrad Böhm: PReVENT: A European program to improve active safety, http://www.prevent-ip.org
4. Car2Car Communication Consortium. http://www.car-2-car.org/
5. "All about the intelligent car", Nikkei Electronics July 2006
6. DARPA, http://www.darpa.mil/GRANDCHALLENGE/index.asp
7. IEEE 802.15, Working Group for Wireless Personal Area Networks (WPANs) http://www.ieee802.org/15/

CONTACT

Tae-Kyung Moon (hinano@mando.com)

Advanced Electronics System Team MANDO Corporation Central R&D Center Giheung-Eub, Youngin-Si, Kyonggi-Do, Korea

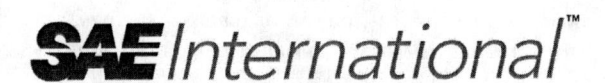

Enabling Safety and Mobility through Connectivity

2010-01-2318
Published
10/19/2010

Chris Domin
Ricardo Inc.

Copyright © 2010 SAE International

ABSTRACT

Vehicle-to-Vehicle (V2V) and Vehicle-to-Infrastructure (V2I) networks within the Intelligent Transportation System (ITS) lead to safety and mobility improvements in vehicle road traffic. This paper presents case studies that support the realization of the ITS architecture as an evolutionary process, beginning with driver information systems for enhancing feedback to the users, semi-autonomous control systems for improved vehicle system management, and fully autonomous control for improving vehicle cooperation and management. The paper will also demonstrate how the automotive, telecom, and data and service providers are working together to develop new ITS technologies.

INTRODUCTION

A primary goal of ITS is to provide substantial benefits in real world fuel economy, road congestion, and general road safety. ITS has its roots in leveraging leading edge technologies, beginning with driver-focused applications, building towards semi-automatic operations, and ultimately arriving at autonomous operations.

FROM SIMPLE FEEDBACK TO AUTONOMOUS SYSTEMS

A number of passive information systems are available to drivers today. Nowadays it is common for basic driver information systems to provide some kind of vehicle status relative to the environment. For instance, a basic collision warning system can alert the driver if there is an impending rear-end collision or if it detects a pedestrian obstacle. Similarly, a travel information system can alert the driver for upcoming road obstructions. By combining information from the immediate environment with longer-range environment, higher degrees of fuel economy and safety are achieved.

As an example of fuel-economy and safety improvements, Ricardo UK Limited with its academic, business, and ITS committee partners, have developed in-vehicle applications to provide feedback to the driver about fuel-economy and safety conditions. Currently available technologies, such as GPS, cell phone, back-office systems, are used. Two recent programs of note are

• Foot-LITE - A smart electronic co-pilot provides fuel economy information and impending economy-changing situations to the user. A small portable display unit is connected to ITS infrastructure and provides the vehicle driver real-time feedback about actual driving behaviour vs. ideal fuel-efficient behaviour. A web-based service provides historical trends for the driver and allows information sharing with other users.

• Co-Driver - An electronic hazard warning system provides situational awareness of road safety conditions and impending hazards. An in-vehicle unit processes hazard information such as steep grade, sharp curves, obstructions, etc., and provides advanced warning of the potential hazard. Co-Driver also indicates the degree of urgency the hazard presents to the user, for instance, a fallen tree across road versus routine road construction markers. Vehicle passengers can easily enter information back to the system in order to alert other drivers of transient hazards, such as obstacles in the road or accidents.

Applications such as these help the driver make decisions about driving habits and navigation. Reduced fuel costs, reduced CO_2 emissions, and safer driving are direct benefits of these technologies. In addition, even further advances are possible by utilizing semi-autonomous and autonomous technology.

CASE STUDY #1: SEMI-AUTONOMOUS "SENTIENCE"

Fuel efficiency can be improved by integrating topographical and geophysical data with automatic vehicle control subsystems. *Sentience* is a recently completed 2 ½ year collaborative R&D program that was co-funded by innovITS[1]. It was jointly developed with six European partners: Ricardo, innovITS, Jaguar/Land Rover/Ford, Ordnance Survey, Orange, and TRL. The overall achievement of this program is the identification and development of a system to improve the fuel efficiency of vehicles using "electronic horizon" data collected with V2I communications. *Sentience* performs intelligent speed adaptation based on situational awareness.

Sentience is built using a web-based server and mobile client application. The server environment includes the telecommunications infrastructure, GPS satellites, weather data, ITS traffic data, historical traffic trend data, and the *Sentience* application web-server. The server translates data from the environment, categorizes them, and communicates them to the client using V2I.

The *Sentience* client application resides in a smart-phone mobile device that is part of the *Sentience* on-board system. The *Sentience* on-board system includes the GPS receiver, the mobile device (cell phone), and real-time supervisory controller unit (SCU). The vehicle interface software and supervisory control algorithms execute on the SCU. The SCU software communicates directly with the acceleration/braking subsystem electronics and over dedicated Ethernet with the client software on the mobile. The SCU software is responsible for optimization of regenerative-braking, air-condition boosting, and EV mode operations.

The team selected a Ford Hybrid Escape as the target vehicle for the prototype system. A hybrid vehicle presents several opportunities not available on conventional vehicles. The Ford Escape is a full hybrid vehicle and can operate in several modes: full electric, conventional combustion, and mixed (parallel) mode. It also utilizes regenerative braking.

SENTIENCE REQUIREMENTS AND SIMULATION

Phase one of the project focused on simulation and requirements specification. Ordnance Survey, Orange, and TRL focused on defining the vehicle routes and supporting data. Ricardo focused on simulation, control strategies, and prototype architecture. The team created and validated a vehicle model to assess baseline vehicle performance.

The primary opportunities found for energy savings are regenerative braking, EV usage during acceleration, and air-conditioning usage. Through measurement and analysis, the team found when it was best to run the electrical motor and when it was best to charge the battery based on road conditions and vehicle characteristics. This analysis allowed selection of optimum tradeoff points between electrical drive and conventional drive for vehicle speed and wheel torque.

U.S. EPA cycles were used to validate the model. Subsequent simulations included different route profiles and varying drive conditions, such as level or hilly routes, constant or varying speeds, with/without air-conditioning, head/tail wind, etc.

- Flat 12.4 mi route, 60mph

- 12.4 mi route (0.6mi flat, 3.7 mi uphill, 7.5 mi downhill, 0.6 mi flat), 60mph

- Flat 12.4 mi route, 30mph, air-conditioning turned on

- Flat 12.4 mi route, variable speed (multiple discrete target speeds) with average of 30mph

- 12.4 mi route (0.6 mi flat, 11.2 mi of repeated alternate 0.6 mi uphill, 0.6 mi downhill gradients, 0.6 mi flat to end), 30mph

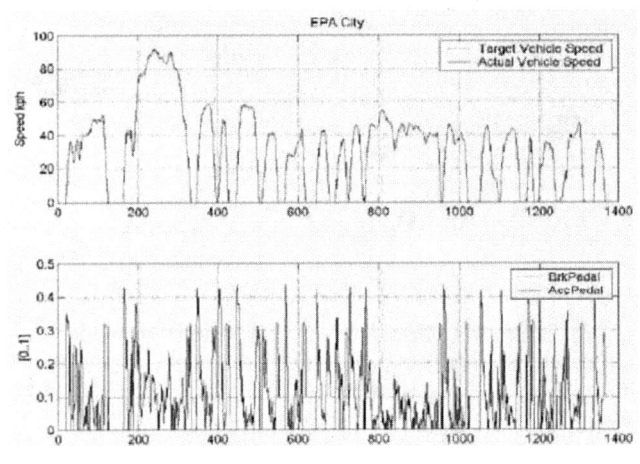

Figure 1. EPA Test Cycle

The on-board *Sentience* architecture incorporates V2I communications to access electronic horizon data, such as topographical, geographical, and traffic data, from the *Sentience* web server. The team performed a sensitivity analysis of the look-ahead algorithm to characterize how deep the queue of traffic/map data must be in order to maximize efficiency of the algorithms and to account for temporary interruptions of service. As a result, the on-board *Sentience*

[1,2] innovITS is the UK Centre of Excellence for sustainable mobility and intelligent transport systems. See http://www.innovITS.co.uk Simulink® is a registered trademark of The MathWorks™.

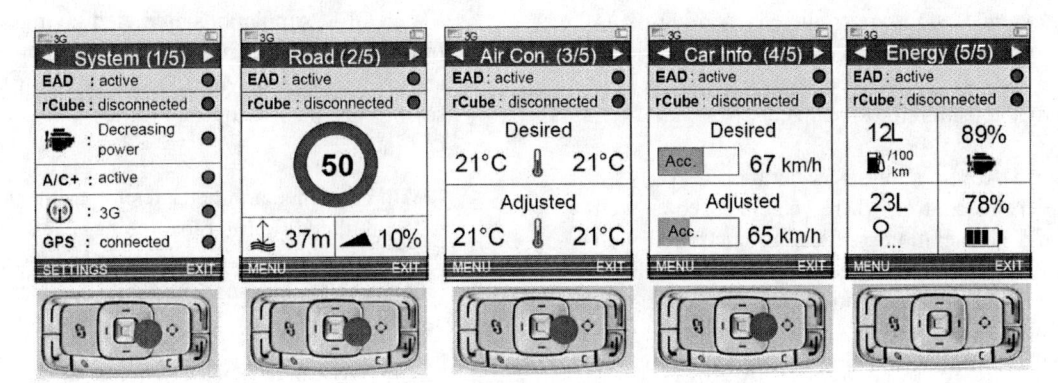

Figure 2. Sentience HMI sample

system views an electronic horizon of up to 3 miles to calculate optimum acceleration/regeneration potential.

A key requirement of the on-board system was a safe implementation with minimal cost. For this reason, a Ricardo rapid prototyping system was selected for the SCU. The SCU intercepts and overrides controls for cruise control and air conditioning. It communicates with the powertrain controller for hybrid, engine, and safety functions. A safety cut-off function for the enhanced acceleration and deceleration was identified as being required during the preliminary safety analysis.

The team, with input from Ford, assessed vehicle systems for suitability, and concluded that a small amount of additional hardware was required to ensure the vehicle system did not raise faults against the cruise control or air conditioning switchgear.

SENTIENCE DEVELOPMENT AND ASSESSMENT

Phase two of the project focused on development, integration, and assessment.

A Nokia N95 cellular phone served as the mobile communications device and human-machine interface (HMI) for the on-board system. For convenience, an external GPS was connected to the phone to provide location information. Ricardo and Orange defined and implemented a telecommunications protocol for communication between the phone and the SCU. Ordnance Survey data provided the historical traffic data. For future use, the *Sentience* architecture supports the of real-time traffic data from ITS infrastructure sources. *Sentience* focuses on three main areas of system operation to optimize energy storage and transfers: engine loading, air-conditioning, and acceleration/braking. See discussions on OEL, EAC, and EAD below.

The *Sentience* HMI on the mobile device displays road information as well as *Sentience* status information, e.g,

• Road speed limit, height and gradient

• Enhanced Air-Conditioning level desired and adjusted temperature set-point

• Enhanced Acceleration/ Deceleration level desired and adjusted vehicle speed

• Energy status information such as battery state of charge

Sentience detects when the vehicle is approaching significant changes in driving conditions due to traffic or geography, and displays pop-ups on the HMI. *A* configuration screen allows the user to select the desired features. The user can selectively enable and disable both pop-ups and audio messages by feature.

Sentience subsystem components are discretely installed in the vehicle under the passenger seat and in the dash. *Sentience* components include a custom harness, a modified A/C control unit, a custom cruise-control unit, the SCU, a wireless router & GPS receiver, and a CAN data logger device.

Sentience Optimized Engine Loading (OEL) executes on the SCU and optimizes the efficiency of the hybrid powertrain through intelligent management of electric, mixed and combustion modes of operation. The OEL algorithm communicates to the powertrain controller via the CAN network and provides supervisory control. Advanced knowledge of opportunities to recharge the battery system allows more flexibility in EV use, e.g., the battery state of charge limits are adjusted with the vehicle operation utilizing these limits. Ricardo developed supervisory control strategies in Simulink® for execution on the *Sentience* SCU. With the current OEL strategies, a 4-9% improvement in fuel consumption is realized.

Sentience Enhanced Air Conditioning (EAC) executes on the SCU and optimizes the A/C operation in order to reduce the CO_2 emissions from the vehicle. Because the combustion engine drives the A/C compressor directly, a specialized strategy was developed to keep the passengers comfortable during extended vehicle stops. EAC overrides the A/C switchgear signals and "pre-cools" the interior 1 or 2 degrees cooler whenever extended stops are predicted. By minimizing the amount of occurrences when the engine runs exclusively for cooling the vehicle interior, a result of a 2-10% improvement in fuel consumption is realized.

Sentience Enhanced Acceleration / Deceleration (EAD) is a form of adaptive cruise control, where vehicle speed as well as acceleration and deceleration profiles are controlled with the knowledge of future traffic and geography features. EAD augments the existing cruise control strategy. *Sentience* automatically controls the speed at a more optimal rate than might be expected through normal driving, allowing potentially significant savings in fuel. Speed set points are a combination of fixed-feature speed limits and probabilistic-feature speed limits. Fixed-features include actual speed limits, bends, roundabouts, speed bumps and stop signs. Probabilistic features include traffic lights, junctions, traffic conditions and pedestrian crossings. EAD slows or accelerates the vehicle at an optimum rate to match legal or safe speeds. The driver can manually override EAD at any time for safety or convenience. Depending on traffic conditions, EAD may have an impact on journey time; the driver therefore could make an informed decision as to whether the trade-off with increased comfort and fuel efficiency is acceptable on that occasion.

Figure 3. OEL Hilly Terrain Optimization

SENTIENCE VALIDATION AND CONCLUSIONS

1. Three new control systems were added to those on a production hybrid vehicle. OEL for enhanced hybrid system efficiency is useful under any mode of operation. EAC can be used whenever air conditioning is turned on. EAD can be used whenever cruise control is active.

2. Track and road testing results indicated significant savings. Improvements in fuel consumption on the order of 5%-10% can be obtained for a low implementation cost.

 ○ With OEL, simulations predicted savings of 4-9%; Initial track test measurements show savings of approximately 2%, with speculation of higher percentage savings on specific routes. Further analysis and testing continues.

 ○ With EAC, dynamometer-testing using an NEDC cycle with simulated sunlight loading has been performed. Over 9% improvement in fuel consumption was seen on an NEDC cycle under moderate mild summer weather conditions.

 ○ With EAD, initial measurements show an average saving ranging between 5% and 24%, with an average of 12% during track testing. Scaling this data to average vehicle usage on real roads gives a total estimated fuel saving of nearly 14%. Initial real world road tests have already shown a fuel savings of over 5%.

3. Implementation costs for *Sentience* using a production system is not restrictive. Typical 3G mobile phones come with GPS capabilities and can be acquired for low cost. Memory and CPU requirements for OEL, EAC, and EAD functions do not prevent those features from being co-resident with software in production ECUs. Suitable sources of traffic data are required, but these traffic data can be easily supported by future ITS infrastructure.

CASE STUDY #2: AUTONOMOUS "SARTRE"

Both fuel efficiency and safety are improved by integrating V2V communications and automatic vehicle control subsystems. *SARTRE* is a Ricardo-led program that shares situational awareness data between vehicles using V2I and V2V communications, thus enabling autonomous vehicle coordination and the creation of "road trains".

The *SARTRE* project began in September 2009 and is scheduled to complete in August 2012. It is being jointly developed with seven European partners in the UK, Sweden, Spain, and Germany: Ricardo, IDIADA Automotive Technology, Institute for Automotive Engineering (ika) of RWTH Aachen University, SP Sveriges Tekniska Forskningsinstitut, TECNALIA Robotiker, Volvo Car Company, and Volvo Technology. The concept behind *SARTRE* is that vehicle platoons improve fuel consumption, increase safety, and reduce congestion on freeways.

Since human driver errors contribute to well over 85% of road fatalities, it is expected that safety will improve dramatically by using autonomous control to remove distractions and errors in judgment. Because autonomous systems can process data much more quickly than a human can, congestion can be reduced automatically by optimizing gaps between vehicles, minimizing traffic dynamics and delaying traffic collapse. Fuel economy can be improved by

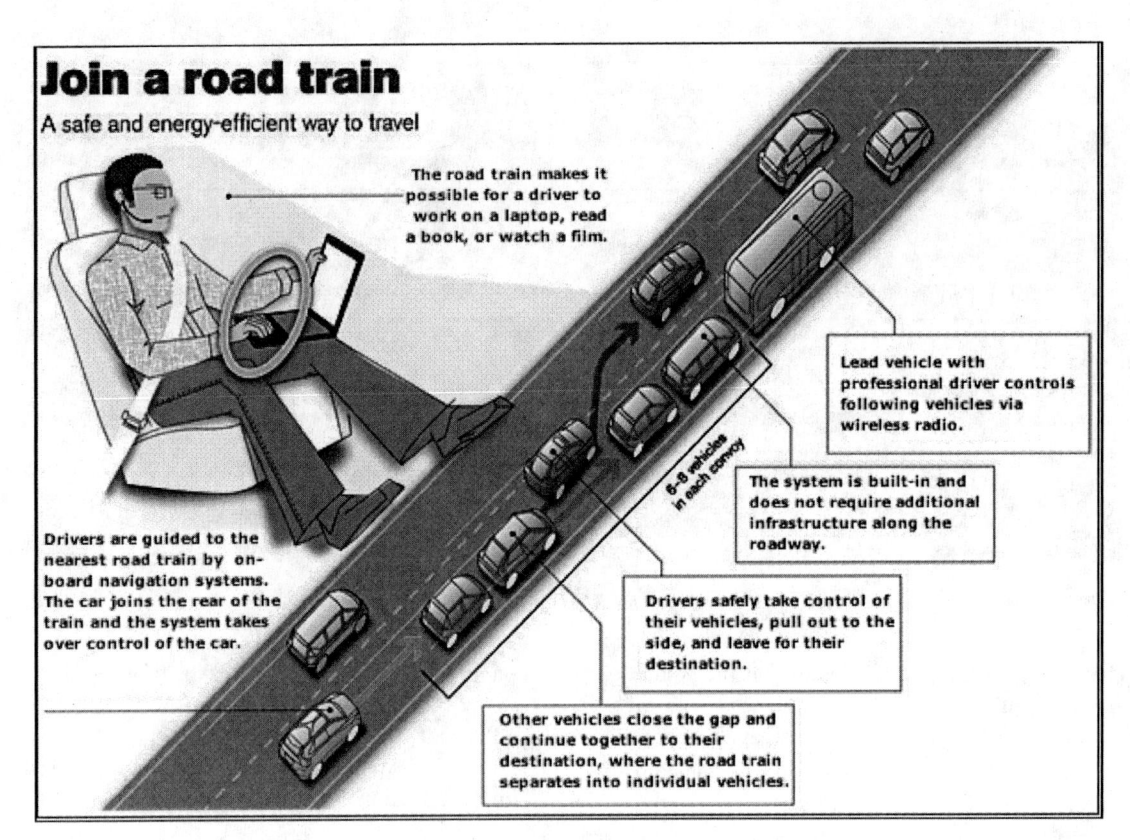

Figure 4. The SARTRE concept

reducing aerodynamic drag, due to drafting in each vehicle's slipstream.

Each road train will consist of up to eight vehicles. Each road train or platoon has a lead vehicle that drives exactly as normal, with human control over the various functions. This lead vehicle is controlled by an experienced driver who is familiar with the route. For instance, the lead may be taken by a taxi, a bus or a truck. A driver approaching the convoy requests entry into the convoy using a human-machine interface. The convoy accepts the vehicle and the vehicle automatically enters the convoy, after which it is completely under autonomous control. A driver approaching his destination leaves the convoy by exiting off to the side and then continues on his own to his destination under his own control. The other vehicles in the road train automatically close the gap and continue on their way until the convoy splits up.

The advantage of such road trains is that all the other drivers in the convoy have time to perform other business while on the road, e.g., talking on the phone, eating, working on a computer, etc. The road trains increase safety and reduce environmental impact thanks to lower fuel consumption compared with cars being driven individually. The reason is that the cars in the train are close to each other, exploiting the resultant lower air drag. Simulation results show the energy saving to be in the region of 20 %: Road capacity is utilized more efficiently by minimizing distance between vehicles.

Researchers see road trains primarily as a major benefit to commuters who cover long distances by motorway every day, but they will also be of potential benefit to trucks, buses, coaches, vans and other commercial vehicle types. As the participants meet, each vehicle's navigation system is used to join the convoy, after which the autonomous driving program then takes over. As the road train approaches its final destination, the various participants can each disconnect from the convoy and continue to drive as usual to their individual destinations.

SARTRE REQUIREMENTS AND SIMULATION

Phase one of the project considered scenarios and constraints during interaction with other road users. A use-case analysis was performed with an emphasis on the human factors. Modeling of the use cases focused on creating a combination of vehicle and traffic specific models, taking into consideration all interchanges occurring between driver, vehicle and other traffic.

An important constraint for *SARTRE* is that the architecture and implementation has to be feasible and use available production components and subsystems. So the team performed additional analysis to understand business

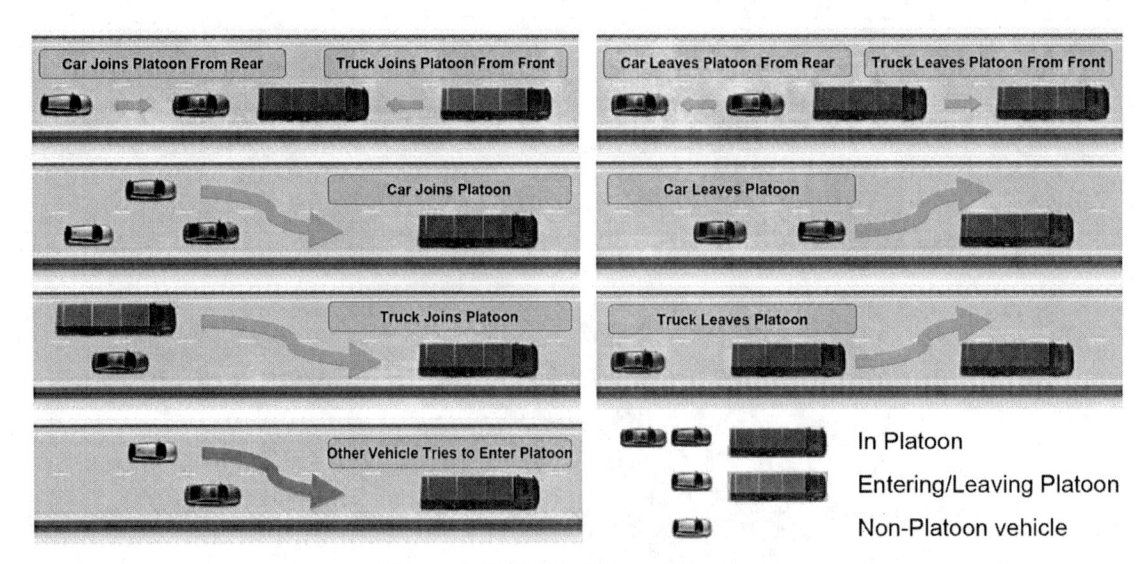

Figure 5. SARTRE Platoon Use-Cases

requirements, usability, risk, and safety, as well as the system itself. As the concept solutions were balanced against available technology, they were rationalized against draft ISO/DIS 26262 using InnovITS[2] Framework Architecture and Classification for ITS (FACITS) process.

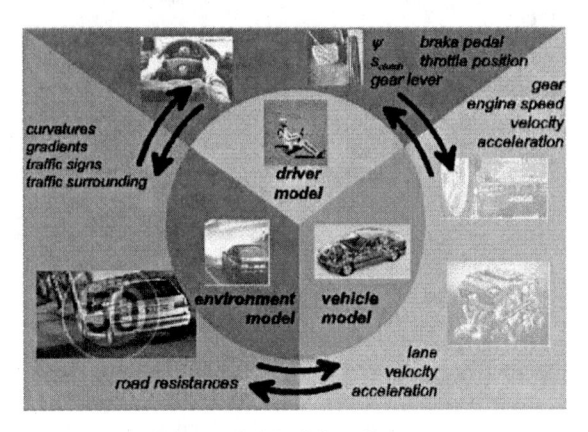

Figure 6. Modeling Process

Use-cases (see Figure 5) needed to take into account a significant number of factors, including, performance/failure of vehicles, braking/acceleration/turning procedures, other vehicles, platoon size, and gap length, and human behaviors, among others. Example use cases are:

• authorized car/truck enters platoon from rear or joins middle of platoon

• unauthorized other car/truck enters platoon or leaves platoon from middle

• authorized car/truck leaves from rear or middle

• authorized leader joins or leaves from front

After all the primary modeling, analysis, and concept generation were complete, the team focused on concept implementation.

SARTRE CONCEPT IMPLEMENTATION

Phase two of the project involves concept selection and implementation. Since intellectual property is being developed by partners to support the implementation of *SARTRE*, only a general discussion of the architecture is given here. Each vehicle is equipped with a dedicated short-range communications (DSRC) radio, an active safety control module, short-range radar, vision systems, active cruise control system, actuators, and supervisory control unit (SCU). DSRC is used to communicate platoon information among all vehicles in the platoon. Once in the platoon, V2V communications, V2I communications, and other active subsystems in each vehicle support autonomous behavior.

To date, the project partners have reached agreements on the factors necessary to proceed with implementation and the concept implementation is underway. Transport behaviour modelling and platoon strategies continue in parallel with human behaviour studies and safety studies. Development of lead vehicles and following vehicles has started. Track studies will soon be performed with the *SARTRE* road-train using three cars and two trucks.

Some of the areas for continued research and refinement are in the areas of

---

[1,2] InnovITS is the UK Centre of Excellence for sustainable mobility and intelligent transport systems. See http://www.innovITS.co.uk Simulink® is a registered trademark of The MathWorks™.

- Number of vehicles in a *SARTRE* platoon and the mix of vehicles (cars/trucks)
- Specification and architecture updates.
- Safety requirements and analysis
- Updates to V2V Communications (DSRC)
- Inputs regarding V2I findings to infrastructure organizations
- Sensor Fusion Systems
- Actuator Systems
- Human Machine Interfaces
- Autonomous Control System
- Platoon Management System

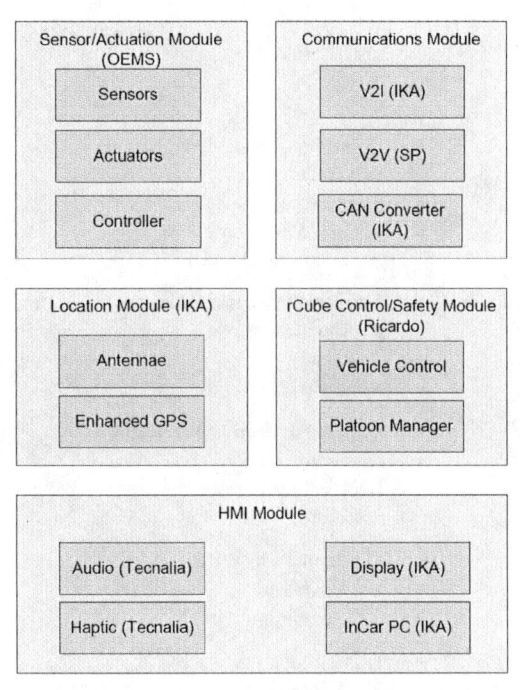

Figure 7. SARTRE Concept Architecture

VALIDATION AND ASSESSMENTS

At the time of the writing of this paper, validation of the systems has not been completed. The plan is to validate the on-vehicle systems, the remote systems, end-to end systems, and fuel consumption claims. Once validation is completed, results of studies will be made available that include assessments of the commercial viability of *SARTRE*, the net impact on infrastructure and vehicles, and potential policy impacts.

SUMMARY/CONCLUSIONS

V2V and V2I communications are changing the ways that people interact with their vehicles. Driver assistance systems are making way for semiautonomous mobility improvements in fuel economy and safety. Future automotive systems will leverage V2I and V2V in order to allow drivers to select semiautonomous and autonomous behaviors, with net gains in safety and mobility. Partnerships between science researchers, policy makers, academia, infrastructure manufacturers, and automotive manufacturers will change the landscape of automotive transportation to a more efficient and safer experience for drivers and passengers.

THE PARTNERSHIPS

The *Sentience* program included each of the following organizations.

Organization	Organization Overview	Role
Ricardo	Automotive Engineering Consultants with expertise in Control Systems, Vehicle Systems, and ITS	• Project Management • Rapid Prototyping Electonics • Vehicle control algorithms
innovITS	UK Dept of Business Enterprise Reform (formerly DTI)	• Promotes UK Telematics/ITS • Funding/Coordination • Program goals
Jaguar/LandRover/ Ford	Prestige UK Vehicle OEM	• Base Hybrid Vehicle • Vehicle Data and Interfaces
Ordnance Survey	UK mapping organization	• Enhanced map data • Traffic congestion data
Orange	Telecoms company	• Telecoms engineering expertise and equipment
TRL	Transport Research and testing Lab	• Vehicle system testing • Test facilities

The *SARTE* program included each of the following organizations.

Organization	Organization Overview	Role
Ricardo (UK)	High value engineering services to the automotive, ITS and clean energy communities	• Coordinator and Management WP leader • Safety Analysis • Platoon Management & Autonomous Control
IDIADA Automotive Technology (Spain)	World-leading company for automotive testing and demonstration	• Validation/Assessment work package leader • Test lead • Road trial lead
Institute for Automotive Engineering (ika) of RWTH Aachen University (Germany)	Leading university in automotive technology	• Concept definition WP leader • Modelling lead • Back office and organisation assistant
SP Sveriges Tekniska Forskningsinstitut (Sweden)	Research institute experienced in automotive safety and communication	• Dissemination WP leader • Use case lead • V2V communications
TECNALIA Robotiker (Spain)	Expert technology centre specialising in ICT	• Human factors assessment • HMI design and implementation
Volvo Car Corporation (Sweden)	Major passengar car OEM	• Implementation WP leader • Following vehicle lead • Following vehicle (car) sensor fusion
Volvo Technology (Sweden)	Major trucks, buses and construction equipment OEM	• Lead vehicle lead • Lead/following vehicle (truck) sensor fusion

CONTACT INFORMATION

Chris Domin
Intelligent & Autonomous Systems
Ricardo, Inc.
40000 Ricardo Drive
Van Buren Twp., MI 48111 USA
Phone: (734) 394-4155
Fax: (734) 397-6677
chris.domin@ricardo.com

ACKNOWLEDGMENTS

Ricardo, Inc. and Ricardo UK Limited would like to thank its partners for support of Foot-LITE, Co-Driver, *Sentience*, and *SARTRE* programs. Chris Domin would like to thank Tom Robinson and Jonathon Hunt from the UK facility for their support while writing this paper.

DEFINITIONS/ABBREVIATIONS

A/C
    Air-conditioning unit

CPU
    Central Processing Unit, a micro controller

DSRC
    Digital Short Range Communications

EAC
    Enhanced air conditioning

EAD
    Enhanced acceleration/deceleration

ECU
    Embedded Control Unit

EV
    Electric Vehicle

HMI
    Human-machine interface or display

ITS
    Intelligent transportation systems

OEL
    Optimised engine loading

SARTRE
    EU Program: Safe Road Trains for the Environment

SCU
    Supervisory control unit

Sentience
    EU Program: Using Electronic Horizon Data to Improve Vehicle Efficiency

V2V
    Vehicle to vehicle

V2I
    Vehicle to infrastructure

Time Determinism and Semantics Preservation in the Implementation of Distributed Functions over FlexRay

2010-01-0452
Published
04/12/2010

Marco Di Natale
Scuola S. Anna

Haibo Zeng
General Motors

Copyright © 2010 SAE International

ABSTRACT

Future automobiles are required to support an increasing number of complex, distributed functions such as active safety and X-by-wire. Because of safety concerns and the need to deliver correct designs in a short time, system properties should be verified in advance on function models, by simulation or model checking. To ensure that the properties still hold for the final deployed system, the implementation of the models into tasks and communication messages should preserve properties of the model, or in general, its semantics. FlexRay offers the possibility of deterministic communication and can be used to define distributed implementations that are provably equivalent to synchronous reactive models like those created from Simulink. However, the low level communication layers and the FlexRay schedule must be carefully designed to ensure the preservation of communication flows and functional outputs. In this paper, we provide a discussion and an analysis of the aforementioned issues and we present possible solutions to the problem of defining FlexRay schedulers that support deterministic communication delays. The aforementioned scheduling options are applied to an X-by-wire case study to highlight tradeoffs between schedulability and additional functional delays in the controls.

INTRODUCTION

Future automobiles are required to support an ever increasing number of additional complex, distributed and interdependent functions. Consider, for example, the set of active safety and X-by-wire functions planned for deployment in future cars.

Functions can be refined top-down or constructed bottom-up from component models or, more likely, the process would be a meet-in-the-middle approach. Today the models of the functional components are typically developed using tools like Simulink from The Mathworks. Safety concerns, the complexity and the interdependency of these functions require that system properties be verified in advance, using models of the functions, by simulation or model checking. The automotive industry together with the avionic industry was the first to embrace model-based design, as a tool to simulate the system functions, verify some properties of interest, remove coding errors and speed up the software development process.

Correctness of systems should be demonstrated by formal reasoning upon the models and their desired properties, provided that models are built on solid mathematical foundations (Model of Computation or MoC) and properties are expressed by logic propositions [1]. When exhaustive proof of correctness cannot be achieved, the modeling language should support simulation and automated testing. Formal methods can be used to guide the generation of the test suite and guarantee some degree of coverage. Different design targets or types of analysis may require different levels of abstractions or viewpoints of the system and its components. All functional and non-functional constraints and properties that are captured by the system-level and component-level models must be propagated at each refinement step in addition to derived requirements. At all levels, the designer should have control over the conditions that guarantee the preservation of the model semantics into its refinement or what are the restrictions that the refinement imposes on the set of behaviors (and of course ensure that no unintended behavior is added). The definition of such a process, with the corresponding models, methods and tools, that goes from system-level to component models, eventually into automatically generated code, is a quite challenging task.

Compared with previous communication standards like CAN, FlexRay offers the possibility of a deterministic communication and can be used to define distributed implementations that are provably equivalent to synchronous reactive models like those created from Simulink. However, the low level communication layers and the FlexRay schedule must be carefully designed to ensure the preservation of model semantics, especially when models include subsystems executing at different rates.

In this paper we provide a discussion and an analysis of the issues that need to be faced when defining a FlexRay communication schedule in a model-based design flow, where tasks and messages provide a distributed implementation of a synchronous model. We present possible solutions to the problem of defining FlexRay schedulers that preserve the semantics of synchronous reactive models including those generated using Simulink and Stateflow. When the model behavior cannot be preserved, the schedule can be defined to at least add known and deterministic delays to the communication links and allow verification of the system behavior by simulation. The next section provides an introduction to synchronous models of computation and Simulink. Section 3 defines the FlexRay scheduling problem and the following Section discusses the consequences of scheduling decisions on the functional behavior of the system. Different scheduling options with deterministic communication delays are then examined for a case study consisting of an X-by-wire system.

STATE OF THE ART

Scheduling techniques for the FlexRay static segment have been developed by extending the work for scheduling messages in a TDMA bus [2] [5]. In [4], the authors consider the case of a hard real-time application implemented on a FlexRay system. Messages are scheduled in the static segment only. In [3], the authors present a timing analysis of applications communicating over FlexRay, using the static and the dynamic segment. The authors first present a static cyclic scheduling technique for messages transmitted in the static segment. Then, they develop a worst-case response time analysis for event-based transmissions in the dynamic segment. Message analysis is integrated in a holistic method that computes the worst-case response times of all tasks and messages.

Synchronous Reactive (SR) models [6] are used for modeling control-dominated embedded applications. SR models are characterized by the "synchronous assumption" or "logical time execution", which requires that the system completes the reaction to an event before the occurrence of any other event. Synchronous languages are implemented in the SCADE commercial tool [7], but also the very popular MATLAB/Simulink [8] tool chain from The Mathworks which allows modeling and simulation of the system according to a synchronous MoC. Both commercial toolsets (Scade and MATLAB/Simulink) offer simulators, automatic code generators and an interface to verification tools such as the plug-in from Prover [9] for SCADE and Design Verifier for Simulink (also based on Prover technology).

Several research papers have defined a possible formal approach to the problem of semantics preservation upon mapping of function onto architecture, at least in the case in which the functional model is a synchronous model. Synchronous models are based on the assumption that the system reaction to any event completes before the next event arrives. In some cases, the mapping of functions into an architecture may require preservation of this property, which refers to the time at which events are defined for the system and the time available for the reaction of the system. Another property of the functional model that can be preserved by its implementation is flow preservation, that is, guaranteeing that the implementation operates on the same values of the input data streams as the model. In both cases, the simplest solution is to restrict the functional model to react to periodic events only and to select for its implementation time-triggered execution platforms. This approach is followed by the Time Triggered Architecture [10]. Techniques for generating semantic-preserving implementations of synchronous models on TTA have been studied in [11].

Methods for desynchronization in distributed implementations have been studied and presented in [12, 13] and a more general approach consists of a intermediate mapping of synchronous models into Kahn Process Networks [14], for which a correct implementation in a non-synchronized architecture platform can be found more easily [15] (albeit, very likely at the price of additional overhead and pessimism in the time performance).

SYNCHRONOUS MODELS AND SIMULINK

In Simulink, the system is defined as a network of communicating blocks. Each block operates on a set of input signals and produces a set of output signals, according to its specifications. Formally, a block transforms an input function (of time) into an output function. The input function's domain can be a set of discrete points (discrete-time signal) or it can be a function defined on a continuous time interval (continuous-time signal). Continuous blocks have a nominal sample time of zero, but in practice, they are implemented by a solver, executing at the base rate. Eventually, every block has a sample time, with the restriction that the discrete part is executed at the same rate or at an integer fraction of the base rate.

Simulink computes for each block, at each step, the set of outputs, as a function of the current inputs and the block state, and then, it updates the block state. A cyclic dependency among blocks where output values are instantaneously produced based on the inputs results in a fixed point problem and possibly inconsistency. A fundamental part of the model semantics are the rules dictating the evaluation order of the blocks. Any block whose output is directly dependent on its input (i.e., any block with direct feedthrough) cannot execute until the block driving its input has executed. Some blocks set their outputs based on values acquired in a previous time step or from initial conditions specified as a block parameter. The output of such a block is determined by a value stored in memory, which can be updated independently of its input. The set of topological dependencies implied by direct feedthrough blocks defines a partial order expressed as a set of precedence constraints among pairs of blocks. The partial order must be accounted for in the simulation and in the run-time execution of the model.

Before Simulink simulates a model, it orders all blocks based upon their topological dependencies. This includes expanding subsystems into the individual blocks they contain and flattening the entire model into a single list. The tool (arbitrarily) chooses one total order in the execution of blocks that is compatible with the partial order imposed by the model semantics. Once this step is complete, the virtual time is initialized at zero. The simulator engine scans the precedence list in order and executes all the blocks for which the value of the virtual time is an integer multiple of the period of their inputs.

Executing a block means computing the output function, followed by the state update function. When the execution of all the blocks that need to be triggered at the current instant of the virtual time is completed, the simulator advances the virtual clock by one base rate cycle and resumes scanning the block list.

In an example multi-rate system, represented in Figure 1, characterized by oversampling of the communication, a possible order of execution of the blocks at simulation time would be the one represented in the lower part of the figure. Block C is executed at the base rate and, because of the feedthrough dependencies, it must follow both A and B.

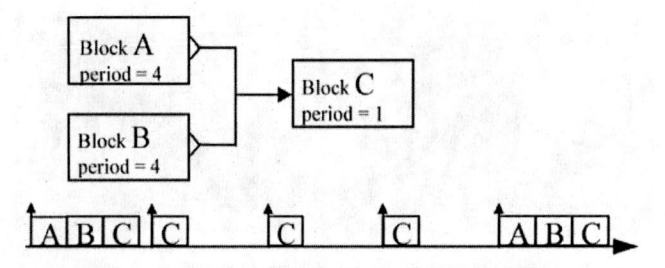

Figure 1. An example of simulation-time execution order.

Figure 2 shows an example of a very simple system in Simulink, in which two source blocks: a free counter modulo 16 and a repeating sequence generator, which produces as output the sequence [4, 2, 1] are executed with period 4 and feed a multiplier and a comparator executing with period 1. We assume that the system is implemented in a distributed platform, where data communications between the source blocks and the multiplier/comparator need to be transmitted (and scheduled) over a bus. In the figure, this is represented by the dashed red line over the communication links. The example is of course very simple for illustrative purposes, but representative of more complex systems in which communication occurs between subsystems.

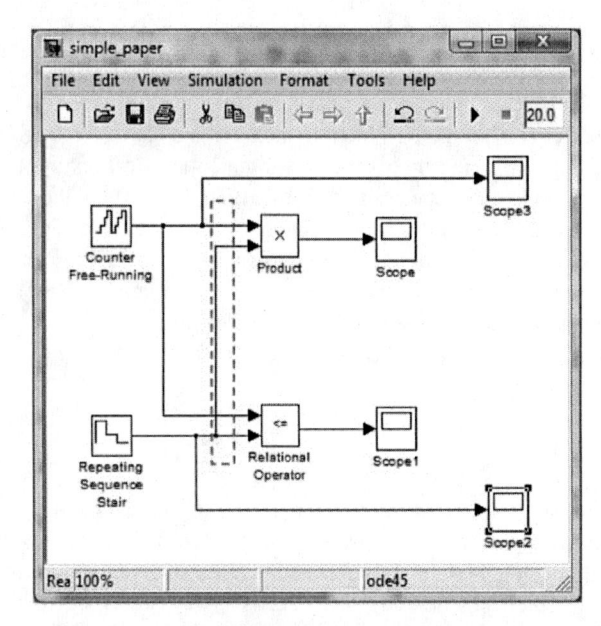

Figure 2. A simple Simulink example with blocks executing at different rates.

Figure 3 shows the results of the simulation of the example model, with the source data shown on the left side, and the output data on the right side.

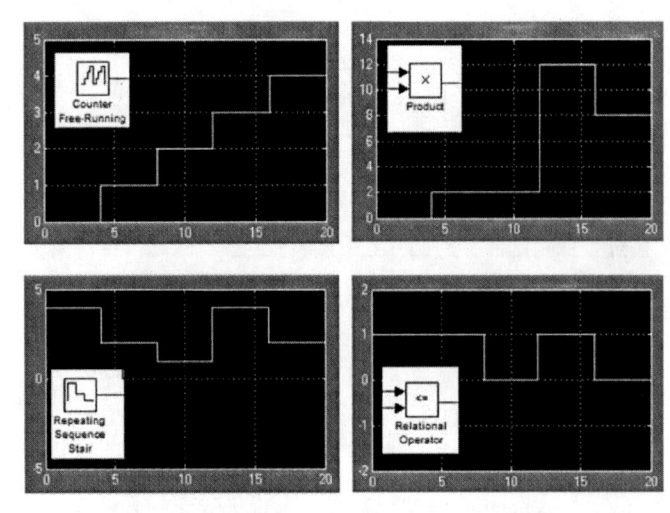

Figure 3. The result of the simulation of the previous example showing the output of the counter source (top-left), the sequence generator (bottom-left), the multiplier (top-right) and the comparator (bottom-right).

The output may be subject to constraints. For example, we might be interested in detecting if the result of the comparator is always equal to one in a given time interval, or if the output of the multiplier ever exceeds a given value.

SCHEDULING IN A FLEXRAY SYSTEM

FlexRay is a modern communication standard for highly deterministic and high speed communication. In Flexray, the communication speed is defined at 10 Mb/s, and the bus bandwidth is assigned according to a time-triggered pattern. The available bandwidth is divided in communication cycles and each cycle contains up to four segments (Static, Dynamic, Symbol and Network idle time - Nit). Clock synchronization is embedded in the standard, using part of the Nit segment.

<figure 4 here>

The static part of the communication cycle enables the transmission of time critical messages according to a periodic pattern or schedule, in which a time slot, of fixed length and in a given position in the cycle, is always reserved to the same node.

Dataflows are a common model used to represent applications to be scheduled on FlexRay. The vertices represent the tasks and the edges represent the data signals communicated among them.

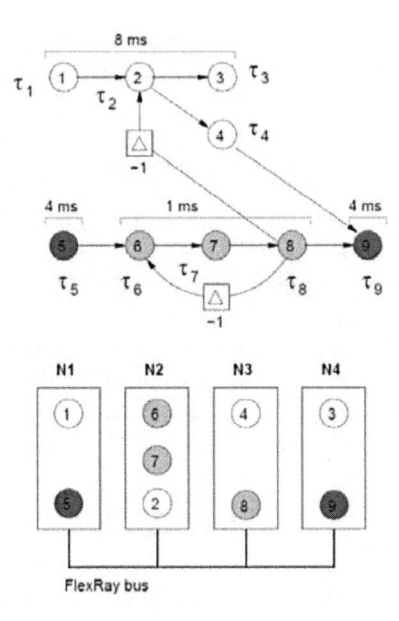

Figure 5. A task model with a unit-delay communication and allocation of tasks

Each periodic task will run an infinite sequence of instances or jobs. Each task is assumed to read its input at the beginning of its execution and writes its results at the end. Each signal may optionally be delivered with a unit delay. Each signal also carries a precedence constraint in the execution of the sender and receiver tasks. If the signal is delivered without a unit delay, the successor must be executed after the sender instance activated immediately before it. However, the successor must read its inputs before the writer executes again, otherwise it will use a signal value generated with a delay equal to the period of the writer. The application cycle or hyperperiod H is defined as the least common multiple (lcm) of the periods of all tasks.

Inside the hyperperiod, each job is considered as an individual scheduling entity. The scheduling problem consists of planning the execution of jobs and the transmission of signals into the available slots inside H.

The set of all the task instances transmitted in the application cycle defines the Application instance graph, as in <u>Figure 6</u>. The FlexRay communication stack may transmit multiple signals in the data content of a single message m_i in a communication slot, upon condition that they are transmitted by tasks allocated to the same node.

<figure 6 here>

The scheduling of FlexRay communication consists of the mapping of the tasks and signals defined in the application cycle into a set of communication cycle instances. This mapping can be performed in different ways, according to the selection of the communication cycle length, the size of the static segment, the slot size and correspondingly the number of static slots in each communication cycle. It is practically impossible to encode all the above into an integrated problem formulation to be solved by an optimization framework without having an exceedingly large search space.

Hence, in our previous research work [16], we investigated a two-step approach. Starting from the design specification, we assumed a FlexRay bus configuration ($l_{app}, l_{comm}, n_{slot}, l_{slot}, b_{slot}$) is given, where l_{app} is the length of the application cycle (the least common multiple of the task periods), l_{comm} is the length of the FlexRay communication cycle, n_{slot} is the number of slots in the static segment of the communication cycle, l_{slot} is the length of the slot in time, and b_{slot} is the size of the slot in bits. Based on this configuration, we apply a mathematical programming framework to encode the problem and synthesize other variables such as slot ownership, signal to slot mapping, message and task scheduling. The two-step approach is consistent with the typical design flows in use by the automotive industry. The application cycle is based on the application on hand, the communication cycle and the slot size are defined based on the need to reuse legacy components and standardize configurations.

In [16] we formulated our problem in the general framework of mathematical programming (MP), where the system is represented with parameters, decision variables, and constraints over the parameters and decision variables. An objective function, defined over the same set of variables, characterizes the optimal solution. The FlexRay scheduling problem allows a mixed integer linear programming (MILP) formulation that is amenable to automatic processing. The example in the previous section is a case of communication with oversampling. A FlexRay schedule that preserves the execution order defined by the model semantics and used to validate the system behavior at simulation time should execute as in the following Figure 7.

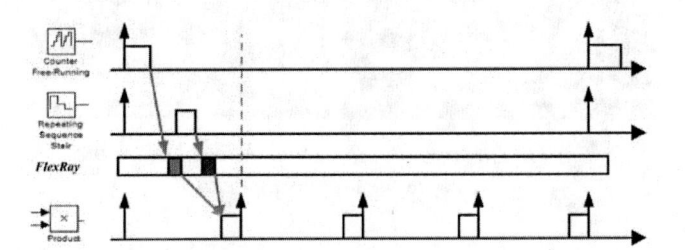

Figure 7. Scheduling the tasks and the communication of the example without delays.

As is clear from the figure, this can be a very tightly constrained scheduling problem. Both senders must be executed before the FlexRay slots allocated for communication of their output data. Those slots, in turn, should be allocated so that they precede the scheduling of the receiver task, which needs to complete before the end of its period. In conclusion, the entire chain must be scheduled before the deadline shown as a dotted line in the figure.

Two observations apply here. In a development flow, the system designer and the software engineers might be presented with a scheduling problem that only defines the execution rate constraints. That is, the scheduler would be requested to execute both senders within their period of 4 units and, following the sender task, to assign the communication slots on the bus for transmission. Receivers would be scheduled in such a way that one instance is executed every unit. The requirement that it is the first instance of the receivers that should process the data (in every cycle of 4) could be dropped, not because of negligence, but because the execution order specification is "implied" in the model (and in the simulator) semantics, and designers might not be aware of it. Otherwise, even when the execution order constraints are known and made explicit in the scheduling problem, because of the tight deadlines, the designer could be tempted to release the execution constraints to ease schedulability.

In both cases, there are tradeoffs between ease of schedulability and functional performance, of which the system architect and the designers of the data communication scheduler should be aware.

IMPACT OF SCHEDULING ON FUNCTIONAL BEHAVIOR

As shown in the previous section, when tasks executed at different rates communicate in FlexRay (such as in the case of oversampling for our example), the scheduling problem may be simplified if we can select the receiver instance that reads the data produced by the sender. This choice, however, is not neutral to the behavior of the functions that are executing the controls algorithms.

Consider the case in which one of the communication paths is delayed allowing communication with the third instance instead of the first one. In this case one task and one slot can be scheduled later easing feasibility of the system-level schedule (including the FlexRay scheduling)

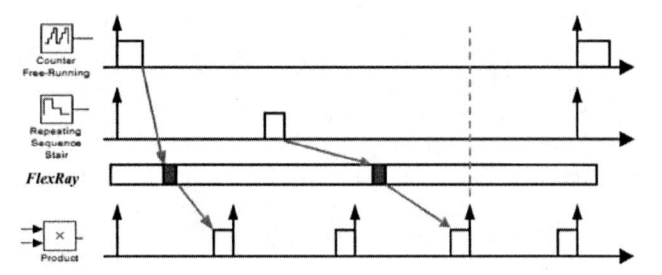

Figure 8. Scheduling the tasks and the communication of the example with a deterministic delay of two receiver periods on one of the communication links.

If such a communication schedule selects in a deterministic way the instance of the receiving task, the behavior of the system can still be represented by a Simulink model (in this case, the model of Figure 9). Of course, the behavior of the functions is going to be affected by these delays. In our simple example, when the delay on one of the communication links is increased from, respectively 0 to 3 receiver periods, the outputs change as represented, respectively in Figures 10 and 11. The output of the multiplier changes significantly. In the case without delays, the output never exceeds 12, in the selected execution window. With the additional delays, the output goes up to 16 for a few cycles. Also, the shape of the output signal is clearly affected. The same happens for the output of the comparator, where the shape of the output function is severely affected. While this is clearly nothing more than a simple example, it is representative of a situation in which large subsystems exchange data over links, computing complex control functions.

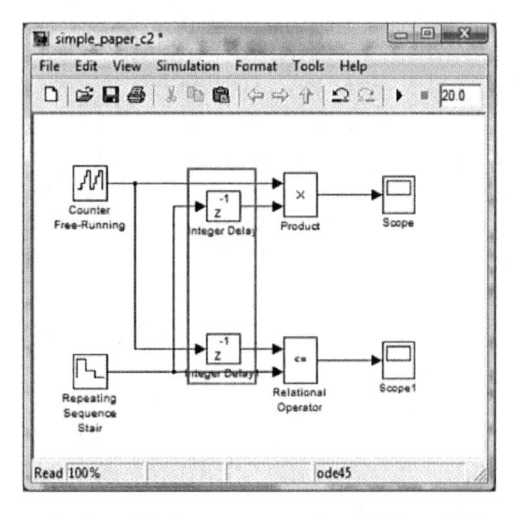

Figure 9. A model that accounts for additional delays on (selected) communication links

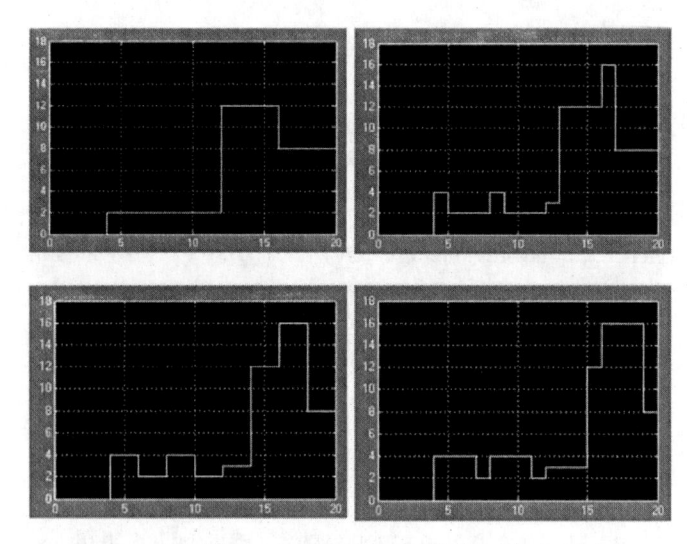

Figure 10. The output of the multiplier block of the example for increasing delays.

The situation can be even worse if even deterministic delays are not guaranteed by the scheduler. If the schedule is generated by only looking at the periods of the communicating tasks, with the guarantee that the execution of each task and the transmission of each message occur only once during the period, then even the simulation of the resulting system becomes more complicated and the resulting behavior is less predictable. Such a situation is unfortunately all too common, and should be avoided altogether.

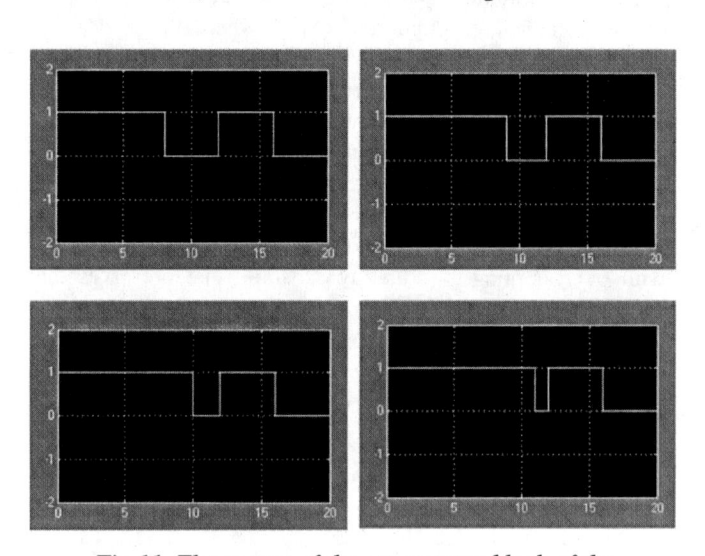

Fig 11. The output of the comparator block of the example for increasing delays.

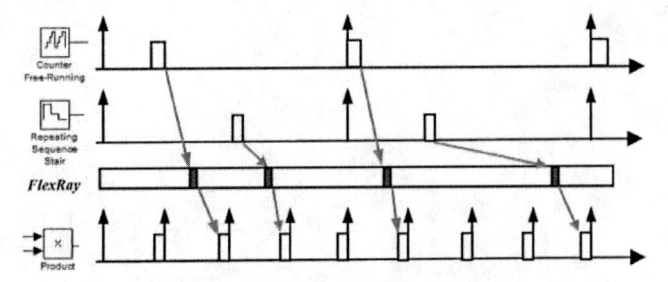

Figure 12. An example of a schedule with nondeterministic delays.

SCHEDULING FEASIBILITY VS. FUNCTIONAL DELAYS: A CASE STUDY

We developed an analysis of a case study to analyze to what degree the selection of the receiver instance, in an oversampling scenario or of the sender instance in an undersampling scenario can affect the feasibility of the schedule. It would be nice to explore also the impact on the functional behavior, but this part of the study is for now beyond our possibilities.

The application configuration of Tables 1 and 2 is obtained from a prototypical X-by-Wire application from General Motors (C_i indicates the worst-case execution time of tasks). The application has 10 ECUs interconnected by a single FlexRay bus. There are a total of 56 tasks, with periods of 1ms and 8 ms respectively, and 132 signals. Tables 1 and 2 show periods and worst case execution time of tasks, in microseconds and the size of each signal, in bits. No end-to-end delay constraints are defined for this case, but the objective is to find the schedule with the minimum number of used slots.

<tables 1, 2 here>

We compare the ease of schedulability for two scenarios in the case of oversampling communications from sender tasks with period 8000ms to receiver tasks with period 1000 ms. If we choose to schedule the first instance of the receivers to read the data, the FlexRay scheduling problem is not feasible, which is proved in the presolve phase, with a runtime = 0.1s. If we ease the communication paths to allow communication with the eighth instance of the receivers instead of the first one, an optimal solution with 13 used slots is found in 223 seconds.

SUMMARY/CONCLUSIONS

FlexRay offers the possibility of a deterministic communication and can be used to define distributed implementations that are provably equivalent to synchronous reactive models like those created from Simulink. However, the low level communication layers and the FlexRay schedule must be carefully designed to ensure the preservation of model semantics, especially when models include subsystems executing at different rates. In this paper we provided a discussion and an analysis of the issues that need to be faced when defining a FlexRay communication schedule in a model-based design flow, where tasks and messages provide a distributed implementation of a synchronous model.

REFERENCES

1. Clarke E. M. and Wing J. M., Formal methods: State of the art and future directions, Tech. Rep. CMU-CS-96-178, Carnegie Mellon University (CMU), Sept. 1996.

2. Pop P, Eles P, Peng Z, Schedulability-driven communication synthesis for time-triggered embedded systems. Real-Time Systems Journal 24:297-325 2004

3. Pop T., Pop P., Eles P., Peng Z., Andrei A., Timing analysis of the FlexRay communication protocol IEEE Transactions 2007

4. Ding S., Murakami N., Tomiyama H., Takada H., A GA-based scheduling method for FlexRay systems. Proceedings of EMSOFT 2005

5. Hamann A., Ernst R., TDMA time slot and turn optimization with evolutionary search techniques. Proceedings of the DATE Conference, pp 312-317 2005

6. Benveniste A., Caspi P., Edwards S. A., Halbwachs N., Le Guernic P., de Simone R., The synchronous languages 12 years later, Proceedings of the IEEE, Vol. 91-1, pp. 64-83, Jan 2003.

7. SCADE Suite Product web page: http://www.esterel-technologies.com/products/scade-suite/

8. Mathworks. The Mathworks Simulink and StateFlow User's Manuals. web page: http://www.mathworks.com.

9. Prover Technology, http://www.prover.com/.

10. Kopetz Hermann and Bauer Günther, The Time-Triggered Architecture, Proceedings of the IEEE, Special Issue on Modeling and Design of Embedded Software, Oct. 2001.

11. Caspi P., Curic A., Maignan A., Sofronis C., Tripakis S., and Niebert P.. From Simulink to SCADE/Lustre to TTA: a layered approach for distributed embedded applications. In Languages, Compilers, and Tools for Embedded Systems (LCTES03). ACM, 2003.

12. Caspi Paul and Benveniste Albert. Time-robust discrete control over networked loosely time-triggered architectures. IEEE Control and Decision Conference, December 2008.

13. Potop-Butucaru Dumitru, Caillaud Benoît, and Benveniste Albert. Concurrency in synchronous systems. Formal Methods in System Design, 28(2):111130, 2006.

14. Caspi Paul and Pouzet Marc. Synchronous kahn networks. ICFP Conference, pages 226238, 1996.

15. Tripakis S., Pinello C., Benveniste A., Sangiovanni-Vincentelli A., Caspi P. and Di Natale M. Implementing Synchronous models on Loosely Time-Triggered Architectures IEEE Transactions on Computer, October 2008, Volume 57 Issue 10.

16. Zeng Haibo, Zheng Wei, Di Natale Marco, Giusto Paolo, Ghosal Arkadeb, Alberto Sangiovanni-Vincentelli, Scheduling the FlexRay bus using optimization techniques. In Proceedings of the 46th ACM/IEEE Design Automation Conference (DAC), July 2009.

CONTACT INFORMATION

Marco Di Natale is associate professor at the Scuola Superiore S. Anna in Pisa, Italy. He can be reached at marco@sssup.it

Haibo Zeng is a researcher at General Motors R&D. He can be reached at haibo.zeng@gm.com

ACKNOWLEDGMENTS

The authors wish to thank Paolo Giusto and Arkadeb Ghosal for valuable discussion and insight.

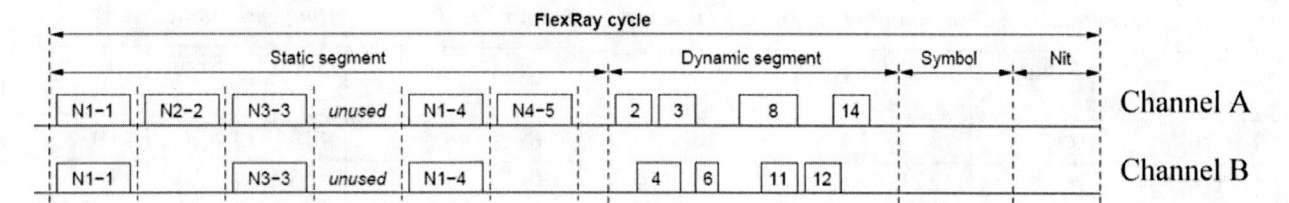

Figure 4. the FlexRay Communication cycle and its four segments

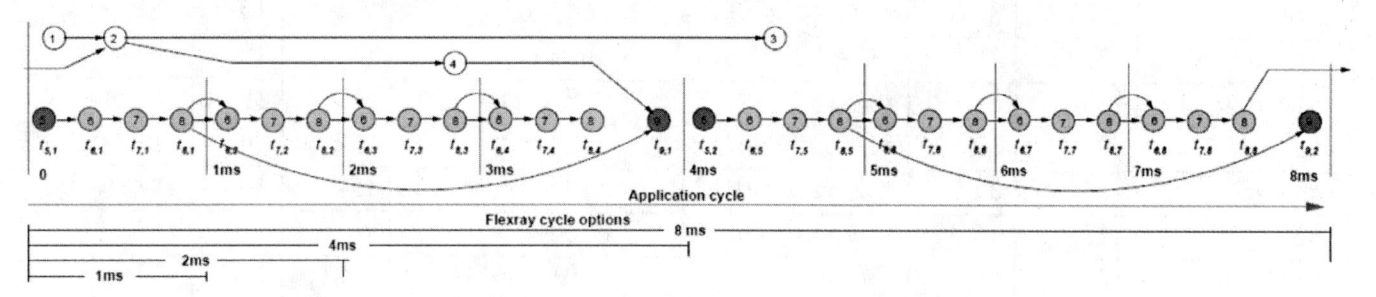

Figure 6. Unrolling the task model in task instances in the application cycle.

Table 1. The definition of the tasks in our case study.

Task	ECU	Period	Ci	Task	ECU	Period	Ci
τ8	e9	8000	810	τ21	e5	1000	25
τ9	e9	8000	550	τ22/ τ26/ τ30/ τ34	e5/e6/e7/e8	1000	60
τ11	e9	8000	100	τ23/ τ27/ τ31/ τ35	e5/e6/e7/e8	1000	40
τ12	e9	8000	200	τ24/ τ28/ τ32/ τ36	e5/e6/e7/e8	1000	20
τ13	e9	8000	200	τ25/ τ29/ τ33	e6/e7/e8	1000	30
τ14	e9	8000	110	τ37/ τ42/ τ47/ τ52	e1/e2/e3/e4	8000	1000
τ15	e9	8000	455	τ38/ τ43/ τ48/ τ53	e1/e2/e3/e4	8000	500
τ16	e10	8000	780	τ39/ τ44/ τ49/ τ54	e1/e2/e3/e4	8000	1500
τ10	e10	8000	510	τ40/ τ45/ τ50/ τ55	e1/e2/e3/e4	8000	1300
τ17	e10	8000	190	τ41/ τ46/ τ51/ τ56	e1/e2/e3/e4	8000	350
τ18	e10	8000	260	τ20	e10	8000	230
τ19	e10	8000	100				

We analyzed the system using our scheduling tool.

Table 2. The definition of the communication signals for our FlexRay scheduling case study.

Signal	Send	Size	Recv	Signal	Send	Size	Recv
σ1 to σ4	τ15	32	τ22/ τ26/ τ30/ τ34	σ48	τ22	32	τ16
σ5 to σ8	τ20	32	τ22/ τ26/ τ30/ τ34	σ49	τ26	32	τ16
σ9 to σ11	τ23	32	τ27/ τ31/ τ35	σ50 to σ53	τ26	16	τ8/ τ16
σ12, σ13	τ21	32	τ22/ τ26/ τ30/ τ34	σ54 to σ63	τ39	16	τ12
σ14	τ12	8	τ39/ τ44/ τ49/ τ54	σ64, σ65	τ42	16	τ12
σ15 to σ18	τ22	16	τ8/ τ16	σ66 to σ77	τ44	16	τ12
σ19, σ20	τ23	8	τ8/ τ16	σ78 to σ87	τ49	16	τ12
σ21 to σ23	τ27	32	τ23/ τ31/ τ35	σ88, σ89	τ52	16	τ12
σ24, σ25	τ25	32	τ22/ τ26/ τ30/ τ34	σ90 to σ99	τ54	16	τ12
σ26 to σ32	τ12	16	τ39/ τ44/ τ49/ τ54	σ100	τ8	1	τ17
σ33, σ34	τ27	8	τ8/ τ16	σ101 to σ116	τ12	16	τ17
σ35 to σ37	τ31	32	τ23/ τ27/ τ35	σ117 to σ124	τ12	16	τ17/ τ18
σ38, σ39	τ29	32	τ22/ τ26/ τ30/ τ34	σ125	τ30	1	τ8/ τ16
σ40, σ41	τ30	16	τ8/ τ16	σ126 to σ128	τ34	16	τ8/ τ16
σ42 to σ44	τ35	32	τ23/ τ27/ τ31	σ129, σ130	τ37	16	τ12
σ45 to σ47	τ33	32	τ22/ τ26/ τ30/ τ34	σ131 to σ132	τ35	8	τ8/ τ16

The FlexRay bus is configured as follows: the application cycle is H= 8ms, the communication cycle =1ms with =14, and the slot size =200bits, or 35μs. The problem is modeled in AMPL and solved using CPLEX.

Exploring Application Level Timing Assessment in FlexRay based Systems

2010-01-0456
Published
04/12/2010

Sandeep U. Menon
Electrical & Controls Integration Lab, General Motors R&D

Copyright © 2010 SAE International

ABSTRACT

One of the motivations to adopt the FlexRay communication protocol is the increased need to integrate active safety, time-critical features into the automotive domain. A deterministic communication protocol can only provide time bounds to guarantee data availability at the network level. Exploration of the timing analysis & function allocation options for application features need to consider effects due to software behaviors such as task preemptions and interrupts, which could skew the response times during execution in ECUs. The data processing and reaction time to network data in each ECU needs to be adjusted during the development phase to achieve real-time closed loop control in adherence with the timing requirements of the application.

This paper provides an innovative approach in using FlexRay to harness the time synchronized nature of the protocol and apply its attributes to support timing assessment of distributed application functions in ECUs. A bench environment will be described, which assist in the determination of task execution time and task trigger information over a distributed set of ECUs. Such a capability would be valuable during the design/integration phase to engineers deploying distributed features over FlexRay.

INTRODUCTION

Increased software content in the automotive domain has been instrumental in enhancing the driving experience over the past decade. The key differentiator of innovation for the current high-end, feature-rich automobiles is mostly in the increased software content. There has also been a steady migration of Electronic Control Units (ECUs) from stand-alone autonomous functional units to networked processors supporting a multitude of features using communication protocols like CAN, LIN etc. With the future trends towards integration of active safety and time-critical advanced features, there is a strong need to enhance these networked ECUs to operate as distributed systems with predictable timing and behavioral implementation. Deployment of these next generation features will require enhancements to existing communication protocols to provide increased bus bandwidth, time determinism and fault tolerant behaviors. The FlexRay communication protocol, jointly developed by automotive, semiconductor and electronic system manufacturers is targeted towards assisting a cost-effective and safety-enhanced deployment of these advanced automotive control applications in future vehicle platforms.

To analyze the functional and timing requirements during the design/development for these complex and distributed features, sophisticated tools and methods need to be used. One of the key aspects of system design involves the timing requirements for these features and their verification. This aspect is critical to optimize the computation power, bandwidth and scalability of vehicle platforms on which these features are deployed. In most cases the OEM is responsible, as the system integrator, for the end-to-end verification of timing associated with distributed features. Conversely, all the relevant internal timing behavior/effects and performance of functions allocated to individual ECUs are abstracted from the system integrator and the view is limited to timing assessments at the network communication level. Timing and performance issues identified during testing, may require iterations of control and architecture design which involve changes to network design/scheduling, activation rates, reallocation among multiple ECUs. Use of system timing models and scheduling analysis [6, 7, 8] are important during early design phases to improve confidence level for deployment of increasingly complex systems. These tools and methods are instrumental in allowing for optimal partitioning and distribution of system features. The V-cycle of the development process for automotive will include incremental development/refinement activities [2] and feature enhancements. For the system integrator to verify the timing contracts (defined in the requirements phase) during the development/integration phase and perform appropriate system level compensations, it is extremely important to not only have visibility at the network communication level but also be able to verify timing behaviors at the ECU application level as well. This aspect is more profound in FlexRay systems than CAN, where minor variation to functionality may require significant adjustments to system configuration, FlexRay schedule and time synchronization across all ECUs [6].

This paper provides a method to assist system designers and integrators with timing assessment values to verify functional allocation, network architecture design and hardware/software requirements as defined in timing models are satisfied for FlexRay based systems. Some of the important application level timing attributes of interest will be considered. Additionally, the FlexRay protocol and the notion of a synchronized global time base along with its timing hierarchy are explained. The use of a time stamping mechanism that builds on the FlexRay global time and its use cases to support system level tracing and optimization of function allocation will be discussed. Use of this method could provide a simplified validation mechanism for timing assessment in FlexRay cluster both at a system and a component level. A case study will be described that demonstrates the utilization of the time stamping concept and its relevance to E/E platform design.

APPLICATION LEVEL TIMING ATTRIBUTES

Vehicle level ECS (Electronic Controls & Software) architecture consists of a set of system features that can be considered as a high level description of the system's capability. Individual system features can be decomposed into software functions that capture a specific behavior which needs to be deployed in one or more ECUs in the vehicle. The software function blocks are allocated to tasks, which are computational units that execute concurrently based on triggers from hardware/network data or from an internal micro-controller clock [7]. Most of the current software tasks associated with function blocks of a distributed feature in the automotive domain are unsynchronized in terms of their control behavior. The rationale for this is the lack of a common notion of time between function blocks operating in different ECUs and the inability of the underlying network systems to provide a common notion of time. Furthermore the variation in hardware and software designs for ECUs within the same communication network makes it difficult for distributed function blocks to operate in a synchronized manner. Figure 1 illustrates a set of features 1, 2 and 3 distributed across multiple ECUs as function blocks interacting via the FlexRay bus. Use of the FlexRay protocol ensures that the underlying network data transmission and reception is deterministic.

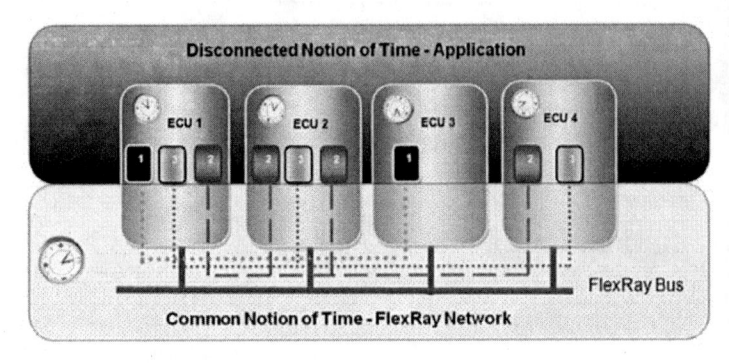

Figure 1. Application Feature Distribution

Defining individual task time budgets to optimize overall time usage at the system level is difficult. Task execution times are subject to change during development [3], as the minimum, average and worst case execution time could vary with modifications to data content and program flow. In addition, the interrupt loading and other higher priority tasks may significantly impact the task initiation time and/or task completion time during the development cycle. As seen in Figure 2 for ECU1, the completion time and start time of a Task 1 scheduled for time t=0 executing on an ECU is dependent on other high priority tasks and interrupts which may statically or dynamically execute in the system. As illustrated in Figure 2, the start of execution of Task1 and its completion time is highly dependent on the state of other tasks, interrupts which execute within ECU1 at that instance. Seq 1 through Seq 5 shown in Figure 2 describe the possible sequences that impact the execution of Task 1 scheduled at time t=0. Current implementations assign periodic rates to tasks to define arrival time and improve predictability. When these options are applied to time critical tasks there would be a need to move away from periodic tasks to event triggered tasks to reduce delays due to timing jitter.

It is extremely rare for individual tasks or communication messages on the network to create performance issues at the system level. Bottlenecks and performance degradation will most likely occur during the integration of these tasks or messages on the networked system. These problems may also manifest when existing system is enhanced, modified, or optimized. As illustrated in Figure 1, a feature consists of multiple software components (that have their own control loop) interacting with each other either locally or via the network communication medium. Integration of these features on multiple ECUs becomes more demanding when the software components have to be implemented by different suppliers [13]. With the introduction of event triggered tasks, use of more traditional worst case analysis needs to be complemented with integration time verification/validation to optimize the system. It would be beneficial for system integrators to be aware of the degrees of freedom available in terms of timing for any give function/task executing in an ECU during initial development phases. This information can be used for system level timing optimization decisions and also to confirm the nominal response time results satisfies the timing/scheduling analysis expectations.

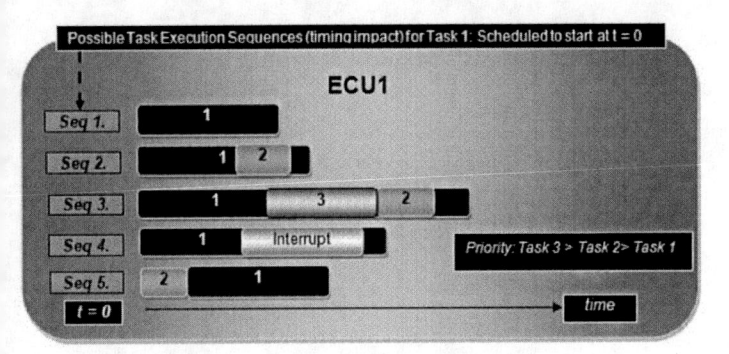

Figure 2. Task Timing Impacts in ECU

In addition to timing impacts at the ECU level from other application tasks, there are communication transmit and receive tasks (part of the communication protocol stack) that are periodic in nature. Existing protocols like CAN, LIN etc are limited in their capabilities for time synchronization and in providing guaranteed bus access. The time spent by the transmitter of a message is also impacted by operating system delays like context switches, OS system calls. At the receiver the same delays apply before application task eventually receives the network messages to perform the desired computational activities.

Currently, OEMs are required to address increasing number of real time problems caused by the integration of networked applications. Even though the OEMs are responsible only for the network design, which is the basis for system integration, the network timing attributes are dependent not only on the communication protocol but also on driver hardware and software which are out of the OEM's scope of responsibility and control [12]. There is a need for reliable verification of system level timing aspects for development of time critical distributed controls. Use of the same time deterministic communication medium, like FlexRay, for not only control but also to perform run time verification activities (using the notion of synchronized time) would be a efficient solution for system integrators. Systems that incorporate time-critical control / actuation can utilize the synchronized time in two ways. Firstly, a control system's decision to activate could be triggered by knowledge from data fusion algorithms that determines the occurrence of one or more network events relative to each other. The second use could be for coordinated actuation, where more than one ECU is required to support an actuation for a given feature; the actions in this case often need to be coordinated with each other in time.

These requirements would greatly benefit from a common reference of global time.

FLEXRAY TIME SYNCHRONIZATION

Since the FlexRay protocol is implemented on concepts of Time Division Multiple Access (TDMA), there is a requirement for a common notion of time between different ECUs in the FlexRay cluster to allocate the appropriate time quanta to individual ECUs for data transmission.

<figure 3 here>

The FlexRay protocol consists of a hierarchical definition of time units as shown in Figure 3. As described in [1] at an ECU level the smallest time unit defined for FlexRay is the Microtick which is derived from the FlexRay communication controller's(CC) oscillator's clock tick. The next level of time unit is the Macrotick which will be an integer multiple of Microticks for any given ECU. The number of Microticks which makeup a Macrotick is however subject to change for any given Macrotick within an ECU. The Macrotick value should however be the same across all synchronized ECUs in the FlexRay cluster. A Cycle is the next level in timing hierarchy of FlexRay and consists of integer number of Macroticks. The number of Macroticks in a cycle will be the same for all 64 cycles[1] defined for FlexRay across all nodes in the system. Each FlexRay cycle is composed of a static segment (time-triggered), a dynamic segment (event triggered), a symbol window and a network idle time (NIT) as shown in Figure 3. The static segment uses a TDMA scheme while the dynamic segment uses a scheme referred to as Flexible TDMA for data transmission. Since FlexRay adheres to a TDMA pattern with equally sized slots in the static segment, the point of time is exactly known when a frame is transmitted on the channel. In the dynamic segment, if the ECU needs to transmit a message, a transmission window (called a minislot) assigned to an ECU expands into a message transmission. If there is no message to be sent, the transmission window elapses unused as a short idle period.

The FlexRay protocol supports a distributed clock synchronization mechanism defined in [1] to allow individual ECUs to synchronize themselves to the cluster. The clock synchronization algorithm enables the time differences between the ECUs in the distributed system to stay within bounds of the desired precision. The timing mismatches between ECUs are basically the offset differences and rate differences due to oscillator clocks. The FlexRay specification defines a mechanism to perform both rate correction and offset correction as part of the time synchronization. The time units described in Figure 3 are used to define a shared "global" notion of time in FlexRay. This global time defined in FlexRay is in effect the local view of each ECU's perception of global time. Every node uses the

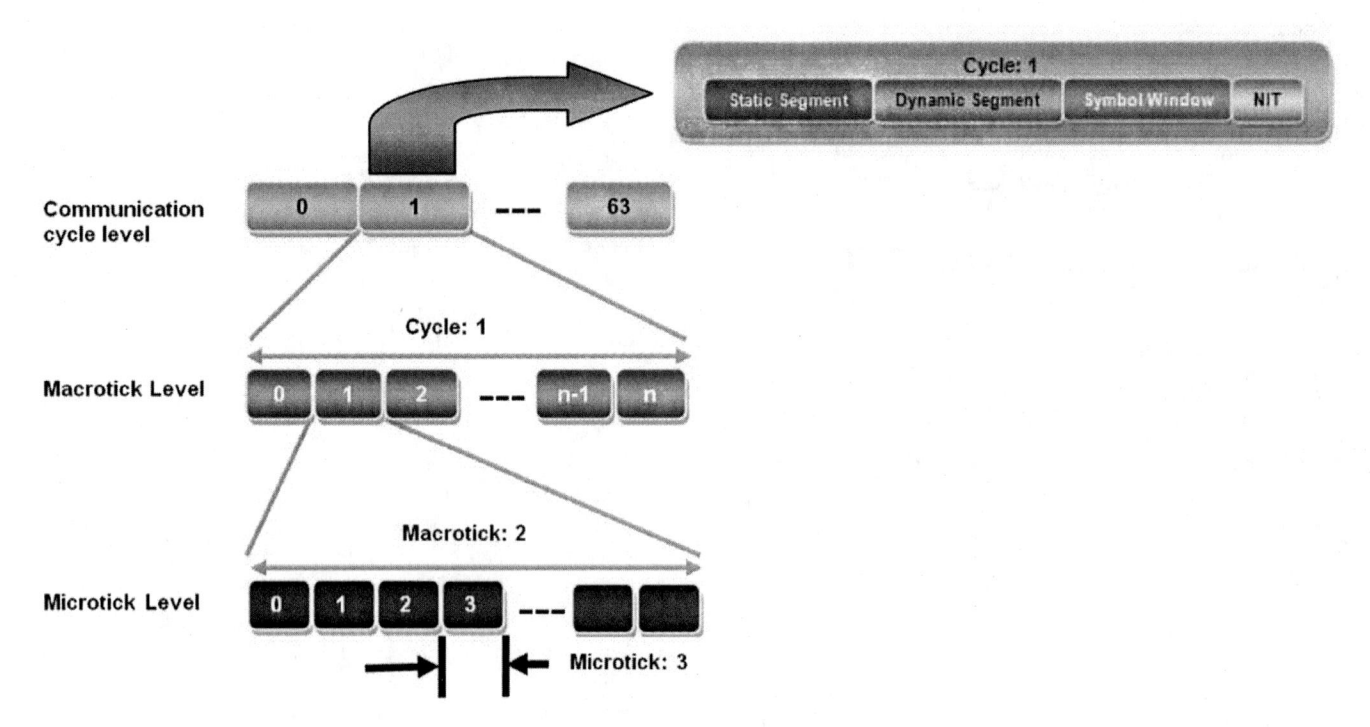

Figure 3. Timing Hierarchy in FlexRay

clock synchronization algorithm to attempt to adapt its local view of time to the global time. The global time [1] is defined using the pair of "cycle count" and "macrotick" values ≪vCycleCounter, vMacrotick≫. For any given time, all ECUs on the cluster are expected to have the same values for vCycleCounter and vMacrotick. Specific ECUs (called synchronization nodes) connected to the FlexRay network are tasked with transmission of synch message frames in the static (guaranteed) segment to negotiate the global time contract for all ECUs on the FlexRay cluster.

All clocks inevitably drift from absolute time and from each other due to their intrinsic frequency error of the hardware oscillator. The intrinsic frequency error in the oscillator will result in linear clock drift on the order of several parts per million (PPM) during the normal course of operation. Therefore, clock drift affects the precision of the relative clock synchronization. The error bound achieved by a clock synchronization method is linked to both the error inherent in the method itself, and the stability of the clock's oscillator standards. The common notion of time shared between different ECUs has a set tolerance limit designed into the FlexRay system which will determine its precision. The accuracy of the time base in a FlexRay system could range from 0.5 to 10 μs.

TIME STAMPING OPTIONS IN FLEXRAY

Software based time stamping in distributed systems helps to provide time interval measurements for real-world lower and upper bounds of task execution, albeit with some limited accuracies. Techniques like time-stamping are relevant for use at the software application level during the design/development stages to compensate for variations to design and also to improve overall system robustness. Use of software time stamping techniques in the past were not possible in the automotive domain due to the desynchronized nature of the communication protocols like CAN, LIN etc and the need for significant additional overhead to implement a system wide time synchronization protocol over existing communication buses. Due to the built in notion of global time in FlexRay network, time stamping activities at multiple ECUs would be an innovative method to apply towards feature verification as part of system integration. This option is also useful to cross reference the observed time bounds to satisfy the expected worst case timing analysis that are based on conservative assumptions. This capability permits making adjustments at a system level to ensure control feature requirements are satisfied. The functional and timing requirements captured using function allocation frameworks as seen in [4], could serve as input to define the scope of applying time stamping in the system.

For any given network architecture, the notion of global time in FlexRay could be used locally by an ECU to timestamp internal executions based on external events. It is extremely important to ensure the actual performance of a control feature meets the design requirements for time to market. If the actual results for timing for the feature vary from timing/schedule analysis expectations, there could be degradation of the control performance. At early development phases, it is useful to correlate time stamp logs from different ECUs of function execution to better understand the system level behavior of a feature. This information can then be transmitted back onto the FlexRay bus to allow system designers to monitor the timing associated with the distributed functions executing on multiple ECUs. Use of time stamps will support the optimization of control behavior and to also fine tune system design parameters, hardware/software resources and task activation frequencies. It is extremely difficult to identify the causality or to reconstruct the sequence of events if the data trace captured does not use a common reference time.

The timestamp information from the FlexRay bus could be used in many ways to assist system and ECU level development and verification activities.

• Time stamping data logs can be used for post processing analysis by system designers to improve overall system performance, justify function allocation and optimize and improve timing requirements for individual features. Any timing contract violations can be analyzed by the system integrator after the application's execution and alterations can be made to accommodate the variations observed.

• Time Stamp information can also be used at run time to improve closed loop real time control multiple ECUs to meet a fixed time deadline. The time-stamp along with signal data identification information could be used by ECUs to synchronize their execution from a specific event. Any timing contract violation in this case can be detected by the application software for necessary corrective measures.

• Time stamping options could be enabled for individual function blocks by means of calibration flags. This allows development activities to proceed with minimal software rebuild/ configuration to evaluate different time critical features on multiple ECUs. Individual timing parameters of interest like functions/task execution time or trigger time can be enabled / disabled as shown in Table 1 using calibration flags. This option could be useful when limited FlexRay message frames are available for verification purposes.

Table 1. Calibration Table for Timing Capture

Function	Function Trigger Time	Function Execution Time
Function 1	0 (Disable)	1 (Enable)
Function 2	1 (Enable)	1 (Enable)
Function 3	0 (Disable)	0 (Disable)
Function 4	0 (Disable)	0 (Disable)
:	:	:
:	:	:

• Time stamping options related to start/execution time of a function block triggered on reception of FlexRay messages in the Dynamic Segment (Figure 3) could be another area of interest. Since the availability of network data on the dynamic segment is impacted by both the event-triggered application data and message delays due to other higher priority dynamic segment messages active in the system. Therefore timing impacts to tasks which activate based on dynamic segment messages maybe more pronounced.

• Individual execution times for software functionality which is partitioned across a chain of ECU tasks can also be traced using time stamping options. In Figure 4, the individual time spent by each task from reception of Message X to its conversion to Message Y can be traced w.r.t FlexRay global time. Use of time stamping for complex interactions at an ECU level could also be beneficial for the ECU development team during in vehicle field evaluations where use of in-circuit emulators and debug tools are not available for timing studies.

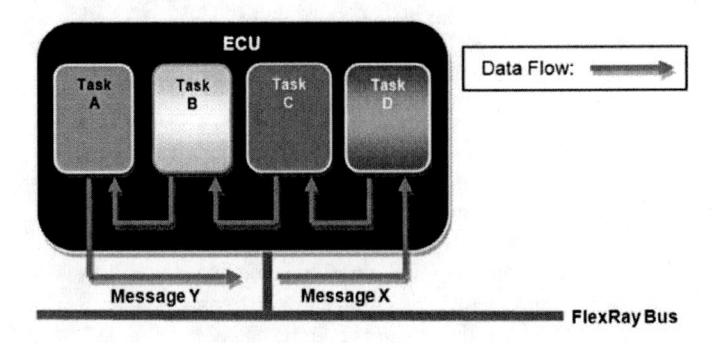

Figure 4. Timing trace for a multi-path data flow in an ECU

• FlexRay Global Time could be used to build a low accuracy multi-network system clock for time stamping activities across CAN, FlexRay systems. As shown in Figure 5, a system wide synchronized time (SST) derived from FGT, could be transmitted by the FlexRay based Gateway ECU on CAN1 and CAN2 as a periodic CAN Message. In order to improve accuracy of this network clock in CAN based systems, the CAN ID for the message has to be a very high priority (if not the highest) to account for bus arbitration on CAN. The software time stamp generation for the periodic message needs to be performed at the CAN Driver level of the protocol stack. It may be possible to achieve 2ms - 5ms accurate system clock for use in CAN networks with such techniques. It would also be necessary to handle the reception of the periodic SST CAN message at the CAN driver layers to avoid skewing of timing information.

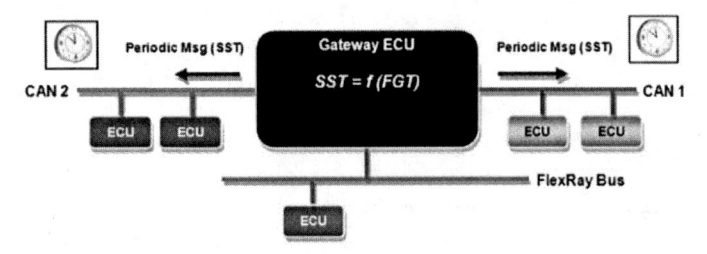

Figure 5. Creating a system wide global time base from FlexRay global time

• To allow time stamping for extended time durations (exceed 64 FlexRay cycle time) the SST time value can be modeled as a multiple of FGT. This implementation could now allow time stamp operations to be done over extended duration across multiple network domains. For FlexRay based gateway nodes, which may be connected to other communication buses it is also possible to evaluate, cross domain interactions and impacts to gateway application software from not only FlexRay messages but also CAN or LIN based messages. System level diagnostics for failures/faults is another use case which could benefit from availability of a cross network global time.

• Use of the FGT information with Universal Measurement and Calibration Protocol (XCP) could be another use case of interest. XCP is extremely useful for developing the driving dynamics of automotive control system [14]. Since the identical input variables and algorithms may need to satisfy different requirements when applied to different vehicle models, ECU level time stamps options on XCP could be enhanced to use FlexRay global time as the time reference for evaluation of function behaviors.

• An option currently in practice is to synchronize all application tasks executing on the ECUs based on the shared notion of the global time provided by FlexRay. When using this approach task priority, interrupts may still influence the start time of an application task scheduled in accordance with global time. However in many cases ECUs (eg: Powertrain systems) connected to more than one communication network or having significant local processing/ actuator timing requirements may choose to remain unsynchronized to the FlexRay global time. Legacy applications and operating system (OS) implementations that are being migrated to FlexRay may also avoid additional redesign and verification to accommodate this need.

When time stamp options are implemented on FlexRay, the time stamp messages that need to be transmitted on the FlexRay bus will require additional bus bandwidth. These messages could be added as low priority dynamic segment messages or as part of any unused static segment messages in FlexRay. Calibration options could be used to limit the number of messages frames used for the verification. The software function used to extract the FlexRay global information needs to be atomic in nature with minimal context switches to support accuracy for the time stamping operation and to avoid introduction of undesirable delays to execution. Implementation of functions like FGT based time stamping may be intrusive towards reusable software components. It may also make sense to limit the granularity of function blocks being analyzed to functions identified in the functional architecture of the vehicle system. Some of the application software implementations may include infrastructure (ECU hardware) function calls which may skew the expected results on function/task level timing impacts when these software functions are migrated to other hardware platforms. Another aspect of importance for time stamping is to retain the context of signal data for which the timing information was captured. For periodic tasks it would be important to verify the signal values used during the task execution to confirm the sequence of events.

The emergence of technologies like AUTOSAR [5, 10] with its layered architecture and hardware abstraction has made it easier to deploy and partition application functions and to support their evaluation. AUTOSAR allows the application software to be abstracted from the infrastructure software there by enabling the reuse of the application software across multiple ECUs and vehicle platforms. The expectation is that with minimal configuration changes, the same software component can be reused across platforms and hardware variants. It would also be possible to predict the behavior of this application software component with respect to time [7]. Instrumentation of time stamping functions for application functions can be applied at the RTE layer, as shown in Figure 6, with minimal intrusions or revisions to application software components. Use of software architectures like AUTOSAR, ensures the capability to verify, compare and isolate software feature level timings and abstract impacts from infrastructure functionalities. This will ensure predictable behavior patterns when a SWC (AUTOSAR Software Component) is migrated to different platforms as part of software reuse or during rapid prototyping activities using pre-existing SWCs. The additional timing overhead from use of such layered software architecture for legacy application software could be another assessment which could be relevant for development and design decisions.

<figure 6 here>

CASE STUDY

To evaluate the concept of time stamping software functions via the FGT, this capability was applied to a simple drive-by-wire model consisting of few sensors, actuators and a control ECU as shown in Figure 7. The scope of the case study was limited to demonstrate the capability for time stamping mechanism in FlexRay and not to perform a comparison between actual timing results with schedule analysis expectations. The system model consisted of two independent sensor ECUs (Sn_1, Sn_2) connected to a CAN based Sensor

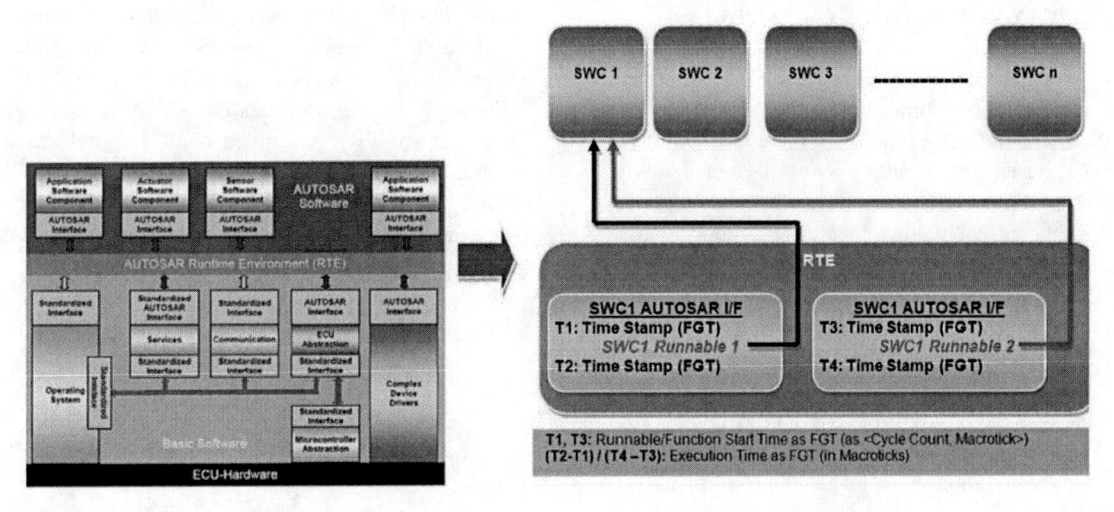

Figure 6. Timing trace insertion in AUTOSAR based ECU

bus. A control ECU (Cn_GW), connected to both CAN and FlexRay buses, performs the sensor fusion and control actions based on sensor input. The output from the control ECU (Cn_GW) were applied to four actuator wheel node ECUs (Ac_FL, Ac_FR, Ac_RL, Ac_RR) that generate the appropriate actuator outputs as RPM values. The sensor modules on CAN and some of the actuators on FlexRay were simulated using Vector.CANoe and are capable of transmitting user controllable sensor/actuator feedback information to the control ECU. The system model, software component allocation, data dictionary generation and application hex file for FlexRay nodes were developed using the AUTOSAR DaVinci tool suite (from Vector CANTech) on a Freescale MPC5567 evaluation board. The FlexRay protocol stack for AUTOSAR 3.0 was used to build the network communication for the actuator and gateway ECUs.

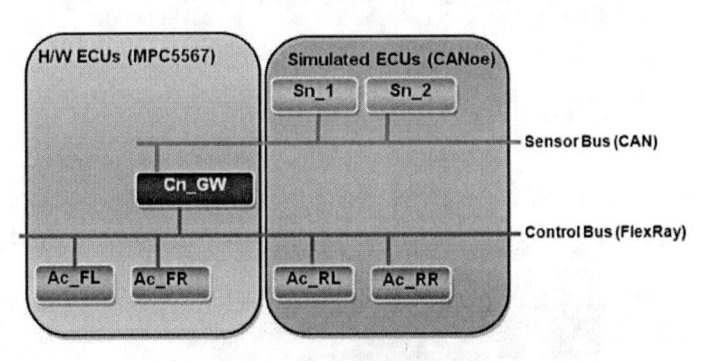

Figure 7. Network Model for Case Study

Some of the FlexRay configurations defined for this system are listed in Table 2 as a reference.

Table 2. FlexRay Configuration for Case Study

FlexRay Configuration	Value
Protocol Specification	2.1A
Channels Used	Channel A only
Baud Rate	10 MBit/s
Cycle Length	5000 µs
Macrotick Duration	1 µs
Static Payload Length	16 bytes
Static Slots	40
Mini Slots	425
Startup Nodes	Cn_GW, Ac_FL
Synchronisation Nodes	Cn_GW, Ac_FR, Ac_FL

The function distribution for this experiment is shown in Table 3. The Cn_GW ECU receives sensor inputs from the CAN system and performs sensor fusion and transmits the appropriate control decisions on the FlexRay bus. The actuator ECUs receive the control signal and performs appropriate actuation along with notification of actuator signal output to the Cn_GW ECU via FlexRay. All functions are configured to be event triggered and will execute on reception of network signal input.

Table 3. Task Breakdown in FlexRay

Functions in FlexRay ECUs		
Cn_GW	Ac_FL	Ac_FR
Sensor Fusion	Wheel Actuation	Wheel Actuation
Control Function		

Using this network model, experiments were performed to illustrate the capability of time stamping using the FGT parameter in a FlexRay network. The experiments listed are used to illustrate a subset of the capabilities which can be achieved using this option and may or may not be relevant to individual feature implementations.

DISTRIBUTED TASK EXECUTION TIMES

The actuation functions are duplicates in terms of behavior (*Wheel Actuation*) for the Ac_FL and Ac_FR nodes and will generate outputs based on a single control signal input transmitted by Cn_GW on FlexRay. These actuator signals need to counterbalance each other to ensure a coordinated control. The execution time (calculated as T2-T1 / T4-T3 in Figure 6) of the *Wheel Actuation* function may need to be monitored on these ECUs as shown in Figure 8, to determine causes for drifts in Actuator signal outputs (approx 2 - 4 rpm observed for the experiment). There are significant differences in the task execution time as seen in Figure 8 for "*Ac_FL Exe Time*" and "*Ac_FR Exe Time*" signal values of the two actuators which maybe relevant for optimization/ control strategies.

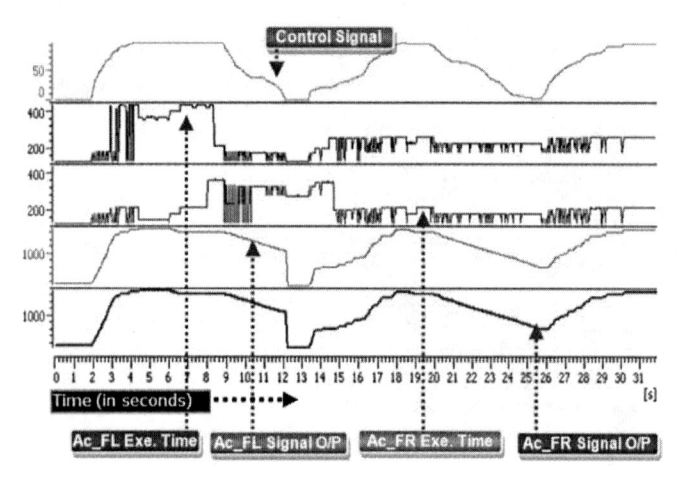

Figure 8. Task Execution Time of Ac_FL and Ac_FR Nodes

Table 4. Parameter Description for Figure 8

Signal	Range (Y axis)
Execution Time (Ac_FL)	190 – 423(µs)
Execution Time (Ac_FR)	190 - 336(µs)
FL Signal O/P	0 – 1800 (rpm)
FR Signal O/P	0 – 1800 (rpm)

DISTRIBUTED TASK TRIGGER TIMES

The start time of actuation tasks "*Wheel Actuation*" on different ECUs based on the same control signal on FlexRay may also be relevant in assisting with system analysis and function partitioning decisions. With the availability of FGT, a common reference time is available to determine triggering of relevant tasks for a given control signal input. The FGT consists of Macrotick count within a specific cycle count (0-63). Capturing both these values at start of task execution (calculated as T1 / T3 in Figure 6) would provide a relative reference point for comparison studies. The time information will assist definition of OS Task configurations/triggers, FlexRay schedule design, load distribution etc. As seen in Figure 9 the trigger times of "*Wheel Actuation*" function on Ac_FL, Ac_FR for a given control signal are captured over time. Since a Macrotick duration corresponds to 1 µs (ref. Table 2), the difference in trigger time from the experiment between the two actuator tasks for the same control input as shown in Table 4 is of the order of 1-2ms.

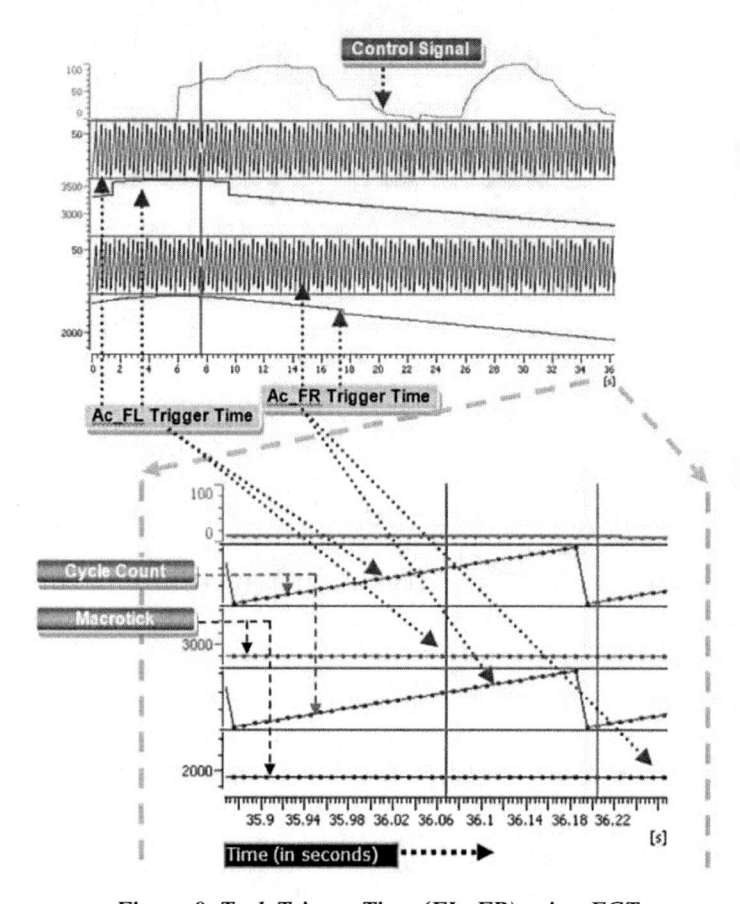

Figure 9. Task Trigger Time (FL, FR) using FGT

Table 5. Trigger Time description for Figure 9

Timing Parameter	Range (Y axis)
Trigger Time (Ac_FL)	
Cycle Count	0 – 63
Macrotick	3700 – 2800
Trigger Time (Ac_FR)	
Cycle Count	0 -63
Macrotick	2900 - 1900

ECU-LEVEL TIMING ASSESSMENT

At an ECU level, some of the relevant timing values could be captured using the FGT option without any additional debugging hardware needs. In-vehicle and system bench development/design activities could benefit from such capabilities during development/ integration sessions to understand overall feature behavior. The Cn_GW ECU consists of multiple tasks (ref. Table 3) which will handle processing of sensor input from CAN and transmit control signals on FlexRay. A possible timing attribute of interest could be the time from reception of the CAN Message to generation of the control signal. Since this timing is distributed across two individual tasks, a timing bound could assist with optimization of task priorities and up-integration of tasks based time constraints at an ECU level. Figure 10 lists the variation in timing from message input to control signal generation at the Cn_GW ECU.

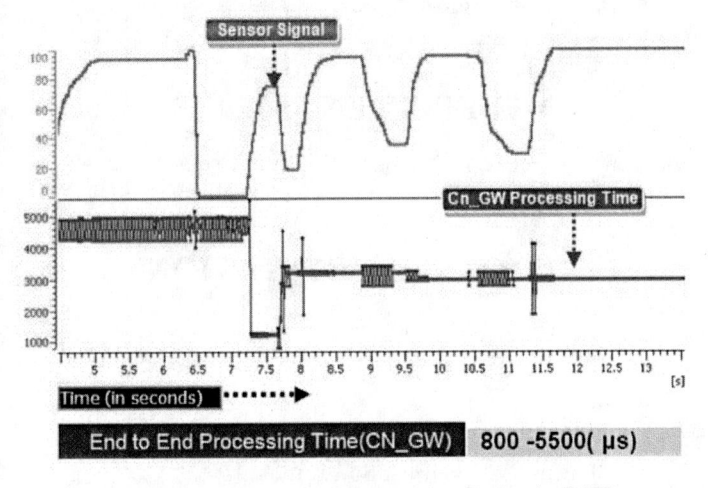

Figure 10. End-to-End Processing Delay in ECU (CN_GW)

Another timing attribute of interest is the time taken from generation of the control signal at the application to the actual transmission of network signal on FlexRay. When using the AUTOSAR stack different configuration options are available at the AUTOSAR COM Layers [5] to define the transmit task execution context as interrupt based or cyclic tasks (Deferred/Immediate). The software architect for an ECU would benefit from time stamping of the message processing delay attribute to identify appropriate configuration settings. Figure 11 lists the processing delay associated with transmit signal updates at the application level. The timing value in Figure 11 is not reset and will transmit the previous value until an update of the information occurs.

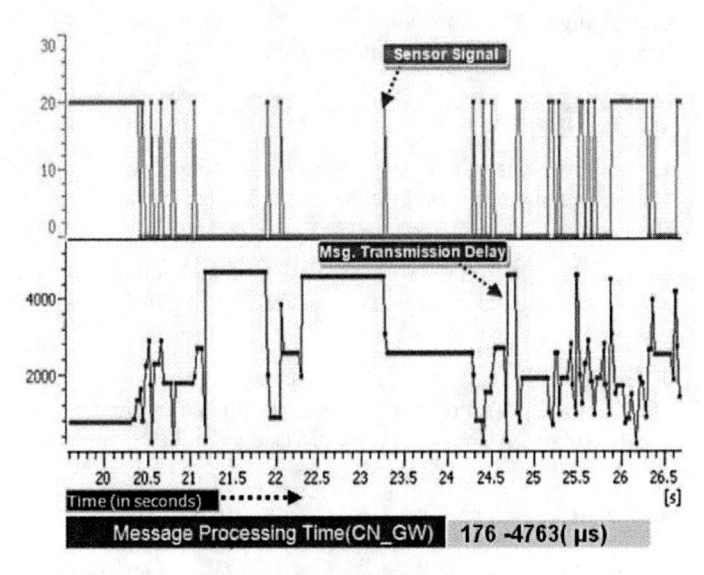

Figure 11. Message Transmission Delay in ECU (CN_GW)

The above evaluations from the case study list possible options available to system designers and architects during the initial design phases to improve/refine system requirements and timing attributes when using FlexRay networks. As in most cases, the timing requirements evolve during the actual development phase and rapid prototyping with use of time stamping could assist with arriving at relatively accurate predictions at an early phase of design.

CONCLUSIONS

The use of FlexRay Global Time for software based time-stamping is an innovative and inexpensive solution to assist network architectures exploration and function allocation studies. This mechanism could be applied to improve overall system timing requirements during the prototyping and development phases. The use of only network monitoring tools without the need for design tools like debuggers on multiple ECUs to study the timing impacts would be an added advantage to this method. The key differentiator for this method is the ability to work with "realistic" timing values to derive or enhance design level requirements. The local notion of global time in an ECU connected to FlexRay could be useful for software design verification/validation for time-critical features in the automotive domain. This approach may influence global time sharing contracts between multiple ECU suppliers on FlexRay systems and also improve message scheduling decisions. Tasks that have execution time on the order of the clock precision of FlexRay may however have limited accuracy when using this method. The allocation of time stamp messages in FlexRay may impact bandwidth of the dynamic segment, so some form of data multiplexing and slot multiplexing strategies may need to be applied to limit the number of FlexRay slot assignments for time stamping purposes in the system.

REFERENCES

1. FlexRay Consortium, www.flexray.com, "FlexRay Communications System Protocol Specification Version 2.1 Revision A".

2. Beck T., "Current trends in the design of automotive electronic systems", Design, Automation and Test in Europe, 2001.

3. Eberhard, D., Schneider, R., Grosshauser F. and Brewerton, S., "Timing Protection in Multifunctional and Safety-Related Automotive Control Systems," SAE Technical Paper 2009-01-0757, 2009.

4. Rambow Thomas, Kiencke Uwe, Schlor Rainer and Seibertz Achim, "A Framework for Optimized Allocation of Control Functions to a Distributed Architecture", SAE 2005 World Congress (2005: Detroit).

5. AUTOSAR GbR "Technical Overview V2.2.1 R3.0 Ver 0001", www.autosar.org.

6. Richter, K., Jersak, M., "OEMs and Supplier Must Cooperate on Timing Analysis when Integrating FlexRay-Based Chassis Systems," SAE Technical Paper 2009-01-0752, 2009.

7. Di Natale Marco, "Virtual Platforms and Timing Analysis: Status, Challenges and Future Directions", Design Automation Conference, 2007. DAC '07. 44th ACM/IEEE.

8. Traub, M., Lauer, V., Becker, J., Jersak, M., Richter, K., Kühl, M., "Using Timing Analysis for Evaluating Communication Behavior and Network Topologies in an Early Design Phase of Automotive Electric/Electronic Architectures," SAE Technical Paper 2009-01-1379, 2009.

9. Hillenbrand Martin, Müller-Glaser K.D., "An approach to supply simulations of the functional environment of ECUs for hardware-in-the-loop test systems based on EE-architectures conform to AUTOSAR", 2009 IEEE/IFIP International Symposium on Rapid System Prototyping.

10. Freund, U., Jaikamal, V., Löchner, J., "Multi-level System Integration of Automotive ECUs based on AUTOSAR," SAE Technical Paper 2009-01-0918, 2009.

11. Montag Pascal, Görzig Steffen, Levi Paul, "Challenges of Timing Verification Tools in the Automotive Domain", ISoLA 2006. Second International Symposium on Leveraging Applications of Formal Methods, Verification and Validation, 2006.

12. Racu Razvan, Hamann Arne, Ernst Rolf, Richter Kai, "Automotive Software Integration", Design Automation Conference, 2007. DAC '07. 44th ACM/IEEE 4-8 June 2007.

13. Hardung B., Koelzow T., and Krueger A., "Reuse of software in distributed embedded automotive systems", In EMSOFT, pages 203-210. ACM Press, 2004.

14. Patzer Andreas, "Optimize ECU parameters with XCP", part of Automotive Design Europe 05/20/09 (in Embedded.com)

CONTACT INFORMATION

Sandeep U. Menon
Electrical & Controls Integration Lab
General Motors R&D
30500 Mound Road
Warren, MI 48090
Tel.: +1 586-298-0261
Fax: +1 586-986-1647
sandeep.menon@gm.com
http://www.gm.com

ACKNOWLEDGMENTS

The author would like to thank Massimo Osella & Lawrence Peruski for their guidance, reviews and comments.

DEFINITIONS/ABBREVIATIONS

FGT
    FlexRay Global Time

ECS
    Electronic Control & Software

TDMA
    Time Division Multiple Access

CAN
    Controller Area Network

ECU
    Electronic Control Unit

AUTOSAR
    AUTomotive Open System ARchitecture

RPM
    Revolutions per minute

RTE
    AUTOSAR Runtime Environment

SWC
    Software Component (AUTOSAR)

XCP
    Universal Measurement and Calibration Protocol

MARKET AND CONSUMER PREFERENCES

2009-01-0784

Driver's Attitudes Toward the Safety of In-Vehicle Navigation Systems

Andrew Varden and Jonathan Haber
University of Guelph

Copyright © 2009 SAE International

ABSTRACT

There is anecdotal evidence of drivers blindly following in-vehicle navigation system (IVNS) commands. IVNSs have shown to be distracting and mishaps with the device have entered popular culture as a source of comedy. Manufactures have reacted by warning drivers of the dangers involved in operating the devices and in some cases prevent address input while moving. While IVNSs are increasingly being used, do drivers perceive their use as distracting, potentially misleading, and thus dangerous?

We conducted an online survey of over 200 drivers to determine their attitudes toward safety while using these devices. This was followed by a series of interviews with an additional 20 drivers to provide more in-depth results. Drivers reported that distraction is not a big issue for them when using an IVNS, with only 8% reporting that the device was too distracting at times. Over 90% of respondents believe IVNSs do not have a harmful or potentially injurious effect and they are not wary of the device. They also placed more trust in directions from IVNSs than from people. There is a discrepancy between drives' attitudes towards safety and potential dangers of using an IVNS. Drivers may be unaware of how distraction affects their driving. Some did not feel using an IVNS was dangerous at all because they are ultimately responsible for any incidents while driving.

INTRODUCTION

IVNS combine information from GPS (Global Positioning System) satellites and electronic maps to provide turn-by-turn directions. Both sources of information are fairly reliable, but not perfect. The systems seem to perform well enough that in recent years the popularity of IVNS has increased significantly. The increased use of IVNSs has led to some concerns about safety. There are concerns about drivers operating the devices while driving. Entering a destination takes significantly longer than dialing a 10 digit cell phone number [1]. In order to avoid this from happening some IVNS manufacturers lock out certain functions when the car is moving, especially destination entry, but others, especially portable systems, do not. There is also concern about drivers becoming distracted by the IVNS while driving and drivers following IVNS directions that are incorrect [2].

There are safety concerns when using an IVNS, including the potential for distraction. Even though an IVNS is a tool designed to help drivers there is still the possibility that they will be misused.

The following is an excerpt from a product safety information sheet produced by an IVNS manufacturer:

When navigating, carefully compare information displayed on the unit to all navigation sources, including information from street signs, visual sightings, and maps. For safety, always resolve any discrepancies or questions before continuing navigation and defer to posted road signs.

Do not become distracted by the unit while driving, and always be fully aware of all driving conditions. *Minimize the amount of time spent viewing the unit's screen while driving and use voice prompts when possible. Do not input destinations, change settings, or access any functions requiring prolonged use of the unit's controls while driving. Pull over in a safe and legal manner before attempting such operations.*

The unit is designed to provide route suggestions. It is not designed to replace the need for driver attentiveness regarding road closures or road conditions, traffic conditions, weather conditions, or other factors that may affect safety of driving [3].

There are three safety issues being identified: be aware of all navigation sources, do not become distracted by the IVNS, and take IVNS directions as suggestions – not commands.

The first safety issue encourages comparing IVNS directions with other navigation sources. There is a warning to defer to posted signs when conflicts arise with IVNS. This warning is important because it establishes that conflicts do exist and that despite all the technology that goes into IVNS directions they can be wrong.

The second safety issue, highlighted in bold, is a warning not to become distracted by the IVNS. One way to become distracted is to try and operate the device while driving. IVNS users are warned not to do this and to safely pull over to the side of the road if they need to operate the IVNS. The other form of distraction is looking at the IVNS screen while driving. The safety warning seems to contend that glances to the screen are inevitable as it instructs people to minimize, and not eliminate, this behaviour. The screen provides useful navigation information making it a tempting distraction. This information may aid a driver but if a glance turns into a stare then the results could be dangerous. Distraction is not a trivial matter as it is estimated to play a factor in anywhere from 25 – 80% of all crashes [4].

The third safety issue concerns IVNS short comings regarding traffic, construction, weather, or any road conditions that can impact driving. This warning reiterates the need to pay attention to the road but also points out that an IVNS is not aware of everything and it is up to the driver to ultimately decide what is best. Instead of route directions, IVNS use the term route suggestions to make the point that is more important to follow the road than the IVNS.

It may seem intuitive that people should pay attention to their surroundings and that IVNS routes can be wrong; however, the media has reported several incidents that illustrate otherwise. An ambulance was supposed to make a 30 minute drive to a hospital 12 miles away in Brentwood, Essex. Instead, the IVNS routed the drivers to Brentwood, Manchester, and it was not until they had driven for four hours before they realized the mistake [5]. Another example involved a pair of car thieves who were apprehended when their IVNS routed them to a border crossing, and right into the hands of law enforcement [6]. It is amazing that the car thieves did not notice that they were approaching the border until they were at the crossing. A twenty year old student, who was borrowing her boyfriend's car, ended up driving onto train tracks where the car was struck by a train [7]. "I put my complete trust in the sat nav," she explained, "and it led me right into the path of a speeding train." This occurred after she approached a gate with a sign that stated: "if the light is green, open the gates and drive through." She opened the gate and drove through. When she got out of her car to close the gate the train struck. Apparently until that point she was not aware that she was at a railroad crossing as it did not show up on the IVNS: "Obviously I had never done the journey before so I was using the sat nav - completely dependent on it." The driver did concede that the IVNS was not completely at fault, and that she had to take some responsibility for what happened:

I can't completely blame the sat nav because up until there, it did get me where I needed to go. If maybe I had been more aware of the situation, I wouldn't have had the accident. But I would be a bit more wary of the sat nav next time because they try to take you the shortest route, and not always the most accessible route and not always the safest route.

Perhaps the most astonishing thing about the above scenario was not an isolated incident. A computer consultant who was driving a rental car made a right turn as advised by an IVNS. This led him onto train tracks where his car became stuck. He was able to get out of the car before a train crashed into it; pushing it more than 30 meters [8].

In all these cases the drivers should have been more aware of where they were going, but they blindly followed the IVNS. There is no indication that the typical IVNS user blindly follows their system, but it happened often enough in the village of St. Hillary, United Kingdom, that the local council erected a sign to warn drivers not to follow their IVNS [9]. A local traffic engineer explained the situation:

We've had a series of problems with drivers getting into trouble by trusting their satnav - and we needed to do something about it. I hope my sign should do the trick.

The problem he was referring to is that the IVNS was directing drivers down a narrow lane that is unsuitable for heavy trucks, with more than a dozen becoming stuck and causing traffic jams for hours.

However, signs are not always enough. The following story illustrates how an IVNS user completely ignored road signs while driving:

A week ago, I saw a car drive the wrong way on a main street in Montreal.

At the corner, another car coming from the correct direction flashed his lights to try to alert the errant motorist, who then started to turn at the intersection, again the wrong way on another one-way street. I waved my arms; he stopped and rolled down his window.

When I explained that both streets were one-way, he replied: "I can't go here? But my GPS system says to go on these streets." I told him, "You have to use your eyes!" and pointed to the one-way sign directly in front of us [10].

It seems that the man was amazed the IVNS could be wrong, that he should take IVNS information as merely route suggestions and not as absolute facts. It is unknown how long he had been using the IVNS, but many of the above incidents involved people who were new to the systems. The ambulance drivers were new to the job, the student who drove onto the train tracks was borrowing the car, and the computer consultant who drove onto the train tracks was in a rental car. In an examination of IVNS usability Nowakowski, Green, and Tsimhoni caution that experienced IVNS users may not encounter the same problems observed by the inexperienced users in their study [11]. Experienced user may still become over reliant on IVNS directions – especially effective systems – according to the IVNS design guidelines created by Stevens, Quimby, Board, Kersloot, and Burns [12]. They predict that the more useful an IVNS is to the driver the more they will adapt their behaviour to become reliant on the system.

Experienced IVNS users may have a good understanding of the abilities and limitations of their IVNS and be able to avoid the pitfalls of the inexperienced users. On the other hand, increased IVNS use may make experienced users even more reliant on their systems. The purpose of our study is to examine experienced users' attitudes toward safety while using an IVNS.

METHOD

A questionnaire was designed to provide insight into measuring IVNS trust, the context for IVNS use, as well understand why IVNS may be used improperly or not used at all. For the purposes of this paper we will focus on why participants used an IVNS.

Demographic questions were asked to understand who was answering the survey and account for difference of age, gender, and experience. Development of the questionnaire started by examining NaviQ an online questionnaire to study IVNS satisfaction [13]. We limited our survey to 44 questions to make it less time consuming for respondents. We asked questions to understand when people use, misuse, and disuse IVNS and have participants rate their trust in IVNSs.

Jian, Bisantz and Drury [14] created a questionnaire to measure trust. In order to create this questionnaire they conducted a word elicitation study where participants generated large set of words related to trust. These words were then rated by a separate group of participants for their relevance with trust or distrust. Lastly, 30 words that were highly rated for trust and distrust were used in a paired comparison study. A factor analysis was then conducted to create 12 clusters of words. We used these questions as the basis for our measurement of trust. In order to examine the specificity of IVNS trust additional trust ratings were collected about IVNS point of interest information.

The entire questionnaire was piloted with four IVNS users known to the investigators to ensure that the directions were clear and the questions were easy to understand. This process led to some questions being reworded and other questions being omitted altogether.

IMPLEMENTATION

Participants were primarily recruited from online communities of IVNS users, while some were also recruited from the local geographic community. Online communities were found by conducting a search at www.google.ca for "forum" and keywords such as "GPS", "car", and "navigation". Forum rules were read to determine if posting a request for survey participants was appropriate. When applicable, forum moderators were contacted to ask for permission. Participants accessed the survey by clicking on a link in the posting. The link directing them to the first page of the survey also acted as the consent form.

Participants were asked to give their consent, confirm that they were at least 18 years of age, and that they had used an IVNS. Participants who answered affirmatively were presented with the remainder of the survey. Participants of the online survey did not receive any compensation.

The questionnaire was also used in a series of interviews that allowed participants to provide more in-depth feedback than in an on-line survey. Interview participants were recruited from southern Ontario. Unlike the online survey where participants were self selected an effort was made to capture representatives from all age groups and genders. Interviews consisted of the same questionnaire as in the electronic survey, but participants were allowed to ask questions and provide additional comments at any time. Interviews took approximately 30 minutes to complete. Participants received $10 in remuneration.

RESULTS

DEMOGRAPHICS

Gender - The respondents were over 90% male, with 190 males participating compared to only 20 females. The amount of male respondents is not surprising given that in a previous IVNS survey [13] where participants were solicited from online forums there were 97% (131 out of 135) male respondents. Although gender disparity in overall Internet use has dissipated [15] males tend to use specialty websites more than females [16]. Another explanation for the gender imbalance could be that males tend to use navigation systems more than females. The 20 female respondents did not show any statistically? significant traits in their responses to differentiate themselves from the males; however, the sample size is too small to make any conclusions.

Age - Age distribution by gender is shown in Figure 1. The survey managed to capture respondents of each gender for all age ranges, although as mentioned in the previous section there were only 20 female respondents. Figure 2 compares the distribution with that of US Internet users [17] and licensed drivers [18]. The comparison with the U.S. Internet Users and Drivers is meant to illustrate the type of age distribution that might be considered normal when responses are obtained through an Internet survey involving drivers. The comparison shows that the survey follows the general bell-shape trend of Internet users, but with fewer respondents on the tail-end, 18-24 and 65 or older categories, than the distribution of Internet users or licensed drivers would suggest. The lower turnout from younger people may be a result of there being fewer IVNS users in that age category due to the cost of an IVNS, the forums targeted may have less people in that age range, or it may reflect an aversion to taking the survey. There may have been few respondents aged 65 years or older because new technology is not designed for older adults [19] and they do not see the need to adopt new technology [20]. Recruiting older drivers for IVNS studies has been a challenge for other researchers [21].

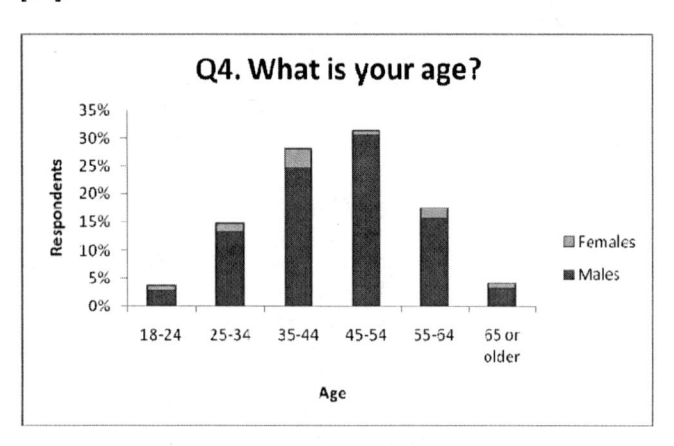

Figure 1 - The age distribution of respondents (n=210).

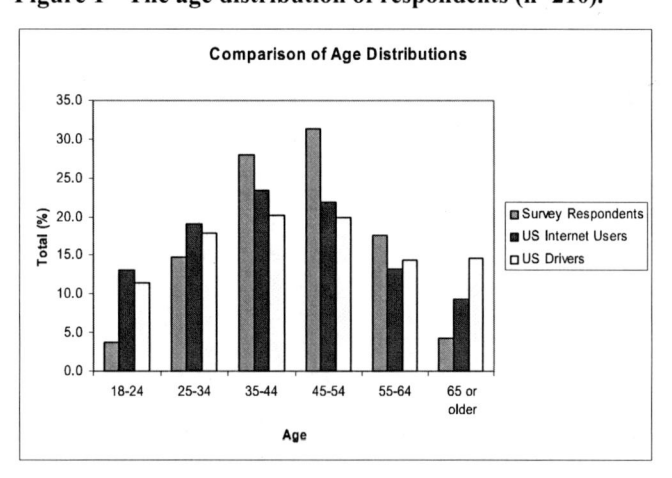

Figure 2 - The age distribution of survey respondents compared with the age distribution for US home Internet users and licensed drivers.

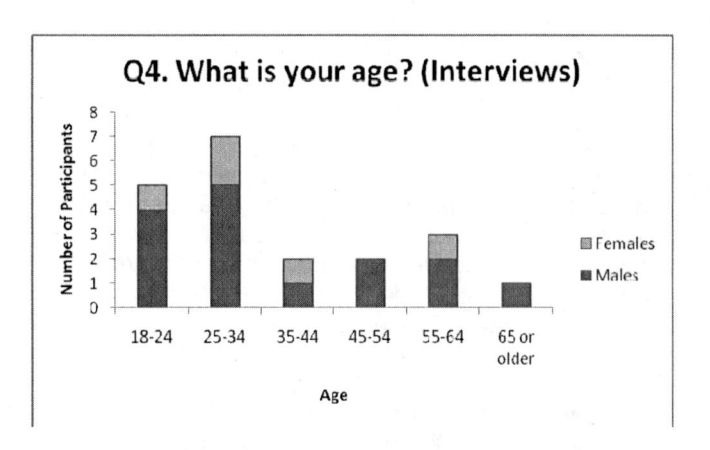

Figure 3 – Age and gender of interview participants.

The age and gender of interview participants is shown in Figure 3. There are representatives from each age range, but there are many more respondents in the 18-24 and 25-34 categories, a reflection of the ease of recruiting participants from a university community compared to older groups.

Experience – In Q5 participants were asked "How often do you use an automotive navigation system?" and were given the choice of "Daily", "At least once a week", "At least once a month", and "Less than once a month". The results shown in Figure 4 are arranged from the most frequent users on the left side of the chart to the least frequent users on the right side of the chart. The majority of respondents are frequent users, with over 60% using an IVNS once a week or more. There is also a good representation from both extreme usage groups, with over 20% of respondents using an IVNS every day and almost 15% using an IVNS less than once a month. All but one driver drove at least once a week providing an opportunity to use an IVNS; however, not all of these drivers were in an unfamiliar area and 90% of respondents who reported driving in an unfamiliar area every day or every week use an IVNS at least once a week.

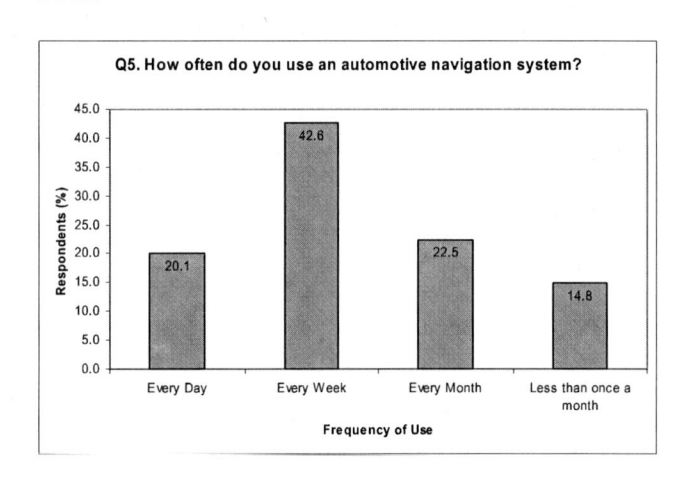

Figure 4 - Frequency of IVNS use (n=209).

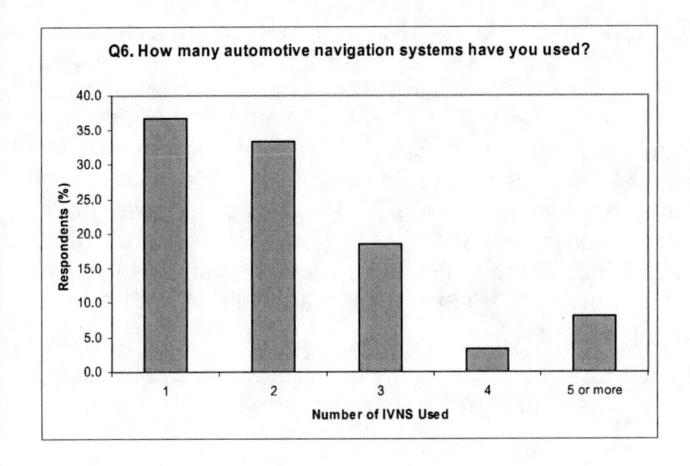

Figure 5 - Number of different IVNS Used (n=210)

It is important to know if respondents' views are based on interacting with only one IVNS or if they have encountered several systems. Experience with different systems gives people a better understanding of IVNS capabilities and should help them use the devices appropriately. The results of Q6 – "How many automotive navigation systems have you used?" shown in Figure 5 indicate that roughly a third have only used one IVNS, another third have used two systems, and the remaining third have used 3 or more systems. Over 8% of respondents fall into a group of very experienced users who have used 5 or more IVNSs. Expert users are less likely to have problems with devices.

Participants were asked how comfortable they are using an IVNS, computers, and driving, and for all three tasks at least 87% reported being comfortable, or very comfortable. The full results are shown in Figure 6 which also reveals that around 9% of participants are also very uncomfortable in all three areas. Using Kendall's tau-b test a substantial positive relationship was found between how comfortable participants are using an IVNS and how comfortable they are using a computer (tau-b = .455, Approx. Sig. = .000, n = 209). There is also a substantial positive relationship between how comfortable participants are using an IVNS and how comfortable they are driving (tau-b = .393, Approx. Sig. = .000, N = 209). This means that the same people who are comfortable with IVNS are comfortable also comfortable driving and using a computer, while those who are uncomfortable using IVNS are uncomfortable driving or using a computer. Those who are uncomfortable are certainly in the minority and that is not surprising given that the method used to recruit participants. Readers of online forums would be familiar with computers, and as previously stated they would tend to be enthusiasts who are more likely to be comfortable using a computer and with IVNS than the general population.

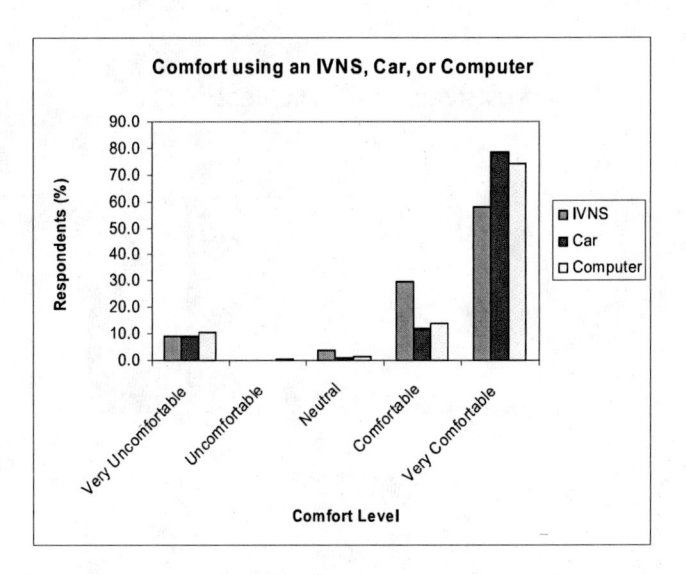

Figure 6 - A comparison of participants' comfort level using an IVNS, car, or computer (n=210).

Participants rated their familiarity with IVNS on a 5 point Likert scale with 1 being not at all familiar, 5 being extremely familiar, and 3 being neutral. As shown in Figure 7 58.5% considered themselves to be extremely familiar with IVNS, while another 33.5% rated themselves as familiar. Kendall's tau-b test found a substantial positive relationship between comfort and familiarity with IVNS (tau-b = .445, Approx. Sig. = .000, n = 188). The fact that two-thirds of respondents have used more than one IVNS also demonstrates that the respondents can be characterized as experienced IVNS users. When examining the rest of the results it is important to keep in mind the results would probably be quite different if our respondents were novice IVNS users.

TRUST

The introduction gave several examples of situations where too much trust in the IVNS resulted in drivers ending up in precarious situations. We measured trust by asking both positive and negatively framed questions. As shown in Figure 7 around 90% of respondents rated the positively framed questions a 4 or 5. The majority of responses for the negatively framed questions were either 1 or 2 as shown in Figure 8, but the percentage varied from as low as 71% for Q33 – "I am suspicious of the navigation system's action or outputs" to 92% for Q35 – "The navigation system's actions will have a harmful or injurious outcome".

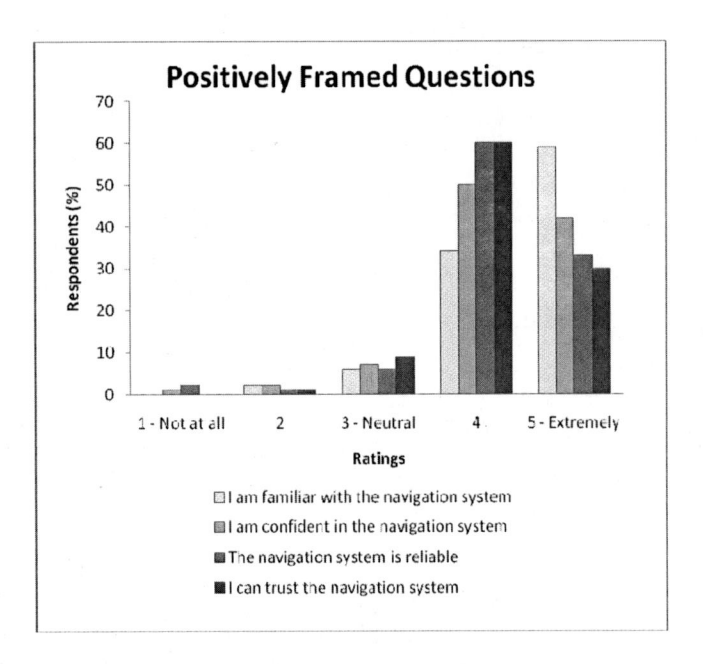

Figure 7 - Positively framed trust questions (n=188).

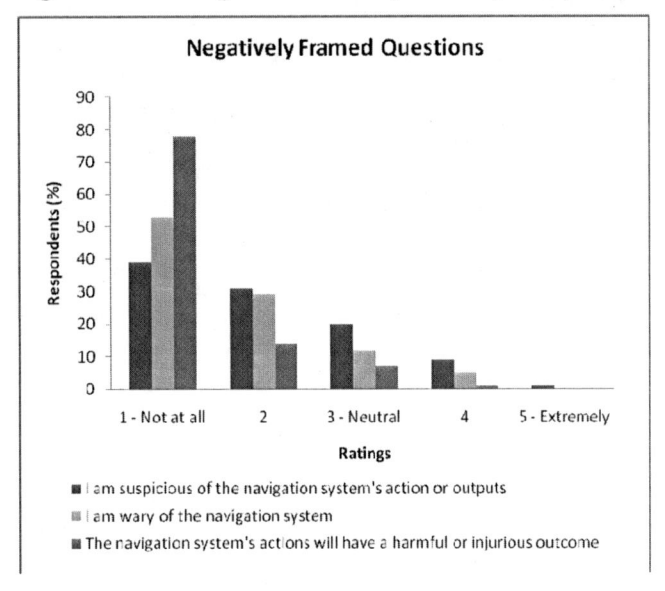

Figure 8 – Negatively framed trust questions (n=188).

Kendall's tau b test was performed on the responses and a very strong positive relationship was found between three of positively framed statements: "I am confident in the navigation system", "The navigation system is reliable", and "I can trust the navigation system" (Kendall's tau-b >= .634, sig. = .000). Both the high ratings for the positively framed statements and the low ratings for the negatively framed statements indicate that the respondents have a high level of trust when using an IVNS.

ATTITUDES TOWARDS SAFETY

A high level of trust does not necessarily mean that too much faith is placed in IVNS directions. In order to address if the amount of trust that respondents have in IVNS is appropriate participants were asked a series of questions about what directions they followed when a conflicts arose between IVNS, people, and road signs. Figure 9 shows that 45.8% of respondents have followed an IVNS when it conflicted with advice from a person, but 41.5% have followed a person when it conflicted with advice from an IVNS. Road signs were more authoritative with 31.9% of respondents having followed an IVNS instead and only 27.4% reported following directions from a person when it conflicted with a road sign.

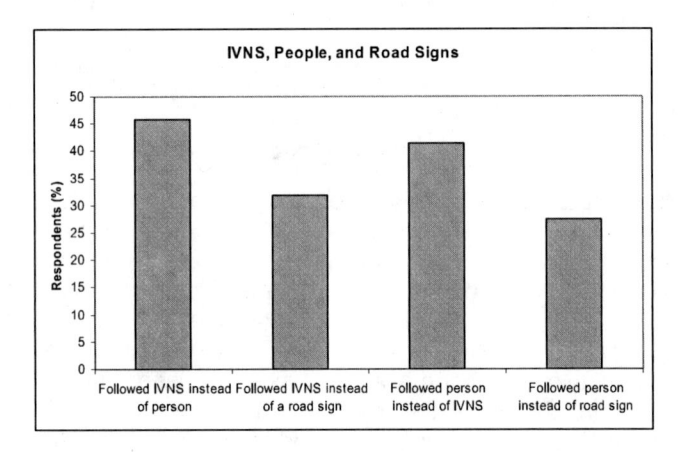

Figure 9 - Respondents were asked if they had followed an IVNS when it conflicted with a person or road signs, and also if they had followed directions from a person when it conflicted with IVNS or a road sign.

Participants also provided examples of when following all three sources turned out to be the right decision and times when they should have done the opposite. A common reason for following an IVNS instead of road signs is that the IVNS would provide a better route:

That was one of the situations that got me into trouble. The highway department had changed the exits so the maps in the unit were incorrect.

Other respondents made it clear that each incident is different and that it is always important to use common sense:

It does make mistakes - it's up to me to check the info I get from it.

One person mentioned serious mistakes that had occurred:

I have seen it tell me to turn when there was a jersey barrier preventing a turn, and a couple other times when turning isn't actually possible.

This participant's awareness that their IVNS might direct them to make an improper turn is good, but if another IVNS user takes an instruction as a command the results could be dire. Respondents provided some indications that they are less attentive to road signs when using their IVNS:

I rarely look at street signs unless the gps tells me to do something illegal. Illegal right turn or something.

While another respondent remarked:

When I am following the nav, I do not give priority to road signs.

Although incidences were mentioned when an IVNS gave illegal or incorrect instructions respondents also said that the device works correctly in most cases. Participants were willing to accept small errors occasionally made by the system:

Roads are always changing, so minor "mis-steps" by a navigation device are not outside reason.

Small errors may even be a good measure at preventing drivers from becoming too reliant on IVNS. Not a single respondent indicated they would blindly follow an IVNS, but some respondents are more inclined to follow the IVNS than other information sources.

<u>Respondents felt there was little/no risk involved</u> – There was very little concern about the possibility of IVNS use being unsafe with 78% of the respondents in our study thought that an IVNS are not at all harmful. In fact, only one respondent felt stronger than neutral about the possible danger of using an IVNS.

Experienced IVNS users feel that they are aware enough not to be led astray by the IVNS directions, and several stated that common sense must be used in all circumstances. This is a different attitude than the student described in the introduction who put her "complete trust" in the IVNS when using it for the first time. Respondents in our study knew, that although their devices are highly accurate they do make mistakes from time to time.

Only 8% of respondents felt that distraction was a problem when using an IVNS. It may be that experienced IVNS users know better than to manually operate the device while driving, pay too much attention to the display screen, or be distracted by the voice commands. Another possibility is that IVNS users are being distracted although they are not aware of it. A study on driver distraction due to cell-phones and perception of performance found that drivers underestimate the negative effects distraction has on their driving [22].

<u>Some people drive differently with an IVNS</u> - Some of our participants mentioned they pay less attention to road signs when using an IVNS. It would be interesting to know how widespread this tendency is among IVNS users. Paying less attention to signs can be seen as a benefit to safe driving or a detriment. The benefit is that drivers spend less time searching intersections for street signs, so that they do not miss a turn, and/or scanning buildings for addresses. They can also spend less time thinking about logistics, debating what exit to take on the highway, As a result, drivers can spend more time with their eyes and attention on the road. The downside is that they may miss signs that are important. Can people filter out the signs that are made redundant by the IVNS and still pay attention to the ones that matter?

It is also possible that some IVNS users are paying less attention to road signs and do not even realize it. This is one problem with self-reporting. If people spend less time looking at signs where they may spend more time looking at the road, the IVNS, their cell phone, or somewhere else entirely. These types of question lend themselves to eye tracker experiments, but even further interview questions on the subject may be fruitful.

Another concern is that extensive use of an IVNS may result in drivers forgetting routes. One participant mentioned how a relative had become completely dependent on their IVNS. When their vehicle with an IVNS was taken in for repair they had trouble finding their way to places they had been driving to for decades. It is possible that by paying less attention to navigational cues with an IVNS, IVNS users forget to pay attention to navigational cues without it, or that without having to call upon routes by memory the information no longer remains fresh and is lost.

FUTURE WORK

Our study elicited responses from experienced IVNS users from many different age groups. There were some groups who were under-represented, mainly females, 18-24 year olds and those over 65 years of age. A simple extension of our study would focus on any of these groups using different recruiting methods. It may also be beneficial to create a new questionnaire or add questions that specifically target an age group or address gender issues. One of the respondents in our study reported that the safety of using an IVNS instead of stopping to ask directions was important to her because she was female. While male respondents also cited safety issues with strangers this may be more important to female IVNS users who were under-represented in our study.

While not explicitly asked in our study, one respondent made a comment about the benefit of knowing when to turn ahead of time:

As I am aging, I appreciate that my GPS helps me better anticipate maneuvers ("in .3 miles exit left").

There may be other aspects of IVNSs that are particularly appealing to older, younger, or female drivers that our study would have missed and there also may be a great difference between experienced and inexperienced IVNS users in these groups.

It is also important to establish how attitudes towards IVNSs develop over time starting with inexperienced users. A long-term study would be able to establish participants' perceptions before ever using an IVNS and examine how those attitudes evolve as participants become familiar with the systems.

CONCLUSION

There is a discrepancy between drives' attitudes towards safety and potential dangers of using an IVNS. Drivers may be unaware of how distraction affects their driving. Some did not feel using an IVNS was dangerous at all because they are ultimately responsible for any incidents that may occur and as one respondent stated: "It's up to me to check the info I get from it." The perceived risk of using an IVNS is low or almost non-existent as respondents are not concerned about distraction or that the IVNS will lead them astray. They understand IVNSs are not perfect and described errors they have encountered. Experienced IVNS users have a great deal of trust in their systems because they find the directions to be more reliable than other sources. Disregard for other navigation is a potentially dangerous habit as drivers could miss important cues, but overall drivers have a good common-sense approach to using an IVNS.

ACKNOWLEDGMENTS

Funding for this research was provided by Auto 21.

REFERENCES

1. Tijerina, L., Parmer, E.B., Goodman, M.J., Individual Differences and In-Vehicle Distraction while Driving: A Test Track Study and Psychometric Evaluation, in Proceedings of the Human Factors and Ergonomics Society Annual Meeting, (2), pp. 982-986, 1999.
2. Green, P., Potential Safety Impacts of Automotive Navigation Systems, presented at the Automotive Land Navigation Conference, June 18, 1997.
3. Important Safety and Product Information Sheet, Garmin, 2007.
4. Neale, V.L., Dingus, T.A., Klauer, S.G., Sudweeks, J., An Overview of the 100-Car Naturalistic Study and Findings, in Proc. Int. Tech. Conf. Enhanced Safety Vehicles, (19), pp. 1-10, 2005.
5. Sun Online, Satnav ambulance 200 miles out, Sun Online, 2 December, 2006, 27 July, 2007. http://www.thesun.co.uk/sol/homepage/news/article73749.ece
6. Wilkes, J., GPS getaway a bum steer, thestar.com, 20 March, 2007, 21 March, 2007. http://www.thestar.com/printArticle/193790
7. Sat nav driver's car hit by train, BBC News, 11 March, 2007, 11 March, 2007. http://news.bbc.co.uk/go/pr/fr/-/2/hi/uk_news/wales/south_west/6646331.stm
8. Associated Press, Man using GPS drives into path of train, MSNBC.com, 3 January, 2008. http://www.msnbc.msn.com/id/22493399/
9. Daily Mail, Council erects sign that tells drivers: Don't trust your satnav, August 27, 2007. http://www.dailymail.co.uk/pages/live/articles/news/news.html?in_article_id=478073&in_page_id=1770
10. Kazenal, S., At the mercy of GPS, The Globe and Mail, Toronto, Ont.: Feb 4, 2008. pg. A.12
11. Nowakowski, C., Green, P., Tsimhoni, O. (2003). Common Automotive Navigation System Usability Problems and a Standard Test Protocol to Identify Them. Transportation Research Institute, (16)
12. Stevens, A., Quimby, A., Board, A., Kersloot, T. Burns, P. (2001). Design Guidelines for Safety of In-Vehicle Information Systems. Transport Research Laboratory, PA, 3721(1)
13. Li, C. NaviQ-A User Satisfaction Questionnaire for In-Vehicle Navigation Systems, M.S. thesis, University of Guelph, Guelph, ON, Canada, 2006.
14. Jian, J.Y., Bisantz, A.M., Drury, C.G. (2000). Foundations for an empirically determined scale of trust in automated systems. International Journal of Cognitive Ergonomics, 4(1), pp. 53-71
15. Ono, H., Zavodny, M. (2003). Gender and the Internet. Social Science Quarterly, pp. 111-121
16. Joiner, R., Gavin, J., Duffield, J., Brosnan, M., Crook, C., Durndell, A., Maras, P., Miller, J., Scott A.J., Lovatt, P. (2005). Gender, Internet identification, and Internet anxiety: correlates of Internet use. Cyberpsychology & Behavior, 8(4), pp. 371-378
17. Day, J.C., Janus, A., Davis, J. (2005) Computer and Internet use in the United States: 2003, US Dept. of Commerce, Economics and Statistics Administration, US Census Bureau. http://www.census.gov/prod/2005pubs/p23-208.pdf
18. Injury Facts, National Safety Council, 2006.
19. Gregor, P., Newell A., Zajicek, M., Designing for dynamic diversity: Interfaces for older people, in Proceedings of the fifth international ACM conference on Assistive technologies, pp. 151-156, 2002.
20. Selwyn, N., Gorard, S. Furlong, J., Madden, L. (2003) Older adults' use of information and communications technology in everyday life, Ageing & Society, (23), pp. 561-582
21. Green, P., Variations in Task Performance Between Younger and Older Drivers: UMTRI Research on Telematics, presented at the Association for the Advancement of Automotive Medicine Conference on Aging and Driving, February 19-20, 2001.
22. Horrey, W.J., Lesch, M.F., Garabet, A., Hopkinton, M.A., Awareness of Performance Decrements Due to Distraction in Younger and Older Drivers, in Proceedings of the Fourth International Driving Symposium on Human Factors in Driver Assessment, Training and Vehicle Design, Stevenson, Washington, pp. 54-60, 2007.

CONTACT

Andrew Varden may be contacted by email: avarden@uoguelph.ca.

SAE International

"Consumer Attitudes and Perceptions about Safety and Their Preferences and Willingness to Pay for Safety"

2010-01-2336
Published
10/19/2010

Veerender Kaul, Sarwant Singh, Krishnasami Rajagopalan and Michael Coury
Frost & Sullivan

Copyright © 2010 SAE International

1. ABSTRACT

The U.S. National Highway Transportation and Safety Agency's (NHTSA) early estimates of Motor Traffic Fatalities in 2009 in the United States [1] show continuing progress on improving traffic safety on the U.S. roadways. The number of total fatalities and the fatality rate per 100 Million Vehicle Miles (MVM), both show continuing declines. In the 10 year period from 1999 through 2009, the total fatalities have dropped from 41,611 to 33,963 and the fatality rate has dropped from 1.5 fatalities per 100MVM to 1.16 fatalities per 100MVM, a compound annual drop of 2.01% and 2.54% respectively.

The large number of traffic fatalities, and the slowing down of the fatality rate decline, compared to the decade before, continues to remain a cause of concern for regulators. The new Corporate Average Fuel Economy (CAFE) standards requiring vehicle manufacturer to meet a fleet wide fuel economy of 35.5 mpg by 2016, has made it even more challenging to maintain the declining rate of fatalities per 100MVM. Automakers' pursuit of vehicle down-sizing and light-weighting strategies works counter to improving safety, as smaller and lighter vehicles feature lower degree of crashworthiness compared to larger and heavier vehicles (greater track width and wheelbase both have positive impact on vehicle stability and safety).

In Europe, the European Transport Safety Council (ETSC) estimates [2] show that 39,000 people lost their lives in road collisions in 2008 across Europe; 15,400 less than in 2001 but still far from the 27,000 deaths limit which the European Union (EU) set for itself in its 2010 Road Safety Target. The average annual progress since 2001 has been 4.4% instead of the 7.2% needed, which could delay the EU in reaching the 2010 target until 2017. In the EU 79 people are killed per million inhabitants in 2008 compared to 113 in 2001. Disparity in road death rates across Europe has decreased since 2001, and in 2008 there was no longer any EU country with more than 150 road deaths per million inhabitants.

The development of passive safety systems has reached near saturation point and now offers limited potential to reduce fatalities. To reduce fatality rates even further, the focus has shifted to active safety systems and advanced driver assistance systems (ADAS). The recent spate of safety recalls involving electronic malfunctioning is likely to lower consumer confidence in advance safety systems in the short-term and impede the adoption of ADAS systems, as consumers are likely to be less willing to pay extra for ADAS systems.

Even as vehicle makers are finding it hard to meet the stricter and contradictory regulations for safety and fuel efficiency; competitive pressures are forcing them to introduce advanced safety systems to achieve highest safety ratings on their vehicles and to differentiate their products. Such systems include Blind Spot Detection (BSD), Lane Departure Warning (LDW), Adaptive Front Lighting (AFL), Night Vision Systems (NVS), Driver Drowsiness Warning (DDW) and Occupant Monitoring systems.

In today's market environment, where demand is weak and margins tight, it is critical for vehicle manufacturers to offer consumers vehicles with features and functions they value most and avoid costly development and consumer dissatisfaction with implementations.

In this paper, Frost & Sullivan analyzes consumer attitudes towards safety and their preferences and willingness to pay for safety features. The analysis is discussed under the following categories.

1. Consumer attitude and concerns for safety

2. Consumer perceptions toward current active and passive safety systems and collision vulnerability

3. Buyer behavior and the influence of safety in the vehicle purchasing process

4. Safety content and system feature preferences

5. Willingness to pay for safety and optimal safety packages

The analysis is based on an online survey of a sample of vehicle owners across vehicle segments and demographic attributes. The study was conducted by Frost & Sullivan in 2008, in the U.S., and in Europe in 2009.

2. INTRODUCTION

Frost & Sullivan conducted separate online survey of the U.S. and the European vehicle owners. While 1,152 consumers participated in the U.S. survey [3] conducted in 2008, 1,938 consumers participated in the European survey [4] conducted in 2009. Respondents were selected based on certain parameters such as age of vehicle owned and future vehicle purchase intentions. Safety features were broadly classified into vehicle stability systems, driver warning and vision systems, collision avoidance systems, occupant protection systems and post collision systems.

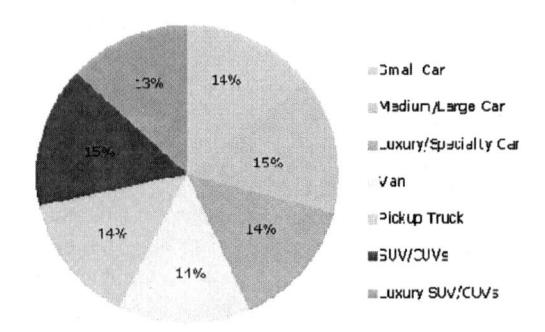

Base, n = 1,152   Source: Frost & Sullivan

Figure 2.a. U.S. Consumer Survey - Respondents Mix by Segments of Vehicle Owned

Vehicle Segment Ownership	Country					
	France	Germany	UK	Spain	Italy	Total
A&B	121	109	102	58	61	451
C	105	109	103	61	62	440
D&E	101	116	101	64	58	440
MPV	87	78	93	39	48	345
SUV	69	68	81	20	24	262
Total	483	480	480	242	253	**1,938**

Source: Frost & Sullivan

Figure 2.b. European Consumer Survey - Respondents Mix by Segments of Vehicle Owned

3. CONSUMER ATTITUDE AND CONCERNS FOR SAFETY

Analysis of U.S. and European consumers' responses clearly highlights the difference between their awareness, attitudes and concerns for vehicle safety. European consumers seem to be more concerned about vehicle safety; have higher level of awareness of advanced safety systems and are more willing to pay extra for additional safety features. U.S. consumers believe they are good drivers; don't require additional safety features in their vehicles and are less willing to pay for additional safety features. This difference is the result of the type of vehicle these consumers drive and the road and driving conditions in the two regions. These findings indicate clear differences in perception of vehicle safety and expectation of safety systems in Europe and the U.S.

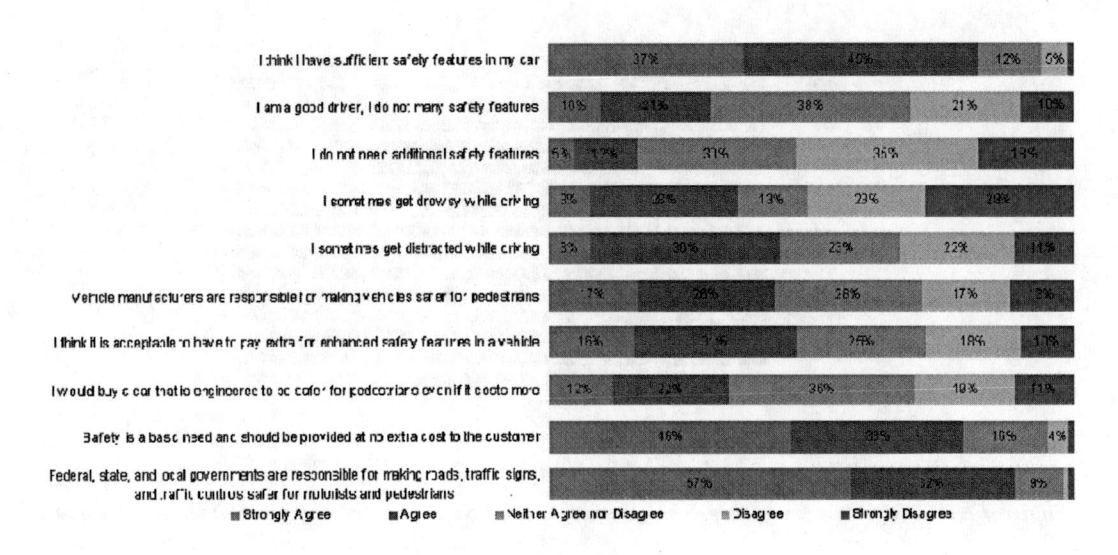

Figure 3.a. U.S. Consumer Awareness and Perception towards Vehicle Safety

Base, n = 1,152   Source: Frost & Sullivan

Figure 3.b. European Consumer Awareness and Perception towards Vehicle Safety

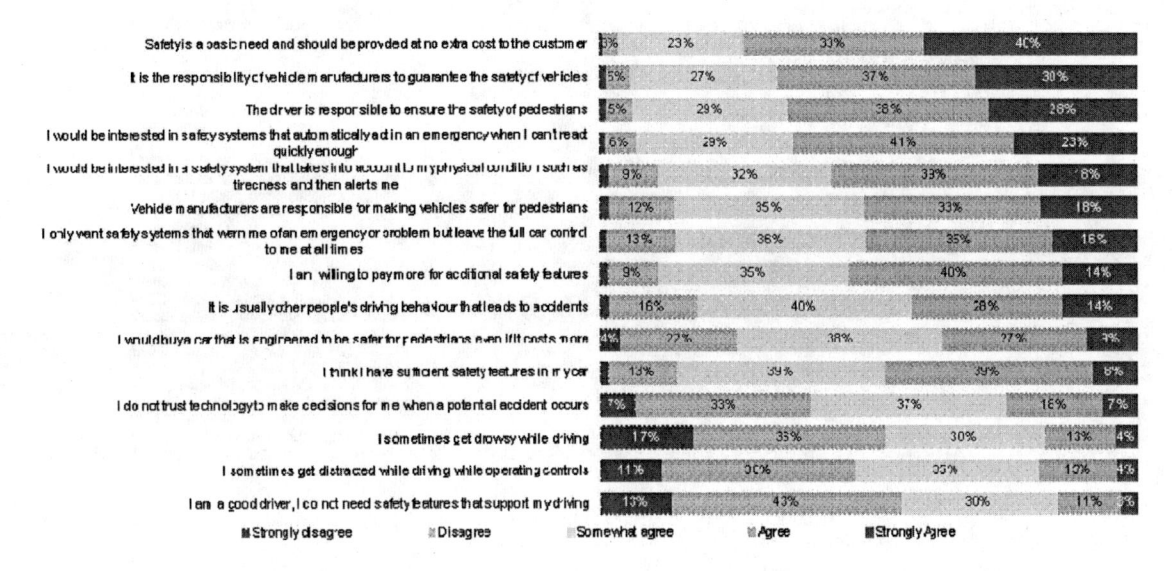

Base, n = 1,938   Source: Frost & Sullivan

4. CONSUMER PERCEPTIONS TOWARD CURRENT ACTIVE AND PASSIVE SAFETY SYSTEMS AND COLLISION VULNERABILITY

U.S. consumers perceive passive safety systems such as seatbelts and airbags to play a very important role when it comes to vehicle safety. While U.S. consumer's strongly associate passive safety features to the concept of vehicle safety, analysis of their responses indicate they are informed and are receptive to newer active safety and advanced driver assistance features. The importance U.S. consumers place on features such as alcohol interlock, BSD, LDW and speed warning systems clearly highlight increasing consumer awareness of advanced safety and driver assistance systems and expectations from vehicle manufacturers to offer these advanced systems.

Though European consumers place high importance on features such as BSD, low speed collision avoidance -partial halt and emergency braking assistance (pre-charging); features such as alcohol interlock, semi-autonomous parking, parking assistance - rear view camera, and traffic sign recognition are not seen as very important.

Base, n = 1,152   Source: Frost & Sullivan

Figure 4.a. U.S. Consumers Perception of Importance of Safety Systems to Overall Vehicle Safety

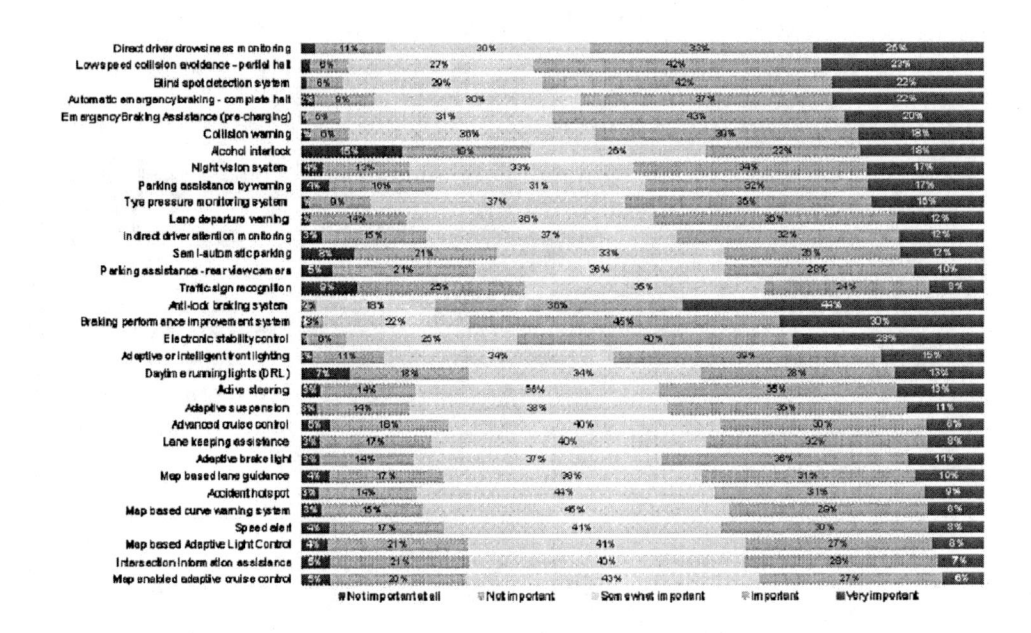

Base, n = 1,938   Source: Frost & Sullivan

Figure 4.b. European Consumers Perception of Importance of Safety Systems to Overall Vehicle Safety

Analysis of consumer responses on vulnerability to collisions highlights that U.S. consumers are more likely to be involved in rear end collisions as compared to other type of collisions. After rear end collisions, it is lateral collisions that U.S. consumers are likely to get involved in. Unlike U.S. consumers, European consumers believe they are more likely to be involved in forward collision, followed by rollover after impact, lateral collision, and then rear end collision.

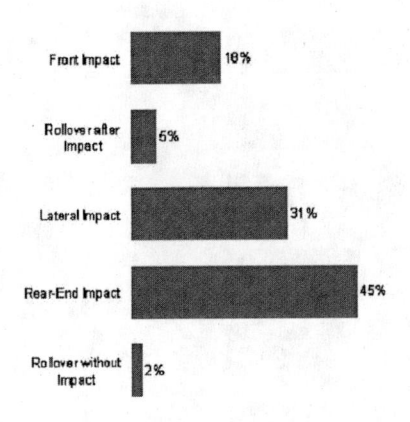

Figure 4.c. U.S. Consumers Feedback on Collision Scenarios Most Vulnerable to When Driving

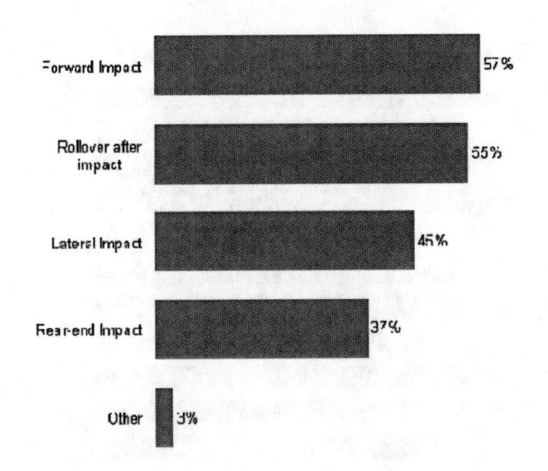

- Front Impact: 18%
- Rollover after Impact: 5%
- Lateral Impact: 31%
- Rear-End Impact: 45%
- Rollover without Impact: 2%

Base, n = 1,152   Source: Frost & Sullivan

Figure 4.d. European Consumers Feedback on Collision Scenarios Most Concerned About When Driving

- Forward Impact: 57%
- Rollover after impact: 55%
- Lateral Impact: 45%
- Rear-end Impact: 37%
- Other: 3%

Base, n = 1,938   Source: Frost & Sullivan

While U.S. consumers tend to perceive a collision scenario where another vehicle hits their vehicle, European consumers tend to perceive it the other way round. The difference in perception is due to the kind of vehicles consumers drive in the two regions. While U.S consumers drive large vehicles with powerful engines, Europeans drive small, low powered vehicles. Also the roadway infrastructure in the two regions plays an important part in consumer perceptions. In the U.S. roads in urban, semi urban and rural areas are wider; relatively less congested and high speed driving is possible when compared to the roads in urban, semi urban and rural areas of major European countries. Analysis of the data also indicate that U.S. consumer perception of vehicle safety is highly influenced by number of safety features offered by vehicle manufacturer in the vehicle; if the vehicle has passed necessary certification tests, and if they are able to see more of the traffic ahead and around their vehicle.

5. BUYER BEHAVIOR AND THE INFLUENCE OF THE SAFETY PACKAGE IN THE VEHICLE PURCHASING PROCESS

U.S. consumers who indicated they were willing to buy a car engineered to be safe for pedestrians (even if it costs more) supported their intentions with a higher budget allocation to technologies that protect them while driving (33%) and minimize injuries to others (24%). Also U.S. consumers allocate most of their budget to safety technologies that protect them while driving, followed by pedestrian protection. Crash compatibility and occupant protection are lower on their priority and budget allocation list, with medium/large car drivers most likely to budget and pay for it, and pick-up truck drivers least likely to budget and pay for it. Proven technology and reputation are key attributes that influence the purchase of safety features. Relative influence of factors like automated driver responses and new, sophisticated technology has increased.

Reliability, safety and comfort lead purchase importance among European consumers when it comes to making a decision on new vehicle purchase. While MPV owner's priority is space, comfort and design & style are more important for D&E and SUV segment vehicle owners. Fuel economy is important for A&B segment vehicle owners. C segment owners see safety significantly more important, which is a reflection of the greater time they spend on the road compared to other segments.

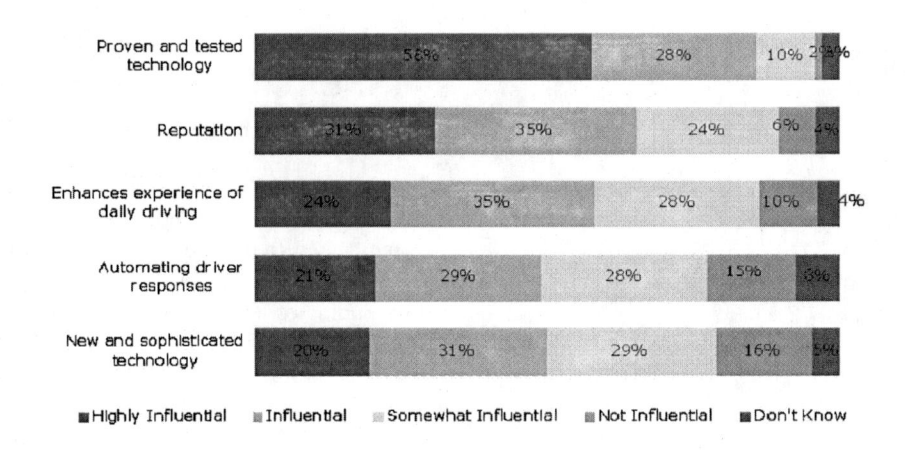

Base, n = 1,152   *Source: Frost & Sullivan*

Figure 5.a. Factors That Influence New Vehicle Purchase Decisions of U.S. Consumers

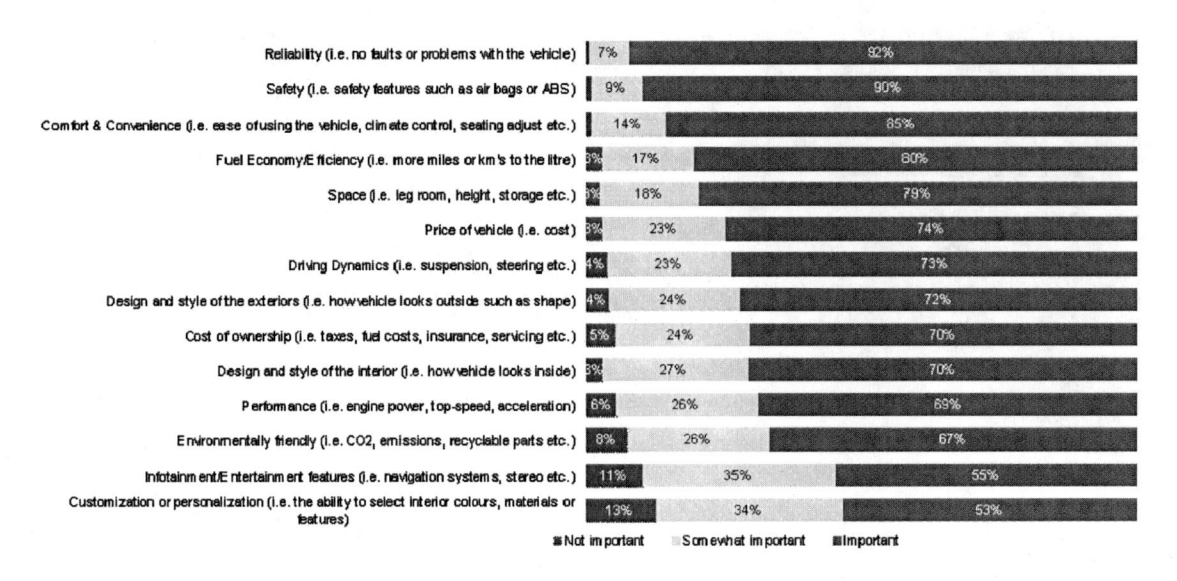

Base, n = 1,938   *Source: Frost & Sullivan*

Figure 5.b. Importance of Factors in New Vehicle Purchase Decisions of European Consumers

6. SAFETY CONTENT AND SYSTEM FEATURE PREFERENCES

U.S. respondents are most likely to purchase occupant protection technologies first, followed by braking or steering technologies. Telematics technologies that relay for assistance after a collision were likely to be purchased last.

When it comes to features that enhance braking, Anti-lock Braking Systems (ABS) remains the most preferred option for U.S. consumers followed by Traction Control System (TCS). The desirability of newer features is lower as consumers prefer tested, proven, and reliable systems/ technologies.

When it comes to features that warn driver of potential risks, technologies that take over driver control (Lane Deviation Control and Driver Drowsiness Control) were the least preferred, with about a fifth of respondents indicating they do not want such features,. A strong preference for these technologies is yet to emerge, highlighting the reluctance of U.S. consumers for such features. However, a third are willing to purchase the vehicle if these features are offered as standard at no extra cost. When it comes to features that protect occupants, with the tendency of respondents to perceive vehicle safety as passive protection, it is not surprising that U.S. respondents continue to desire front and side airbags.

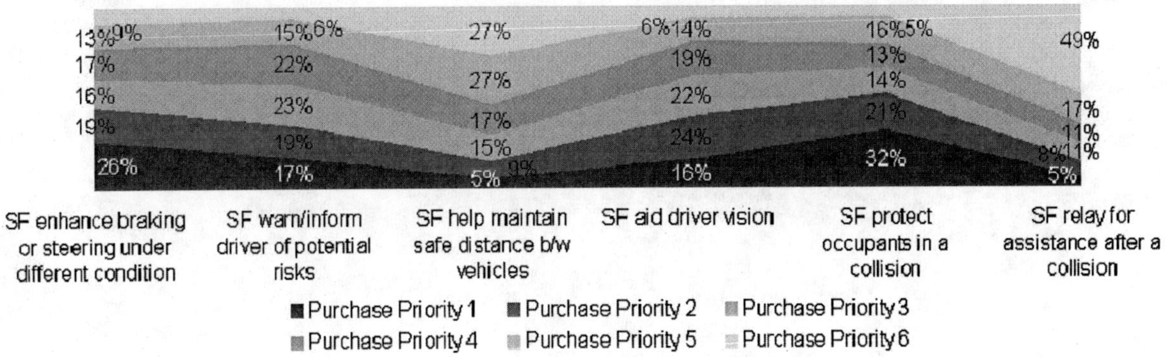

Base, n = 1,152 Source: Frost & Sullivan

Figure 6.a. U.S. Consumers Willingness to Purchase Safety Features if Offered as Optional Add-ons

Safety Group	Safety Systems/Features
Safety features that enhance braking or steering under different condition	Anti-lock Braking System (ABS), Traction Control System (TCS), Electronic Stability Control (ESC), Rollover Protection, Active Steering
Safety Features that warn/inform driver of potential risks	Lane Deviation Warning , Lane Deviation Control, Driver Drowsiness Warning, Driver Drowsiness Control, Park Assist
Safety features that help maintain safe distance between vehicles	Conventional Cruise Control, Adaptive Cruise Control (ACC) with Emergency Stop, ACC with Emergency Stop and Go Functionality
Safety features that aid driver vision	Lane Change Assistant, Collision Warning System, Blind Spot Detection, Night Vision System, Adaptive Front Lighting, Automatic Headlamp Control
Safety features that protect occupants in a collision	Occupant Detection System, Rear Seat Headrests, Curtain Airbags, Side Airbags, Front Airbags
Safety features that relay for assistance after a collision	Automatically notify emergency services, including driver information and medical records, Button to summon emergency services

Source: Frost & Sullivan

Figure 6.b. Safety Features Grouping in U.S. Consumers Survey

From the response of European consumers, it is clearly evident that technologies which help prevent accidents are the most desired by vehicle owners. There are no significant differences by age, by gender, by country or by vehicle segment. Clearly low speed collision avoidance - partial halt and braking performance improvement system, lead overall importance for consumers. Interestingly, while they are the most expensive technologies, automatic emergency braking - complete halt and collision warning, are also highly rated by consumers, almost twice more important than speed alert and parking assistance by warning (ultrasonic park assist).

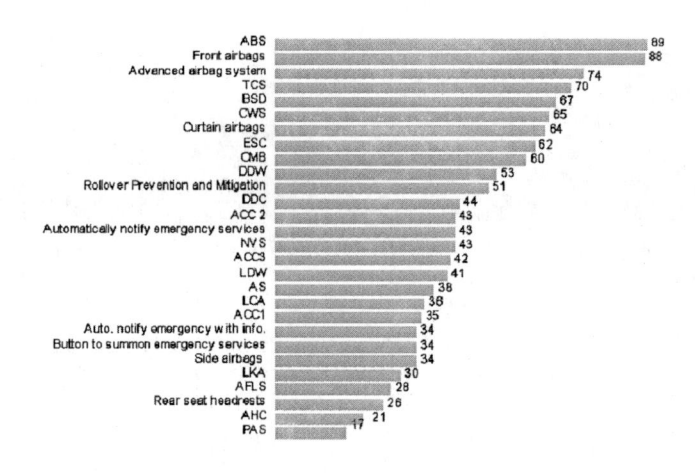

Base, n = 1,152   Source: Frost & Sullivan

Figure 6.c. U.S. Consumers Importance Ranking of Safety Features

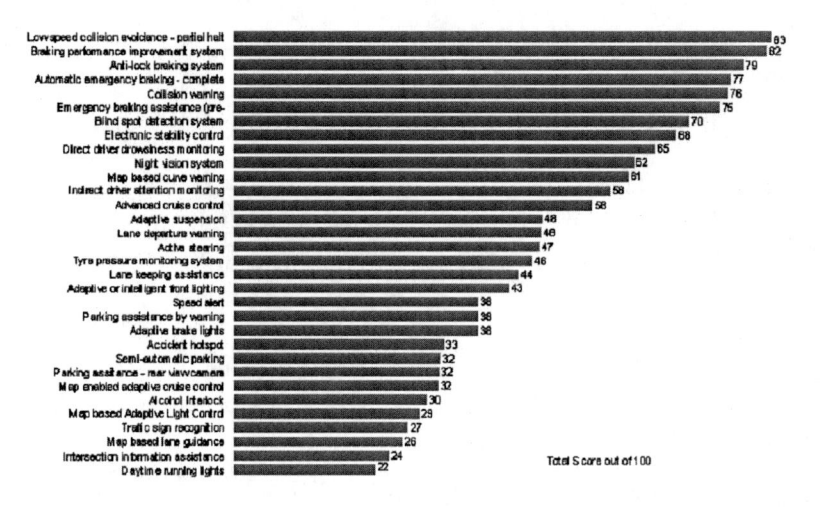

Base, n = 1,938   Source: Frost & Sullivan

Figure 6.d. European Consumers Importance Rating of Safety and ADAS Features

7. WILLINGNESS TO PAY FOR SAFETY AND OPTIMAL SAFETY PACKAGES

Analysis of respondent's feedback clearly indicates that U.S. consumers consider safety features that protect occupants in a collision and features that enhance braking or steering such as ABS, TCS, ESC, airbags, BSD as important features when it comes to new vehicle purchase. Consumers expect safety features that help maintain safe distances between vehicles, and features that enhance braking or steering such as ACC, Collision Mitigation Braking, Collision Warning System and Rollover Mitigation and Prevention to be offered as standard equipment. U.S. consumers expect other safety features such as LDW, AFL, park assist system, NVS to be offered as optional.

Analysis of European respondent ratings indicate that consumers are willing to spend 33% of their budget on safety technologies that protect while driving as against only 18% for post-crash support systems. European consumers had rated safety second in terms of importance, only after reliability. The importance consumers place on high cost (over €1,000) ADAS features indicate that they are willing to pay more for avoidance systems clearly indicating that ADAS systems are likely to buck the trend and attract European consumers to pay for these features.

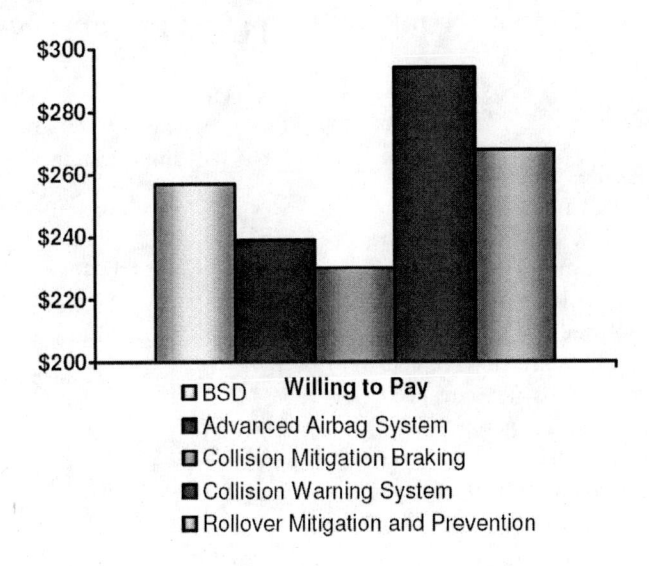

Base, n = 1,152   Prices are in US Dollars   Source: Frost & Sullivan

Figure 7.a. U.S. Consumers Willingness to Pay for Key Safety Features

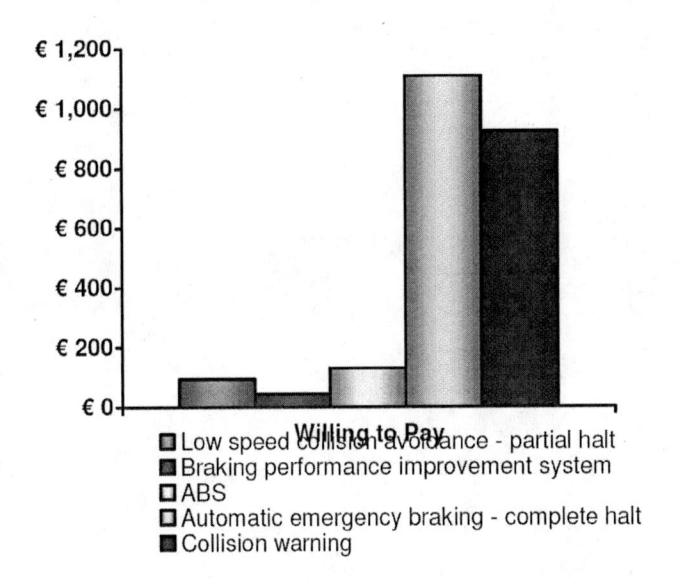

Base, n = 1,938   Source: Frost & Sullivan

Figure 7.b. European Consumers Willingness to Spend on Additional Safety Features on Next Vehicle Purchase

8. SUMMARY/CONCLUSIONS

Analysis of consumer feedback from the U.S. and European surveys clearly illustrate the difference in perception and awareness of U.S. and European consumers when it comes to vehicle safety. It also highlights the difference in their attitude towards active and passive safety systems and advanced driver assistance systems, and willingness to pay for these systems during their next vehicle purchase.

Type of vehicles (based on size, power and road presence) and road infrastructure play a key role in shaping consumer attitudes and perceptions towards vehicle safety. Majority of U.S. consumers believe they are good drivers, they don't require significant additional safety features, and they are likely to get involved in accidents because of the vehicles around them. Whereas majority of European consumers tend to think the other way round; that they require additional safety features to keep them safe and they are likely to be more involved in accidents caused by frontal impact. The difference in perceptions of the U.S. and European consumers is explained by the vehicles they drive and conditions in which they drive. While U.S. consumers drive larger vehicles, European consumers drive compact vehicles primarily in urban roads. Narrower roads and traffic congestions on average are higher in Europe as compared to the U.S.

U.S. consumers perceive passive safety systems such as seatbelts, airbags and occupant protection systems as most important followed by active safety systems such as ABS and TCS. With ABS, TCS and smart airbags already mandated by NHTSA [5]; U.S. consumers perceive BSD, collision warning system, collision mitigation braking and rollover mitigation system as important in order of purchase priority.

European consumers perceive advanced driver assistance systems such as low speed collision avoidance -partial halt, braking performance improvement system, ABS, automatic emergency braking - complete halt, and collision warning as most important. However interesting to note is that both U.S. and European consumers rate occupant protection systems and braking / collision related systems as most important when it comes to safety.

With a spate of recent vehicle recalls raising number of questions about safety of vehicles, consumer perceptions and attitudes towards advanced safety features is expected to be less positive in the short-term. Recent vehicle recalls have also raised questions about the reliability of electronic and autonomous systems that take control away from the driver. As a result, consumers are expected to be less willing to pay for newer safety features compared to tried and tested safety features. While there is a big opportunity for vehicle manufacturers and suppliers to introduce / offer advanced driver assistance systems in the long-term, it is going to be a major challenge to prove the reliability of these systems and convince consumers to pay for these systems in their next vehicle purchase.

9. REFERENCES

1. U.S. Department of Transportation, DOT HS 810 983, "Evaluation Program Plan, 2008 - 2012", (August 2008)

2. ETSC News Release Embargoed until 14 January 2010, "Ambitious Road Safety Targets Needed for 2020", (14 January 2010)

3. Frost & Sullivan Report "U.S. Consumers' Desirability and Willingness to Pay for Safety Systems", (2008)

4. Frost & Sullivan Report "European Consumers' Desirability and Willingness to Pay for Advanced Safety and Driver Assistance Systems", (2009)

5. U.S. Department of Transportation, DOT HS 811 291, NHTSA Traffic Safety Facts Crash Stats - A Brief Statistical Summary, (March 2010)

10. CONTACT INFORMATION

Brian Drake
248-836-82603
brian.drake@frost.com

11. DEFINITIONS/ABBREVIATIONS

NHTSA
National Highway Transportation and Safety Administration

EU
European Union

ETSC
European Transport Safety Council

ADAS
Advanced Driver Assistance Systems

ABS
Anti-lock Braking System

LDW
Lane Departure Warning

BSD
Blind Spot Detection

ESC
Electronic Stability Control

AFL
Adaptive Front Lighting

NVS
Night Vision Systems

DDW
Driver Drowsiness Warning

SAE International

Commercial Business Viability of IntelliDrive℠ Safety Applications

2010-01-2313
Published
10/19/2010

Robert White, Tao Zhang, Paul Tukey and Kevin Lu
Telcordia Technologies

David McNamara
MTS LLC

Copyright © 2010 Telcordia Technologies, Inc.

ABSTRACT

This paper presents modeling, analysis, and results of the business viability of a set of IntelliDrive [1] safety applications in a free market setting. The primary value drivers for motorists to adopt the IntelliDrive system are based on a set of safety applications developed and analyzed by the US DOT. The modeling approach simulates IntelliDrive on-board equipment adoption by motorists based on the value of the safety applications. The simulation model uses parameters that are based on adoption rates in a similar dynamical system from recent history and incorporates feedback loops such as the positive reinforcement of vehicle-to-vehicle applications value due to increased adoption. This approach allows the analysis of alternative IntelliDrive business approaches, deployment scenarios, and policies. The net present value of the IntelliDrive system to the nation is computed under alternative scenarios.

INTRODUCTION

The Promise of IntelliDrive is Improved Safety and Mobility

IntelliDrive is a U.S. Department of Transportation (US DOT) initiative to develop and demonstrate technologies for using wireless communications technologies to improve transportation safety, mobility, and sustainability. The nation-wide deployment represents a significant investment by a diverse set of stakeholders, both public and private. As described by the US DOT outreach website, "IntelliDrive aims to enable safe, interoperable networked wireless communications among vehicles, the infrastructure, and passengers' personal communications devices." IntelliDrive will ultimately enhance the safety, mobility, and quality of life of all Americans, while helping to reduce the environmental impact of surface transportation.

IntelliDrive as the Vehicle Infrastructure Integration (VII) was formally announced at the ITS America's 2003 Annual Meeting. The US DOT gathered stakeholders from state DOTs, automobile manufacturers, and others to create a working group representing the public and private interests and to create a consensus over deployment. Many at that time expected that VII would be a federally funded project of the magnitude of the US Interstate Highway System; it would be a "wireless network" that covered our nation's highways and intersections. Since that time, several successful test beds and demonstrations have been funded, but no consensus or plan for a nation-wide deployment has emerged. Today, IntelliDrive remains largely a federally funded set of research projects.

As IntelliDrive matures beyond the research phase, a crucial issue becomes a viable business case for the nation-wide deployment of IntelliDrive systems and applications. Studies have shown that IntelliDrive applications, when widely deployed, can provide significant economic benefits [1]. However, it is also widely recognized that some important IntelliDrive applications, such as vehicle safety applications based on vehicle-to-vehicle communications, will provide benefits to drivers only after a high percentage of all the vehicles are equipped with the same applications, which can take many years and require heavy investments. Similarly,

---

[1] The IntelliDrive℠ logo is a service mark of the U.S. Department of Transportation (US DOT).

traffic signal phase and timing (SPAT) applications based on Dedicated Short-Range Communications (DSRC) will generate significant benefit only when a large number of the dangerous intersections are equipped with DSRC communications capabilities.

Collaboration and Investment possible with a Viable Business Model

Further complicating matters is the fact that supporting IntelliDrive applications requires collaboration among many industry sectors, private and public. For example, automotive manufacturers need to install communications and applications capabilities on vehicles. Network operators and transportation agencies need to collaborate to deploy roadside network infrastructures such as DSRC equipment at intersections. Network providers, software providers, and automotive manufacturers need to collaborate to establish the enabling infrastructure required for vehicle communications, such as the public key infrastructure for supporting security for vehicle communications. Device makers will deploy IntelliDrive applications on after-market devices. As illustrated by these examples, supporting any set of IntelliDrive applications will require multiple parities to deploy different pieces of an integral system.

Given that many IntelliDrive applications will provide significant value only when either a large number of vehicles or a large infrastructure network is deployed, stakeholders have been reluctant to jump in with their respective investments. Automotive manufacturers want to see roadside infrastructures be deployed before deploying onboard equipment. Parties involved in deploying infrastructure networks don't want to invest in the deployment and wait through a long uncertain period of time before their deployment can generate economic benefits.

The experience of the authors is that before an endeavor of this magnitude is undertaken -- installation and on-going maintenance of a vast transportation system -- key business questions must be addressed:

• Who will benefit?

• Who can invest?

• What and when is the pay back: i.e., what is the business model?

• When and how will the project be launched; i.e., the business plan?

These important questions are typically addressed as part of a business plan which details the cost of deployment and projects the return on investment over-time. We think the current issues with deployment stem from a lack of a viable and credible business plan. The attitude of "build it and they will come" is even more unrealistic in light of the current global financial crisis. An IntelliDrive business plan, including the formal statement of a set of goals, the rationale for why they are attainable, and the financial and operational plans for reaching these goals, is of paramount importance today. Fortunately, many in the community see this need and the dialogue is now underway, with deployment ideas being considered. The authors consider the model detailed by this paper an important element in creating a viable IntelliDrive business plan - assessing and quantifying the benefits in credible financial terms, is the cornerstone.

System Dynamics as a Tool to Model and Quantify Benefits

In this paper, we study the commercial viability of deploying safety applications. We present a business-modeling tool that can be used to answer the fundamental question: driven by commercial markets, what IntelliDrive business models and deployment strategies will be viable and practical?

A major challenge in understanding the business viability of deploying complex systems and applications, such as IntelliDrive systems and applications, is to model the complex interactions of the many factors that impact the business cases. As discussed above, the value to the users depends on the set of applications and how widely they are deployed. Some applications can provide value only after a high percentage of other vehicles are equipped with the same capabilities or when a large roadside network infrastructure is implemented, while other applications may provide value even when a small number of vehicles are equipped with the applications. As the deployment of the IntelliDrive systems and applications grow, the value to the users will grow; recognizing that this user value growth is typically not linear. As the user value grows, more and more users will be motivated to join the system, further increasing the system value. We present in this paper a business case modeling tool that uses system dynamics techniques [2] to model and analyze the interrelations among the many business impacting factors.

The model can support any combination of applications. For the results presented in this paper, we focused on the safety applications in the Volpe study [1] : Signal Violation Warning, Stop Sign Violation Warning, Curve Speed Warning, and Electronic Braking Lights. We further considered the impact of deploying more applications to increase the value provided to the users.

VALUE OF THE INTELLIDRIVE SYSTEM

The John A. Volpe National Transportation Systems Center in the US DOT has published a study that estimated the benefits of a set of VII[2] safety applications based on reduced

Table 1. National benefits and present value per vehicle of VII safety applications from the Volpe study

Safety Application	Type	National Benefit ($B)	Value per Vehicle ($)
Electronic Brake Lights	V2V	14	54
Signal Violation Warning	RSE/Intersection	11	44
Stop Sign Violation Warning	RSE/Intersection	3	11
Curve Speed Warning	RSE or Database	15	58
Total		42	168

crashes and other benefits under the assumption that intersections and vehicles would be equipped over time under an assumed deployment of vehicles [1]. That study calculated a benefit for each of the safety applications as a present value, in which the benefits from reduced future crashes were discounted to the present day as a total for the entire United States. The safety applications and the projected benefits (in billions of 2008 dollars) are shown in Table 1.

In the current analysis, we are interested in studying the effect of the value of the safety applications on the decisions of individual motorists to adopt the system. This requires a value that the motorists would gain by having an On-Board Equipment (OBE) on their vehicles. The average value of the system per vehicle shown in Table 1 is derived by dividing the national benefit (the national present value of the safety applications) by the current number of vehicles in use in the US (250 million).

The "Intersection" type safety applications rely on an IntelliDrive Road-Side Equipment (RSE) being installed in an intersection for an equipped vehicle to receive transmissions that carry warnings and other information. This means that a significant portion of all intersections need to be equipped before equipped vehicles can receive benefits.

"Electronic Brake Lights (EBL) is a Vehicle-to-Vehicle (V2V) application that would provide a warning to the driver in case of the sudden deceleration of a forward vehicle. The OBE of the lead vehicle would send a signal to other vehicles if its longitudinal deceleration exceeds a predetermined threshold, thereby allowing those following drivers to be aware of this deceleration even if their visibility is limited by weather conditions or obstructed by large vehicles."[3] The value of V2V applications depends on the fraction of all vehicles that are equipped to support the application. This is an externality similar to that of a communications network, in which the benefit of joining the network grows as the network size grows.

"The Curve Speed Warning (CSW) application provides an in-vehicle warning to the driver if the vehicle's speed is higher than the recommended speed for the curve. The system can be designed to receive the information from an RSE or to use the OBE and a downloaded navigation map to make an assessment. In the first case, the RSE compares the vehicle speed with the recommended speed and sends a signal to the vehicle if there is a potential danger. In the latter case, the OBE compares the vehicle speed to the recommended speed that is stored with the navigation map data. Road condition data can also be used in this process to fine-tune the speed warning based on weather and other factors."[4]

Other safety and commercial applications envisioned can be added to the system over time. These include electronic payment for tolls as well as goods and services, and private applications that could be installed either by an OEM or in the aftermarket.

The current modeling and analysis envisions the initial OBEs will be available in the aftermarket to provide the safety applications listed in Table 1. Over time, additional applications will become available and increase the value of the system. As the system is adopted by increasing numbers of motorists, OEMs will be incented to offer an IntelliDrive option on new vehicles.

MODEL STRUCTURE

The approach is to model the growth of value to motorists as intersections are deployed with RSE. Additional value will accrue as OBE are deployed on vehicles and the benefits of V2V applications such as EBL increase. Some safety applications such as CSW can be available as soon as the system is initiated. The general model structure is shown in Figure 1.

---

[2] Vehicle Infrastructure Integration (VII) was the prior name given to IntelliDrive by the US DOT.
[3] See reference [1].
[4] See reference [1].

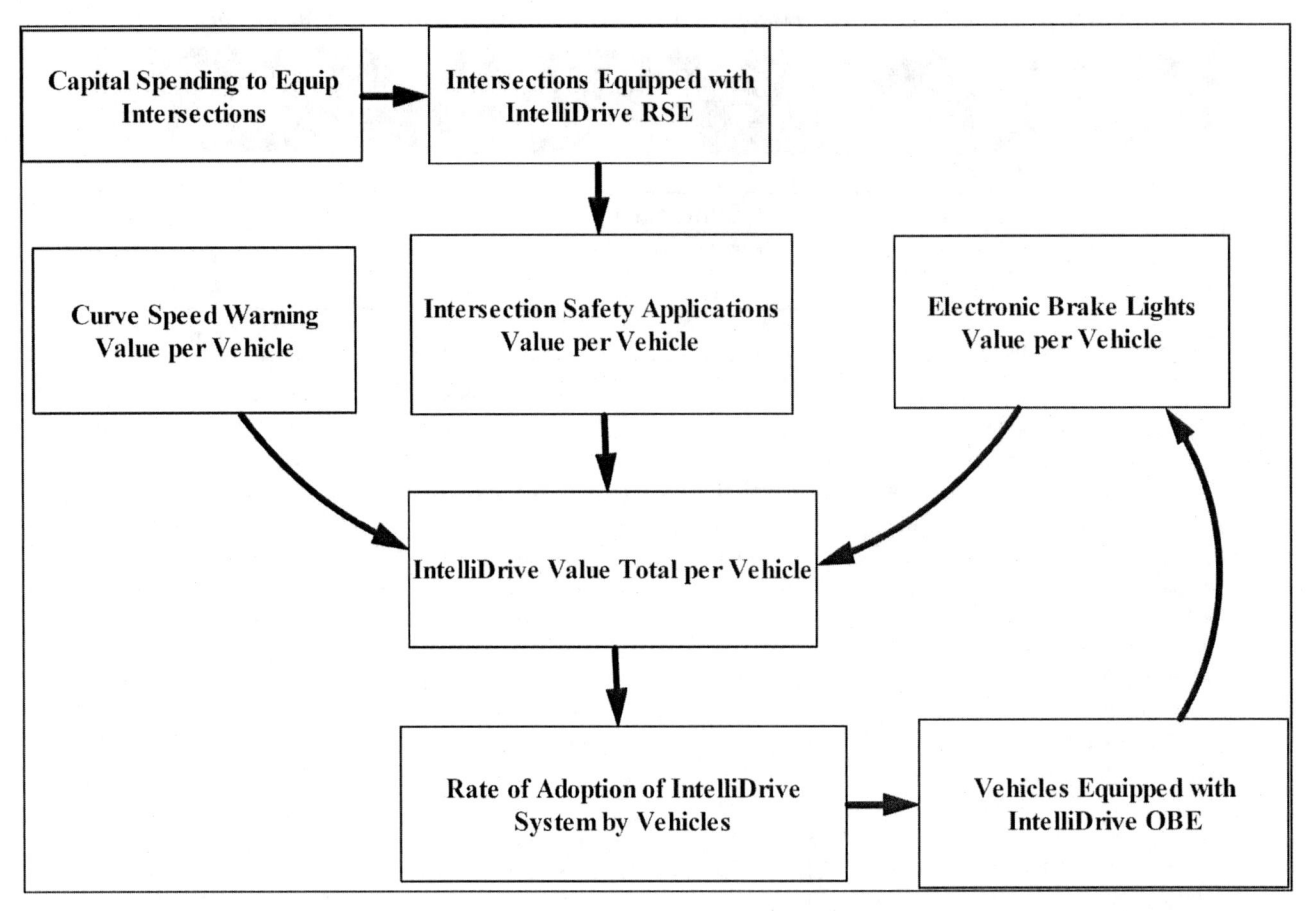

Figure 1. Model influence diagram showing contribution of value flows from safety applications

The model dynamics are based on the underlying assumption that the growth in system adoption is proportional to its average value to the motorist. System value comes from the value of the three groups of safety applications shown in Table 1. The model was implemented in the Vensim system dynamics modeling language [3].

The value of intersection safety applications is driven by the deployment of intersections. Capital spending by the government drives the deployment of RSEs in intersections and other locations. As the number of equipped intersections increases, the value to motorists increases. This value increase, in turn, will motivate more motorists to install the OBE on their vehicles.

The value to motorists of the CSW application can be realized as soon as the system is initialized if this safety application is map driven. The OBE can estimate safe curve speeds from its current position on a map. A warning is issued if the OBE detects that the speed of the vehicle entering the curve exceeds the calculated safe limit. In the current analysis, CSW is an RSE-based application.

The value of V2V safety applications such as EBL is directly tied to the number of other vehicles that are equipped. As the number of equipped vehicles grows, the chances of a vehicle encountering another equipped vehicle increases; this, in turn, increases the value of EBL adding to the value of the entire system.

The question of setting the parameters in the IntelliDrive penetration model is difficult because the system has not yet begun to be deployed and decisions by motorists have not yet been made. Similar systems for which we have a full history of deployment and customer adoption can be used to provide approximate parameters if care is taken in correctly mapping the coefficients. One such system is the E-ZPass electronic tolling system that was originally deployed in New York and New Jersey in the 1990's [4, 5, 6]. We set the functional form in the E-ZPass penetration growth model identically to that in the IntelliDrive penetration growth model. In each case, the penetration growth is proportional to the average value of the system per vehicle. This common structure allows us to use the calibrated E-ZPass coefficients in the IntelliDrive model. By transferring these coefficients we are assuming that motorists' reaction to average value per vehicle is the same

Fraction of Target RSE Locations in Operation by Year

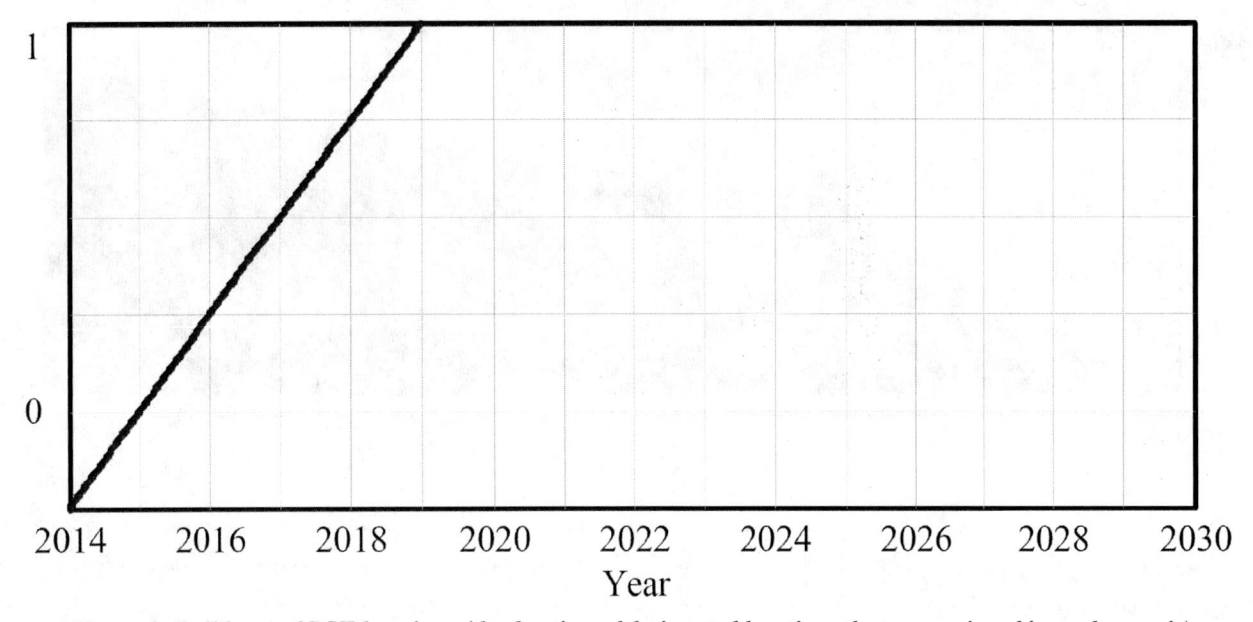

Figure 2. Build-out of RSE locations (the fraction of designated locations that are equipped is on the y-axis)

for both systems. This model calibration is discussed in Appendix 1.

INTELLIDRIVE PENETRATION MODEL BASELINE RESULTS

The baseline scenario for the IntelliDrive Penetration Model (IPM) is based on the Volpe study assumptions in [1]. However, since the business drivers are fundamentally different in the current commercial business model [7] than the model assumed in the Volpe study, some basic assumptions will be different. For example, the Volpe study assumes a deployment schedule of new light vehicles based on a prior US DOT ITS Joint Program Office study [8]. In the current model, the schedule of vehicle OBE adoption is not an input, but is calculated as an output.

Contrasting to the Volpe study that assumed 100% of all vehicles would adopt the system regardless of the value, the ultimate IntelliDrive system penetration in a commercial model cannot be known at this point. Market research can be employed to estimate the market potential as a function of price and features. We have assumed 80% ultimate penetration for the baseline scenario.

The Volpe study assumed a five-year build-out of RSE locations starting in 2014 and continuing through 2018. We interpret this to mean that all 252,000 sites for RSEs identified in the Volpe study are equipped in the five-year rollout. Figure 2 shows the assumed build-out of RSE locations.

The value of the system grows as RSE locations are developed and equipped. The resulting aggregate value from all safety applications is shown in Figure 3. This chart shows value from EBL, which begins to accrue as vehicles are equipped. We included "other" applications in the baseline scenario that add value to the system without identifying these explicitly. Other applications are assumed to be introduced in 2014 and add $5 per vehicle in value per year thereafter.

The value of the RSE safety applications builds as locations are equipped between 2011 and 2016; after 2016 it holds flat at $122 per vehicle. In the baseline analyses, we assume that CSW is an RSE application. Alternate scenarios can consider CSW being based on location and downloaded map data. In the later case, the value of the CSW application would be constant at $59 per vehicle, and becomes available on day one.

The Volpe study [1] uses $50 for the cost of the installed OBE. Volpe notes that they have received comments that this cost may be too low. The initial OBE in the current study is envisioned as an aftermarket unit that motorists will purchase and install. Typical of such products, the initial cost will be significantly higher than the long term cost. In the baseline scenario, we have assumed that the initial cost of the OBE will be $200, dropping over time to $50. The rate at which the cost will drop was taken in this model to be independent of annual sales in order to allow this to be a control variable. The government can influence the cost drop rate through policies that, for example, require government fleets to install OBEs, thereby insuring sales and incenting a cost drop.

IntelliDrive Safety Applications Value per Vehicle ($)

RSE Safety Applications Value ·············· $/vehicle
Elecronic Brake Lights Value ·············· $/vehicle
Other Applications Value ·············· $/vehicle

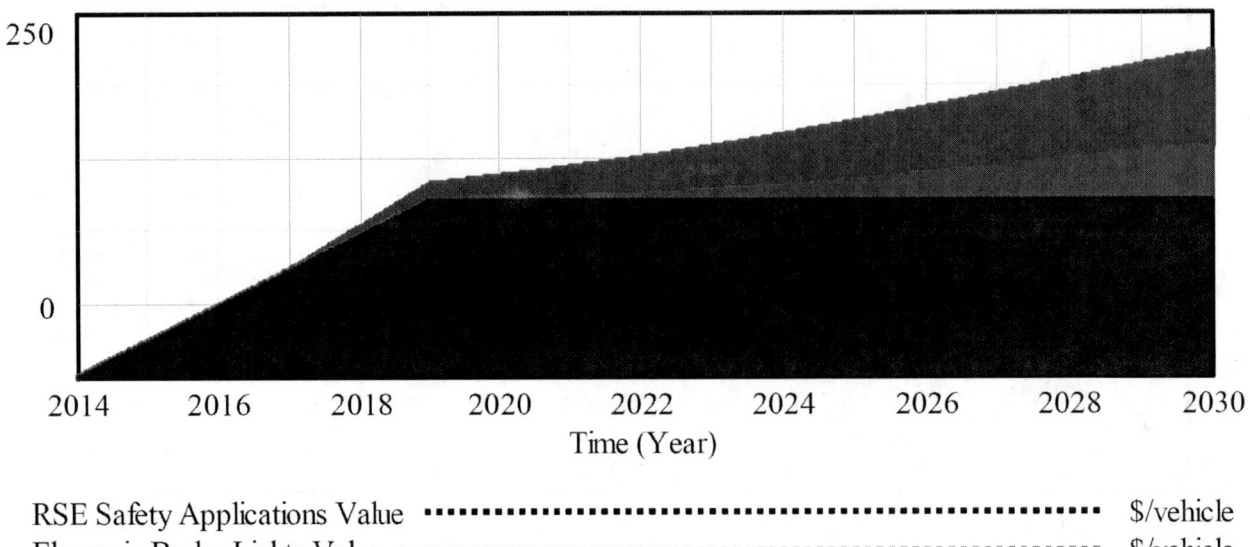

Figure 3. IntelliDrive safety applications value

Total Equipped Vehicles (M)

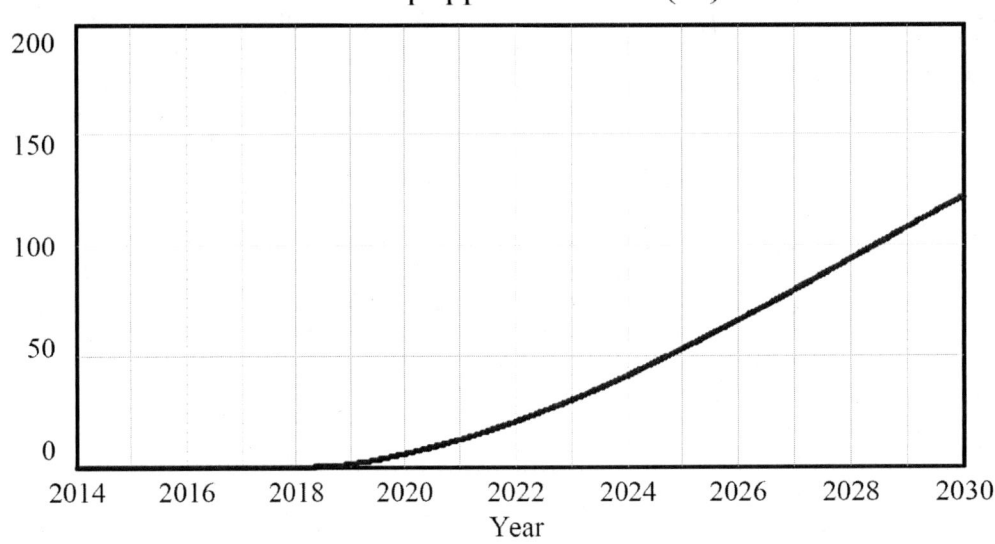

Figure 4. Equipped vehicle growth: total vehicles equipped with OBE in USA

In the current model, motorists will not adopt the system until the value exceeds their cost. Figure 4 shows the growth of vehicles with OBE after the value per vehicle surpassed the cost of the equipment in late 2018.

This equipped vehicle curve shows that the vehicle adoption rate growing through 2030, rising to 80% of all vehicles in the long run.

The costs to build-out and operate the IntelliDrive system are based on the costs in the Volpe study [1]. We assumed that the average cost to build-out an RSE is $15,000. For simplicity, we did not explicitly model the RSE replacement based on an average lifetime; rather, we set the annual maintenance cost to 20% of total imbedded cost to cover RSE replacements, operations, and maintenance. The annual costs to build, operate, and maintain the system is shown in Figure 5.

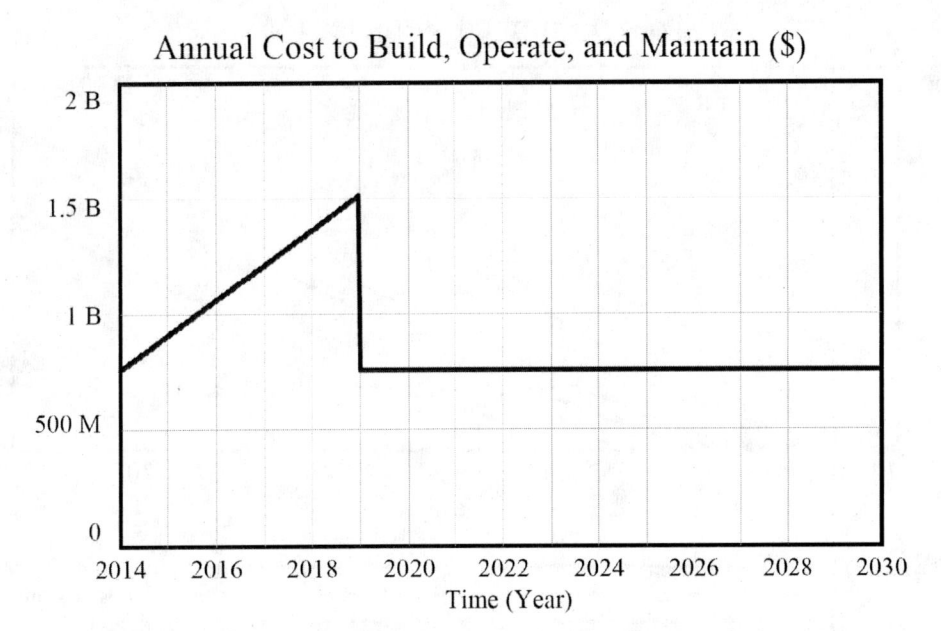

Figure 5. Expenditures to build, operate, and maintain the IntelliDrive system

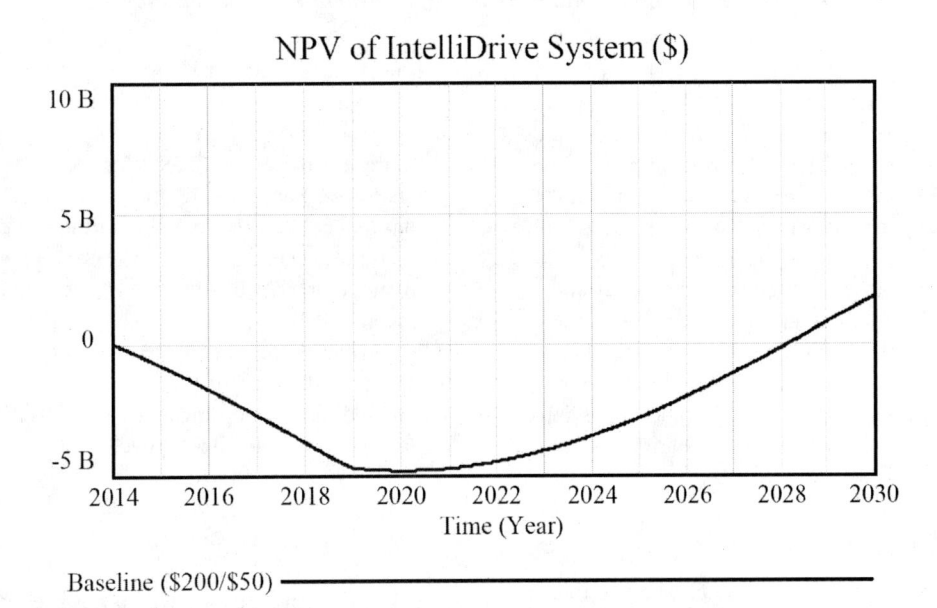

Figure 6. Net present values of IntelliDrive system with baseline scenario assumptions

The cost to build-out the RSE locations during the first five years grow as the cost of operations, maintenance, and replacements grow. In the sixth year, there is no further system build-out, so the annual costs drop to just operations, maintenance, and replacements.

As in the Volpe study [1], we can compute the net present value (NPV) to the nation as a whole by discounting the future benefits and future expenditures to the present day. In the Volpe study, all value came from benefit-producing safety applications. In that case, value was the same as benefit and could be directly compared to cost. In the current case, we are including "other" applications that increase system value, but may not have benefits to the public beyond the motorists that are using these applications, and therefore cannot be included in the NPV calculation. We set 50% of the other-applications' value as providing benefit in the baseline scenario.

Figure 6 shows the NPV of the IntelliDrive system under the baseline scenario assumptions with a 7% discount rate.

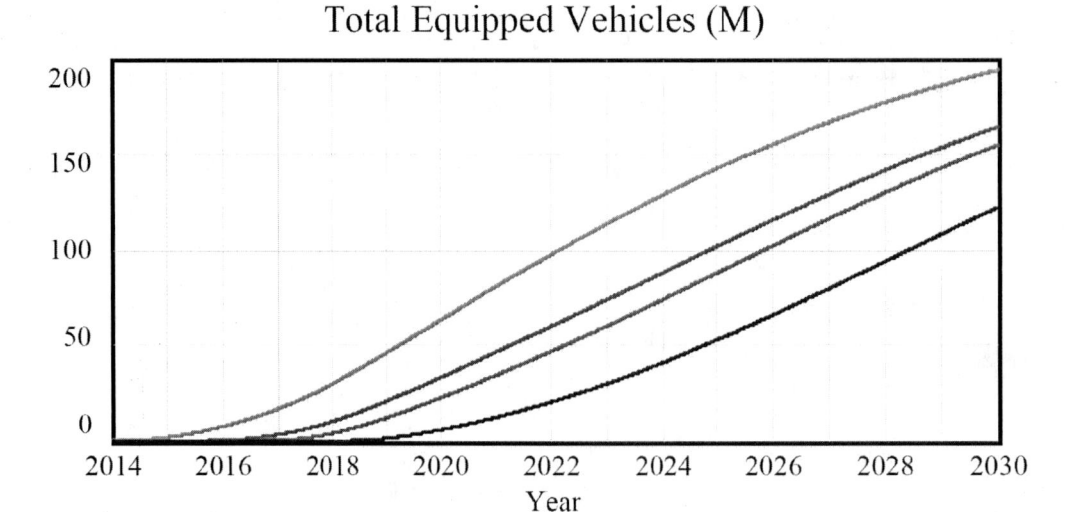

Figure 7. Effect of alternative OBE costs on IntelliDrive penetration

The NPV declines to about -$5B in mid 2019 before enough vehicles have installed OBEs to turn the NPV in a positive direction. The NPV goes positive in early 2028, and eventually reaches almost $20B. This NPV is similar to the Volpe study. This shows that under the baseline scenario, the net discounted payback period is over 14 years.

The downside of this result is the high risk: the government had to invest nearly $5B building out the system without any vehicles adopting the system until the very end of the build-out. This is untenable from a business strategy point of view; therefore, an approach is needed that brings more users into the system earlier. We examine alternative scenarios designed to do this in the next section.

ALTERNATIVE SCENARIOS

The IntelliDrive Penetration Model (IPV) can analyze many alternative scenarios involving alternative assumptions such as the speed of the RSE build-out, the value of safety and other applications, the ultimate potential system adoption level, unit costs, discount rates, and other parameters. For the purposes of this paper, we focus our attention on the cost of the OBE since it has the most impact on the IntelliDrive system NPV.

The primary reason for the delay in user adoption until 2018 in the baseline scenario is that the cost of the OBE is higher than the average system value per vehicle until then. This implies that the IntelliDrive business strategy should focus on bringing down the cost of the OBE as rapidly as possible to bring users into the system earlier, reducing risk and improving NPV.

There are many approaches that can be taken to reduce the OBE cost. For example, the government could adopt policies that reduce the effective cost of the OBE to motorists such as instituting tax credits, or providing subsidies to the manufacturers. In the E-ZPass case, the government appointed a single company, Mark IV Industries, to exclusively manufacture and supply the transponders, which are the equivalent of the OBEs in the IntelliDrive case. This allowed the government to control the OBE quality and distribution. This also allowed the government to set the up-front cost of the OBE to the motorist to $0 from day-one, resulting in immediate net positive value to the motorists, and corresponding uptake in system adoption.

In the current analysis, we do not specify or analyze specific strategies to bring the OBE cost down, although the model is capable of such analyses. Instead, we merely look at the effect of OBE cost reductions on the IntelliDrive system adoption and NPV. We looked at reducing both the initial OBE cost and the long-run OBE cost. The baseline scenario assumed the initial and long-run OBE costs were $200/$50. We now consider alternative scenarios in which these costs are reduced to $100/$50, $50/$50, and $0/$0. The resulting impacts on system adoption are shown in Figure 7.

The effects of lower OBE costs on bringing users into the system earlier are significant. Lower OBE costs mean that the time it takes for the value of the system to exceed these thresholds is reduced so that more vehicles are equipped with an OBE sooner. This leads to improved NPV and reduced payback period as seen in the following chart.

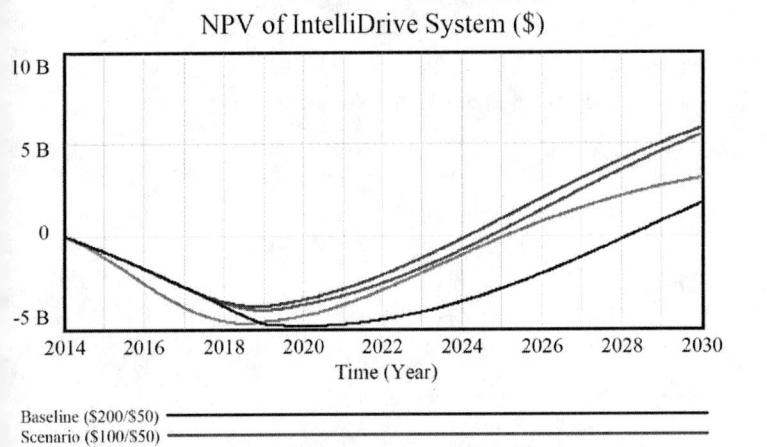

Figure 8. Effect of reduced OBE cost on IntelliDrive system NPV

This result shows an improving business case as the OBE cost is reduced, with the $50/$50 case providing the highest NPV and fastest payback. The $0/$0 OBE case, where the OBE is free to motorists on day-one, shows an immediate uptake in system adoption, however the added costs of paying for the OBEs is too much to overcome and this case is ultimately the worst of all considered scenarios. These results are central findings of the modeling, and suggest that the focus for developing an IntelliDrive business model be on finding a way to launch the system with a low cost OBE.

We also looked at the possibility of slowing the deployment of RSE locations to reduce annual construction budgets, and thereby reducing risk. Unfortunately, this leads to much slower system adoption, markedly extended payback periods, and reduced NPV. Therefore, this strategy is not recommended. This finding leads to considering a geographic deployment strategy of placing the system in localized but sufficiently large areas so that the value to motorists in these areas builds quickly, leading to rapid uptake in system adoption in these areas. More study, modeling, and analysis are needed to explore business strategies along these lines.

SUMMARY/CONCLUSIONS

IntelliDrive is a US DOT initiative to develop and demonstrate technologies for using wireless communications technologies to improve transportation safety, mobility, and sustainability. It envisions vehicles to communicate with each other and with road-side and infrastructure servers to achieve situational awareness in real time to detect and warn drivers of imminent dangers of collisions, and to provide traveler information to the drivers.

As IntelliDrive technologies mature, a crucial issue becomes how to develop a viable business case for widespread deployment of IntelliDrive systems and applications. It is widely recognized that some important IntelliDrive applications, such as vehicle safety applications based on V2V communications, will provide benefits to drivers only after a high percentage of all the vehicles are equipped with the same applications.

This paper studied the viability of a commercial business model for the IntelliDrive system with a dynamic simulation model. This model was calibrated against the market adoption of E-ZPass, which followed a similar commercial business model.

The baseline results are based on assumptions closely aligned with the Volpe study of IntelliDrive safety applications. The IntelliDrive penetration model predicts that a commercial business approach is viable only if a significant reduction in the cost to the motorist of the OBE is achieved. Specifically, the model predicts that a low cost OBE to motorists is required upon system launch to incent a strong enough adoption uptake to sufficiently reduce business risks to an acceptable level.

How can an affordable OBE be provided? One approach is to launch on luxury vehicle first, the "trickle down" adoption model. Luxury vehicles are less cost sensitive and value safety features as important to the brand. The luxury brands make up about 10% of the US fleet and could lead with V2V safety related features. This was the case for other safety related features such as airbags and ABS, with "encouragement" from NHTSA.

Another opportunity to reduce the cost of OBE equipment is through an aftermarket fitment program. At this time there is not a compelling reason for the driving public to purchase an aftermarket device. A program of the magnitude of the federal initiative related to driver distraction is needed to educate drivers about IntelliDrive safety applications and why an aftermarket device is useful. At this point, the driving public is not aware of why they need an aftermarket IntelliDrive device even if it is essentially given away. The US DOT JPO has appropriately reached out to the Consumer Electronics Industry to jointly develop this strategy. There is much work ahead.

The stakeholders who benefit are those who should take the upfront risk of "subsidizing" OBE costs by the strategies discussed. Our modeling tool helps estimate the size of the investment and determines the payback period. We envision public and private stakeholders joining forces to make this investment. Our model indicates that without government incenting significantly lower OBE costs coupled with cultivating driver awareness, it is difficult to predict if, how and when IntelliDrive will become a reality.

REFERENCES

1. Vehicle-Infrastructure Integration (VII) Initiative Benefit-Cost Analysis Version 2.3 (Draft), May 8, 2008, Prepared by: Economic and Industry Analysis Division, RTV-3A, John A. Volpe National Transportation Systems Center, United States Department of Transportation, Cambridge, Massachusetts

2. Sterman, J.D., Business Dynamics: Systems Thinking and Modeling for a Complex World, New York, NY: McGraw-Hill; 2000

3. Vensim, The Vensim Simulation environment, Vensim Professional 32 Version 5.4

4. Evaluating EZPass - Using conjoint analysis to assess consumer response to a new tollway technology, Vavra, Terry C., Green, Paul E., and Krieger, Abba M., Summer 1999,

5. E-ZPass Evaluation Report, Vollmer Associates, LLP, August 2000

6. Operational and Traffic Benefits of E-ZPass to the New Jersey Turnpike (August 2001)

7. Achieving the Vision: From VII to IntelliDrive Policy White Paper, RITA Intelligent Transportation Systems, April 30, 2010

8. Jones, W.S., "VII Life Cycle Cost Estimate," December 2006, updated April 2007, and "A VII Deployment Scenario," December 2005, US DOT ITS Joint Program Office

CONTACT INFORMATION

Robert White
Telcordia Technologies
rwhite@telcordia.com

Tao Zhang
Telcordia Technologies
tao@research.telcordia.com

Paul Tukey
Telcordia Technologies
paul@tukey.org

Kevin Lu
Telcordia Technologies
klu@telcordia.com

David A. McNamara
MTS LLC
coachdavemc@gmail.com

DEFINITIONS/ABBREVIATIONS

CSW
Curve speed warning

DSRC
Dedicated Short-Range Communications

EBL
Electronic brake lights

I2V
Infrastructure to vehicle

IPM
IntelliDrive Penetration Model

JPO
Joint Program Office

OBE
On-board equipment

OEM
Original equipment manufacturer

NPV
Net present value

RSE
Road-side equipment

SPAT
Traffic signal phase and timing

US
DOT United States Department of Transportation

V2V
Vehicle to vehicle

VII
Vehicle Infrastructure Integration

APPENDIX 1

MODEL CALIBRATION

The question of setting the parameters in the IntelliDrive model is difficult because the system has not yet begun to be deployed and decisions by motorists have not yet been made. Similar systems for which we have a full history of deployment and customer adoption can be used to provide approximate parameters if care is taken in correctly mapping the coefficients.

One such system is the E-ZPass electronic tolling system that was originally deployed in New York and New Jersey in the 1990's. Since this time, the E-ZPass system has expanded considerably to the Mid-Atlantic, Mid-West, and New England; it currently is in use in 25 agencies spread across 14 states. We focus attention on the initial E-ZPass deployment because it has been studied extensively and data are readily available.

The initial E-ZPass build-out was started in 1993 and completed in 1997. To reflect this build-out schedule, we set the budget for E-ZPass build-out so that 25% of the construction was completed in each year. E-ZPass was opened for sales in 1995. 80% was taken to be the asymptote of the E-ZPass penetration of rush-hour vehicles in the model.

Figure 9 shows that the E-ZPass Penetration model fits the tracking data very well.

There are clear similarities between the E-ZPass model and the IntelliDrive model. Both are driven by the value to the motorist and adoption is throttled by the construction of the system. There is an externality with E-ZPass as with IntelliDrive, but in the E-ZPass case, this is a negative reinforcement. This is because as drivers adopt E-ZPass, the E-ZPass electronic toll lanes become more congested while the cash toll lanes become less congested; this means that as more motorists adopt E-ZPass, its value is reduced somewhat. This negative externality effect was detected in the model fitting.

The functional form in the E-ZPass penetration growth model is identical to that in the IntelliDrive penetration growth model. In each case, the penetration growth is proportional to the average value of the system per vehicle. This common structure allows us to use the calibrated E-ZPass coefficients in the IntelliDrive penetration model. The implication of transferring these coefficients is the following: we are assuming that motorists will react to the average value per vehicle the same in both systems.

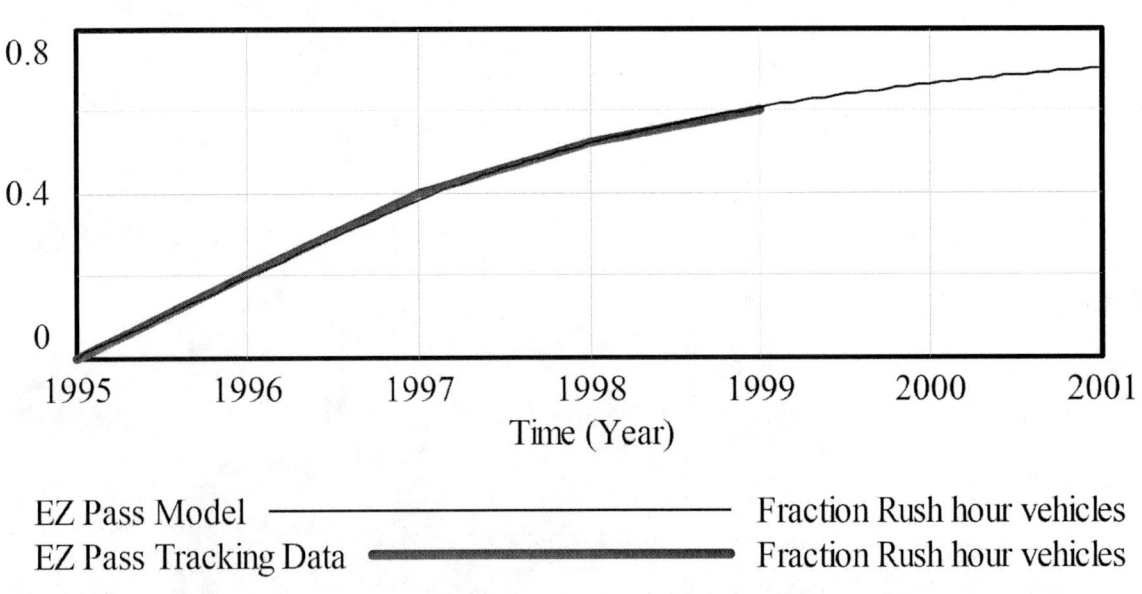

Figure 9. Resulting fit of the E-ZPass Penetration model (blue) against tracking data (red)

EDITOR'S AND SPECIAL CONTRIBUTORS' BIOGRAPHIES

About the Editor

Dr. Andrew Brown, Jr.
Executive Director & Chief Technologist, Delphi Corporation

Dr. Andrew Brown, Jr. is Executive Director & Chief Technologist for Delphi Corporation, and as such he provides leadership on corporate innovation and technology issues to help achieve profitable competitive advantage. Dr. Brown also represents Delphi globally in outside forums on matters of innovation and technology including government and regulatory agencies, customers, alliance partners, vendors, contracting agencies, academia, etc. Prior to this assignment, Dr. Brown had responsibility for common policies, practices, processes and performance across Delphi's 17,000 member technical community globally and its budget of $2.0 billion, including establishing Delphi's global engineering footprint with new centers in Poland, India, China, and Mexico, among others.

In April of 2009, SAE International's Executive Nominating Committee named Dr. Andrew Brown Jr., as its candidate for 2010 SAE International President. He was elected as 2010 SAE International President and Chairman in November of 2009, and was sworn into office in January of 2010.

As an NAE member, Dr. Brown was appointed by the National Research Council (NRC) to serve as chair of the Committee on Fuel Economy of Medium and Heavy Duty Vehicles. The report developed by this group was recently referenced by President Obama in his enhanced efforts on fuel economy improvement.

Dr. Brown joined Delphi coming from the GM Research and Development Center in Warren, Michigan, where he was Director - Research, Administration & Strategic Futures. He also served as a Manager of Saturn Car Facilities from 1985 to 1987. At Saturn, he was on the Site Selection Team and responsible for the conceptual design and engineering of this innovative manufacturing facility.

Dr. Brown began his GM career as a Project Engineer at Manufacturing Development in 1973. He progressed in the engineering field as a Senior Project Engineer, Staff Development Engineer, and Manager of R&D for the Manufacturing Staff. During this period, he worked on manufacturing processes and systems with an emphasis on energy systems, productivity improvement and environmental efficiency. Before joining GM, he supervised process development at Allied-Signal Corporation, now Honeywell, Incorporated in Morristown, New Jersey.

Dr. Brown earned a Bachelor of Science Degree in Chemical Engineering from Wayne State University in 1971. He received a Master of Business Administration in Finance and Marketing from Wayne State in 1975 and Master of Science Degree in Mechanical Engineering focused on energy and environmental engineering from the University of Detroit-Mercy in 1978. He completed the Penn State Executive Management Course in 1979. A registered Professional Engineer, Dr. Brown earned a Doctorate of Engineering in September 1992.

Special Contributors

Dr. Joseph N. Kanianthra
President, Active Safety Engineering LLC

Dr. Joseph N. Kanianthra is the President of Active Safety Engineering LLC, a company set up specifically to provide automobile safety consulting services to the industry and safety community. He retired from the National Highway Traffic Safety Administration (NHTSA) in August 2008 after 32 years of service. There he was the Senior Technical Advisor and an Associate Administrator for the Office of Vehicle Safety Research, where he directed all vehicle safety research related to crashworthiness, crash avoidance, driver-vehicle performance, tire research, driver distraction research, and intelligent technologies research programs. His chief accomplishments in research and safety standards in NHTSA were the development of safety standards based on research for dynamic side-impact protection, roof crush injury prevention, ejection mitigation, head injury protection, and frontal impact protection of occupants in light passenger vehicles. He was also very active in the area of international harmonization of research and safety standards. Until August 2008, he was the Chairman of the International Enhanced Safety of Vehicles Government Focal Points Program. Dr. Kanianthra received several distinguished awards, including the Secretary's Award for meritorious and technological achievements and the Presidential Rank Award for sustained superior achievements. In 2008, he was awarded the Pathfinder Award by the Automotive Occupant Restraint Council and the Award of Merit for his significant contributions in safety from the Association for the Advancement of Automotive Medicine. In June 2009, he received the Award for Engineering Excellence from the U.S. Government Department of Transportation and, in October 2009, in recognition of his outstanding technical achievements in vehicle safety, he was elected to be an SAE Fellow by SAE International's Board of Directors and the SAE Fellows Committee.

Patrick Lepercq
Corporate Vice President Public Affairs, Michelin

Patrick Lepercq has been Corporate Vice President Public Affairs for Michelin since 2002. He started his career leading a consultancy firm providing services in management and communication in Paris. He joined Michelin in 1980, holding several human resource responsibilities at the Michelin Group headquarters in France and serving a four-year assignment as Director of Personnel for Michelin Tires Canada Ltd. He was Managing Director of Michelin Tire Public Ltd Co. in the U.K. from 1996 to 2001. Mr. Lepercq is also Vice President of the European Tyre & Rubber Manufacturing Association (ETRMA), a member of the executive committee of the International Road Federation (IRF), and the Chairman of the Global Road Safety Partnership (GRSP). In France, he is Vice President of the International Development Research Foundation (FERDI), Treasurer of the French Road Union (URF), and a member of the board of the France-China Committee. He has a Master's degree in psychology from the René Descartes Institute at the University V of Paris and a Master's degree in social sciences from the Social Studies Institute of Paris.

Special Contributors

Dinesh Mohan,
Volvo Chair Professor for Biomechanics and Transportation Safety at the Transportation Research and Injury Prevention Programme of the Indian Institute of Technology, Delhi

Dr. Dinesh Mohan is Volvo Chair Professor for Biomechanics and Transportation Safety at the Transportation Research and Injury Prevention Programme of the Indian Institute of Technology, Delhi. He obtained his BTech in mechanical engineering from the Indian Institute of Technology Bombay, followed by a Master's degree in mechanical and aerospace engineering from the University of Delaware and then a Ph.D. in biomechanics from the University of Michigan, Ann Arbor. He started his research career working on vibrations of anisotropic plates and moved on to mechanical properties of human aortic tissue. This was followed by work on head, chest, and femur injury tolerance, injuries in human free falls, effectiveness of helmets and child seats, and the first evaluation of airbags in real world crashes. This background helped him work on epidemiology of road traffic crashes and injuries in rural India; helmet design; pedestrian, bicycle, and motorcycle crash modeling; and technological aids for the disabled. He has co-authored and edited four books on safety. He is the recipient of the Distinguished Alumnus Award of Indian Institute of Technology Bombay, the American Public Health Association International Distinguished Career Award, the Bertil Aldman Award of the International Council on Biomechanics of Impacts, the Association for Advancement of Automotive Medicine's Award of Merit, and the International Association for Accident & Traffic Medicine's International Award and Medal for outstanding achievement in traffic safety.

Richard Harris
Director, Intelligent Transport Systems, Logica plc
English Language Secretary of the World Road Association (PIARC) Technical Committee Network Operations, and Technical Secretary of the PIARC/FISITA Joint Task Force on Intelligent Cooperative Systems

Richard Harris, internationally recognized as a leading expert in Intelligent Transport Systems, is the Director of Intelligent Transport Systems at Logica plc. His experience spans more than 30 years, and he has over 20 years' experience of European Commission research and development, policy support, implementation, and outreach and promotion projects. Mr Harris is the International Director of ITS U.K. and a U.K. representative and the English Language Secretary of the World Road Association (PIARC) Technical Committee B2 "Network Operations" (2008-2011). He is a member of the Joint Task Force on "Intelligent Cooperative Vehicles and Vehicle-Infrastructure Communications and Safety" established by PIARC and FISITA in 2008. He is a founding member of the International Benefits Evaluations and Costs (of ITS) group and serves on the management committee as Chairman. In March 2010 he received the prestigious ITS U.K. Rees Hill Award for outstanding contribution to ITS by a U.K. professional.